Material Analysis in Cultural Heritage

Material Analysis in Cultural Heritage

Editors

Žiga Šmit
Eva Menart

MDPI • Basel • Beijing • Wuhan • Barcelona • Belgrade • Manchester • Tokyo • Cluj • Tianjin

Editors
Žiga Šmit
Faculty of Mathematics and Physics
University of Ljubljana
Jožef Stefan Institute
Ljubljana
Slovenia

Eva Menart
Department of Conservation and Restoration
National Museum of Slovenia
Jožef Stefan Institute
Ljubljana
Slovenia

Editorial Office
MDPI
St. Alban-Anlage 66
4052 Basel, Switzerland

This is a reprint of articles from the Special Issue published online in the open access journal *Materials* (ISSN 1996-1944) (available at: https://www.mdpi.com/journal/materials/special_issues/Material_Analysis_in_Cultural_Heritage).

For citation purposes, cite each article independently as indicated on the article page online and as indicated below:

LastName, A.A.; LastName, B.B.; LastName, C.C. Article Title. *Journal Name* **Year**, *Volume Number*, Page Range.

ISBN 978-3-0365-7488-2 (Hbk)
ISBN 978-3-0365-7489-9 (PDF)

Cover image courtesy of Eva Menart

© 2023 by the authors. Articles in this book are Open Access and distributed under the Creative Commons Attribution (CC BY) license, which allows users to download, copy and build upon published articles, as long as the author and publisher are properly credited, which ensures maximum dissemination and a wider impact of our publications.
The book as a whole is distributed by MDPI under the terms and conditions of the Creative Commons license CC BY-NC-ND.

Contents

About the Editors vii

Preface to "Material Analysis in Cultural Heritage" ix

Žiga Šmit and Eva Menart
Special Issue of "Material Analysis in Cultural Heritage"
Reprinted from: *Materials* **2023**, *16*, 2370, doi:10.3390/ma16062370 1

Roman V. Balvanović and Žiga Šmit
Emerging Glass Industry Patterns in Late Antiquity Balkans and Beyond: New Analytical Findings on Foy 3.2 and Foy 2.1 Glass Types
Reprinted from: *Materials* **2022**, *15*, 1086, doi:10.3390/ma15031086 3

Roxana Bugoi, Alexandra Țârlea, Veronika Szilágyi, Ildikó Harsányi, Laurențiu Cliante, Irina Achim and Zsolt Kasztovszky
Shedding Light on Roman Glass Consumption on the Western Coast of the Black Sea
Reprinted from: *Materials* **2022**, *15*, 403, doi:10.3390/ma15020403 27

Jingyi Shen, Li Li, Ji-Peng Wang, Xiaoxi Li, Dandan Zhang, Juan Ji and Ji-Yuan Luan
Architectural Glazed Tiles Used in Ancient Chinese Screen Walls (15th–18th Century AD): Ceramic Technology, Decay Process and Conservation
Reprinted from: *Materials* **2021**, *14*, 7146, doi:10.3390/ma14237146 43

Philippe Colomban, Michele Gironda, Divine Vangu, Burcu Kırmızı, Bing Zhao and Vincent Cochet
The Technology Transfer from Europe to China in the 17th–18th Centuries: Non-Invasive On-Site XRF and Raman Analyses of Chinese Qing Dynasty Enameled Masterpieces Made Using European Ingredients/Recipes
Reprinted from: *Materials* **2021**, *14*, 7434, doi:10.3390/ma14237434 63

László Szentmiklósi, Boglárka Maróti, Szabolcs Csákvári and Thomas Calligaro
Position-Sensitive Bulk and Surface Element Analysis of Decorated Porcelain Artifacts
Reprinted from: *Materials* **2022**, *15*, 5106, doi:10.3390/ma15155106 89

Minghao Jia, Pei Hu and Gang Hu
Corrosion Layers on Archaeological Cast Iron from Nanhai I
Reprinted from: *Materials* **2022**, *15*, 4980, doi:10.3390/ma15144980 101

Ana Fragata, Carla Candeias, Jorge Ribeiro, Cristina Braga, Luís Fontes, Ana Velosa and Fernando Rocha
Archaeological and Chemical Investigation on Mortars and Bricks from a Necropolis in Braga, Northwest of Portugal
Reprinted from: *Materials* **2021**, *14*, 6290, doi:10.3390/ma14216290 117

Marco Vona
Characterization of In Situ Concrete of Existing RC Constructions
Reprinted from: *Materials* **2022**, *15*, 5549, doi:10.3390/ma15165549 135

Patrizia Santi, Stefano Pagnotta, Vincenzo Palleschi, Maria Perla Colombini and Alberto Renzulli
The Cultural Heritage of "Black Stones" (*Lapis Aequipondus/Martyrum*) of Leopardi's Child Home (Recanati, Italy)
Reprinted from: *Materials* **2022**, *15*, 3828, doi:10.3390/ma15113828 155

Francesca Assunta Pisu, Pier Carlo Ricci, Stefania Porcu, Carlo Maria Carbonaro and Daniele Chiriu
Degradation of CdS Yellow and Orange Pigments: A Preventive Characterization of the Process through Pump–Probe, Reflectance, X-ray Diffraction, and Raman Spectroscopy
Reprinted from: *Materials* **2022**, *15*, 5533, doi:10.3390/ma15165533 . . . . **165**

Mariana Domnica Stanciu, Mircea Mihălcică, Florin Dinulică, Alina Maria Nauncef, Robert Purdoiu, Radu Lăcătușand Ghiorghe Vasile Gliga
X-ray Imaging and Computed Tomography for the Identification of Geometry and Construction Elements in the Structure of Old Violins
Reprinted from: *Materials* **2021**, *14*, 5926, doi:10.3390/ma14205926 . . . . **183**

About the Editors

Žiga Šmit

Žiga Šmit is a retired professor at the Faculty of Mathematics and Physics of the University of Ljubljana. His research work is conducted at the Jožef Stefan Institute in Ljubljana. His general field of interest is atomic collisions with light ions. He studied ionization mechanisms of inner shells by proton impact in semiclassical approximation and proposed corrections beyond the first step models. At the same time, he developed numerical models for practical applications of induced X-ray for chemical analysis and produced several methods for application of PIXE on historic materials, like analysis of archaeological metals, differential PIXE for the analysis of layered samples, a combined PIXE-PIGE algorithm for the analysis of historic glasses. Important historical cases include studies of usewear on flint tools, Roman military equipment metals, glass from almost all historical periods from the Bronze Age to the early 20th century, (semi) precious stones and paint pigments. Currently, he is interested in building a new code for deconvolution of X-ray spectra, modeling of XRF and in further analysis of historic materials.

Eva Menart

Eva Menart is the head of The Department of Conservation and Restoration of the National Museum of Slovenia and is additionally employed as a researcher at the Jožef Stefan Institute. She began her research, which combines science and cultural heritage, during her studies, after which she obtained a doctoral scholarship from University College London and in 2013 a PhD in Heritage Science. She has researched paper degradation and historical inks and worked on electrochemical gas sensors for detecting heritage material degradation. Her research now mainly focuses on the analysis of heritage objects with an emphasis on non-destructive methods (XRF, PIXE) and interpretation of results. She is also engaged in research in the field of preventive where she participates in international and national projects. She is currently the vice-chair of a COST Innovators Grant ENDLESS Metal, focusing on low-cost analytical tools and their dissemination in the metal conservation community.

Preface to "Material Analysis in Cultural Heritage"

This book describes the application of modern analytical techniques in the research of historic materials, ranging from glass and ceramics to metals, architectural objects, lithics, pigments and musical instruments. Its target audience is investigators of cultural heritage objects, such as material scientists, conservators and restorers, but also historians and archaeologists. The published articles are results of collective works of research teams; however, the list of first authors (in alphabetical order) includes R. Balvanović, R. Bugoi, Ph. Colomban, A. Fragata, M. Jia, F.A. Pisu, P. Santi, J. Shen, M.D. Stanciu, L. Szentmiklósi, and M. Vona. The editors are grateful for the collaboration of the authors and the excellent help of MDPI publishers.

Žiga Šmit and Eva Menart
Editors

Editorial

Special Issue of "Material Analysis in Cultural Heritage"

Žiga Šmit [1,2,*] and Eva Menart [2,3]

1 Faculty of Mathematics and Physics, University of Ljubljana, SI-1000 Ljubljana, Slovenia
2 Jožef Stefan Institute, SI-1000 Ljubljana, Slovenia
3 National Museum of Slovenia, Prešernova 20, SI-1000 Ljubljana, Slovenia
* Correspondence: ziga.smit@fmf.uni-lj.si

The objects of cultural heritage represent memories of human activities from the past. Therefore, we pay special attention to their preservation, and study them to learn more about the lives and creativity of our ancestors. The present book offers insight into several analytical techniques applied to selected heritage materials.

Glass is now a leading technological material, especially in the field of optical communications. It was also useful and widespread in the past. In Roman times and late antiquity, the production of glass attained industrial dimensions. Raw glass was produced in Egypt and Palestine, exploiting local siliceous sands and alkalis from dry Egyptian lakes as raw materials. According to the glass composition, several glass types were recognized in an early study [1], which is still valid today [2]. A review paper by Balvanović et al. [3] studies the development and distribution of glass types of Foy Serie 3.2 and Foy Serie 2.1 in the Mediterranean during late antiquity. Glass from the 1st to 4th century from Histria and Tomis, in present-day Romania, was studied through prompt-gamma activation analysis by Bugoi et al. [4], showing both early Roman and later 3.2 and 2.1 glass types.

Ceramic objects are often covered in glazes to improve their visual impression and increase their stability and durability. The composition of glazes used on ancient Chinese screen walls was studied by scanning electron microscopy (SEM) and X-ray diffraction (XRD), and the degradation processes were studied from a conservation point of view by Jingyi Shen et al. [5]. In goldwork masterpieces from the Quing Dynasty, it was found that the cobalt pigment in blue enamel was likely imported from Europe, as demonstrated by a non-destructive investigation through optical microscopy, Raman microspectroscopy, and X-ray microfluorescence spectroscopy by Colomban et al. [6]. For porcelain, it is important to show whether the products were made in a renowned or local workshop. Specifically in painted porcelain, an efficient analytical tool for surface and bulk analysis was found in position-sensitive X-ray fluorescence spectrometry (XRF) and prompt-gamma activation analysis (PGAA) by Szentmiklósi et al. [7]. The preservation of metals is another difficult task for conservation science, as they are prone to corrosion. One of the most challenging problems is the conservation of iron artifacts retrieved from a marine environment, which is the topic of the paper by Minghao Jia et al. [8]. Two papers deal with masonry and architectural remains. Fragata et al. [9] describe bricks and mortars in the Roman city of Bracara Augusta, modern-day Braga in Portugal. Chemical/elemental analysis performed by XRF demonstrates that the composition of the bricks differs according to their functionality, depending on whether they are being used as coatings, floorings, or masonry mortars, and their age of origin between the 4th and 7th centuries. In modern buildings, produced mainly by reinforced concrete since the middle of the 20th century, the deterioration and strength of the concrete need to be constantly monitored. Non-destructive tests with ultrasound are generally preferred, but their reliability needs to be regularly checked by destructive tests. Vona [10] constructed a comprehensive database to assess the reliability of the relationship between destructive and non-destructive methods for in situ concrete testing.

Citation: Šmit, Ž.; Menart, E. Special Issue of "Material Analysis in Cultural Heritage". *Materials* **2023**, *16*, 2370. https://doi.org/10.3390/ma16062370

Received: 9 March 2023
Revised: 14 March 2023
Accepted: 14 March 2023
Published: 16 March 2023

Copyright: © 2023 by the authors. Licensee MDPI, Basel, Switzerland. This article is an open access article distributed under the terms and conditions of the Creative Commons Attribution (CC BY) license (https:// creativecommons.org/licenses/by/ 4.0/).

Studies of lithic materials conclude with a thrilling story of "magic" black stones that have been preserved in the childhood home of famous Italian poet Giacomo Leopardi. The analysis performed by Santi et al. [11] suggests that the stones originate from Tuscany and were used by the Romans as counterweights. However, according to popular belief, they could also have been used as weights for torturing early Christians. Modern analytical techniques are further useful tools for the characterization of paint pigments. During the 19th and 20th centuries, yellow pigments based on cadmium sulfide (CdS) attained popular use. However, the pigment is subject to degradation with time. A study by Pisu et al. [12] employed the methods of SEM-EDS (energy-dispersive spectroscopy), Raman, and XRD on model samples painted with CdS, demonstrating its degradation into cadmium sulfate.

Another important aspect of culture, past or present, is music. Stanciu et al. studied the art of making ancient violins [13]. The authors investigated the structure and density of wood. They performed X-ray imaging and computed tomography on several violins manufactured by famous violin makers.

Author Contributions: Draft, Ž.Š.; final version Ž.Š. and E.M. All authors have read and agreed to the published version of the manuscript.

Funding: This work was partly funded by the Slovenian Research Agency (ARRS), programme P6-0283 "Archaeological Heritage Research".

Conflicts of Interest: The authors declare no conflict of interest.

References

1. Foy, D.; Picon, M.; Vichy, M.; Thirion-Merle, V. Caractérisation des verres de la fin de l'Antiquité en Méditerranée occidentale: l'émergence de nouveaux courants commericiaux. In *Échange et Commerce du Verre dans la Monde Antique*; Actes du colloque de l'Association Française pour l'Archéologie du Verre, Aix-en-Provence et Marseille 2001; Foy, D., Nenna, M.-D., Eds.; Éditions Monique Megoil: Montagnac, France, 2003; pp. 41–85.
2. Cholakova, A.; Rehren, T.; Freestone, I.C. Compositional identification of 6th c AD glass from the Lower Danube. *J. Archaeol. Sci. Rep.* **2016**, *7*, 625–632. [CrossRef]
3. Balvanović, R.; Šmit, Ž. Emerging glass industry patterns in Late Antiquity Balkans and beyond: New analytical findings on Foy 3.2 and Foy 2.1 glass types. *Materials* **2022**, *15*, 1086. [CrossRef] [PubMed]
4. Bugoi, R.; Târlea, A.; Szilágyi, V.; Harsányi, I.; Cliante, L.; Achim, I.; Kasztovszky, Z. Shedding light on Roman glass consumption on the Western Coast of the Black Sea. *Materials* **2022**, *15*, 403. [CrossRef] [PubMed]
5. Shen, J.; Li, L.; Wang, J.-P.; Li, X.; Zhang, D.; Ji, J.; Luan, J.-Y. Architectural glazed tiles used in ancient Chinese screen walls (15th-18th century AD): Ceramic technology, decay process and conservation. *Material* **2021**, *14*, 7146. [CrossRef] [PubMed]
6. Colomban, P.; Gironda, M.; Vangu, D.; Kirmizi, B.; Zhao, B.; Cochet, V. The technology transfer from Europe to China in the 17th–18th Centuries: Non-invasive on-site XRF and Raman analyses of Chinese Qing dynasty enameled masterpieces made using European ingredients/recipes. *Materials* **2021**, *14*, 7434. [CrossRef] [PubMed]
7. Szentmiklósi, L.; Maróti, B.; Csákvári, S.; Calligaro, T. Position-sensitive bulk and surface element analysis of decorated porcelain artifacts. *Materials* **2022**, *15*, 5106. [CrossRef] [PubMed]
8. Jia, M.; Hu, P.; Hu, G. Corrosion Layers on archaeological cast iron from Nanhai I. *Materials* **2022**, *15*, 4980. [CrossRef] [PubMed]
9. Fragata, A.; Candeias, C.; Ribeiro, J.; Braga, C.; Fontes, L.; Velosa, A.; Rocha, F. Archaeological and chemical investigation on mortars and bricks from a necropolis in Braga, northwest of Portugal. *Materials* **2021**, *14*, 6290. [CrossRef] [PubMed]
10. Vona, M. Characterization of in situ concrete of existing RC constructions. *Materials* **2022**, *15*, 5549. [CrossRef] [PubMed]
11. Santi, P.; Pagnotta, S.; Palleschi, V.; Colombini, M.P.; Renzulli, A. The cultural heritage of "black stones" (lapis aequipondus/martyrum) of Leopardi's child home (Recanati, Italy). *Materials* **2022**, *15*, 3828. [CrossRef] [PubMed]
12. Pisu, F.A.; Ricci, P.C.; Carbonaro, C.M.; Chiriu, D. Degradation of CdS yellow and orange pigments: A preventive characterization of the process through pump–probe, reflectance, X-ray diffraction, and Raman spectroscopy. *Materials* **2022**, *15*, 5533. [CrossRef]
13. Stanciu, M.D.; Mihălcică, M.; Dinulică, F.; Nauncef, A.M.; Purdoiu, R.; Lăcătuș, R.; Gliga, G.V. X-ray Imaging and Computed Tomography for the Identification of Geometry and Construction Elements in the Structure of Old Violins. *Materials* **2021**, *14*, 5926. [CrossRef]

Disclaimer/Publisher's Note: The statements, opinions and data contained in all publications are solely those of the individual author(s) and contributor(s) and not of MDPI and/or the editor(s). MDPI and/or the editor(s) disclaim responsibility for any injury to people or property resulting from any ideas, methods, instructions or products referred to in the content.

Review

Emerging Glass Industry Patterns in Late Antiquity Balkans and Beyond: New Analytical Findings on Foy 3.2 and Foy 2.1 Glass Types

Roman V. Balvanović [1,*] and Žiga Šmit [2]

1. Laboratory of Physics, Vinča Institute of Nuclear Sciences, National Institute of the Republic of Serbia, University of Belgrade, P.O. Box 522, 11001 Belgrade, Serbia
2. Faculty of Mathematics and Physics, Jožef Stefan Institute, University of Ljubljana, 1000 Ljubljana, Slovenia; ziga.smit@fmf.uni-lj.si
* Correspondence: broman@vinca.rs

Abstract: Resolving issues posed by our paper describing the late antiquity glass from Jelica (Serbia), we performed a thorough analysis of similar glass, systematically collected from the literature. The analysis showed that Foy 3.2 type evolved gradually from a composition similar to the Roman antimony-decolorized glass to a composition approaching Foy 2.1, lasting longer (second−seventh century AD) and spreading wider than originally described, including large parts of the Balkans, France interior, Germany, and Britain. The center of its distribution seems to be the Balkans and Italy. During the sixth century, Foy 3.2 glasses in the Balkans showed a significant increase of average MgO concentration compared to the earlier period and Foy 3.2 glasses outside the Balkans, implying different sand quarries and perhaps different trade routes for its imports. Recycling criteria for Foy 3.2 glass has been established. Similarly, 125 high-iron Foy 2.1 glasses are selected from the literature. They cluster within two groups regarding iron concentrations, which we term high iron (HI) and very high iron (VHI) Foy 2.1. In addition, there is a low lime subgroup of the VHI group, termed VHILL. The paper offers two possible explanations for the elevated iron, color branding, and different silica sources. High-iron glasses seem relatively evenly spread across the entire Mediterranean and its interior, representing, on average, around a quarter of the local Foy 2.1 assemblages. The percentages of high-iron samples are almost double in manufactured glass compared to raw glass, suggesting that the addition of iron was happening in the secondary workshops, i.e., for color branding. Among the manufactured glass, the proportions were higher in glassware than in windowpane glass. To capture the changing sand exploitation conditions, we propose the term "generic composition/type" or "(geochemical) class".

Keywords: late antiquity; Foy 3.2; Foy 2.1; Fe-rich; color-branding; Balkans

1. Introduction

In a recent study, we described a new glass assemblage from the sixth-century from the Byzantine settlement of Jelica in Serbia [1,2]. This manganese-decolorized glass, typical of late antiquity, is characterized as Foy 3.2 and Foy 2.1 [3]. Foy 3.2 is characterized by smaller average concentrations of oxides derived from sand minerals, especially alumina (1.92%) and, to a smaller extent, lime (6.99%), indicating the use of cleaner glass-producing sands, with less heavy minerals and feldspars. Foy 2.1 is characterized by higher average concentrations of heavy-mineral related oxides like iron (1.35%), titanium (0.16%), and magnesium (1.23%), showing the use of sands with more impurities. Foy 3.2 was originally described in 17 glasses from Southern France (late fifth/early sixth century AD) and in two earlier glasses from Tunisia (late first and the second century). It has recently been recognized in several locations across the Balkans in Kosmaj [4]; Caričin Grad [5] in Serbia; Serdica, Dichin, and Odartsi in Bulgaria [6,7]; Butrint in Albania [8], and, most recently, in Slovenia [9]. Noting the similarity between Foy 3.2 and Foy 2.1 types of glass from Bulgaria, Cholakova and colleagues proposed a hypothesis that these groups are related to each other in terms of origin, but should nevertheless be considered separate primary groups. Similar observations led others to the contrary conclusion and the joint term for both groups, namely Foy 2 [10,11]. Cholakova and Rehren further discussed Foy 3.2 type in contexts from the late fourth to the early seventh century in the Western Mediterranean, Italy, and Balkans, noting a gradual increase in sand impurities over time. New glasses of Foy 3.2 type that are appearing in the literature are dated to ever-wider time-span, and show different compositions and similarities to other types such as Foy 2.1 and Roman antimony-decolorized glass. The paper focuses on the general distribution, duration, geochemical characteristics, and evolution of its composition with time, trying to interpret the reasons for this change.

Our study also found some of Jelica glasses compositionally similar to the neighboring iron-rich subgroup of the Lower Danube composition. The high-iron glass otherwise similar to Foy 2.1 was later termed high-Fe Foy 2 [10]. This glass has since been described in several assemblages [11–13]. A clearer picture of the characteristics and distribution patterns of the Fe-rich Foy 2.1 type is needed. The paper also examines the origin of elevated iron in these glasses.

2. Evolution and Characteristics of Foy 3.2-Type Glass

To obtain a synoptic compositional overview of the considerable diversity of the Roman and the Late Antiquity natron glass types, a large amount of data have been plotted on the principal component analysis (PCA) diagram (Figure 1). The selection of oxides (SiO_2, Na_2O, MgO, Al_2O_3, K_2O, CaO, TiO_2, MnO, and Fe_2O_3) is limited by the published data. Before PCA, power transformation $\hat{x}_{ij} = \tilde{x}_{ij} - \overline{\tilde{x}}_i$, where $\tilde{x}_{ij} = \sqrt{x_{ij}}$ is performed on the data because of their heteroscedasticity (different variances across oxide values). The diagram plots five late antiquity glass groups from the Balkans classified as série 3.2 (Jelica, Kosmaj, and Caričin grad from Serbia; Butrint from Albania; and a Bulgarian fifth century composition), and compares it against several groups of typical Roman natron glass and Foy série 3.2 and série 2.1 as the referential groups. The comparison groups include Roman manganese added glass (group AD/N1), several Roman antimony added glass groups (AD/N2, Roman Sb glass from Carthage, and colorless group CL1/2), and Roman blue-green glass group Ic1a.

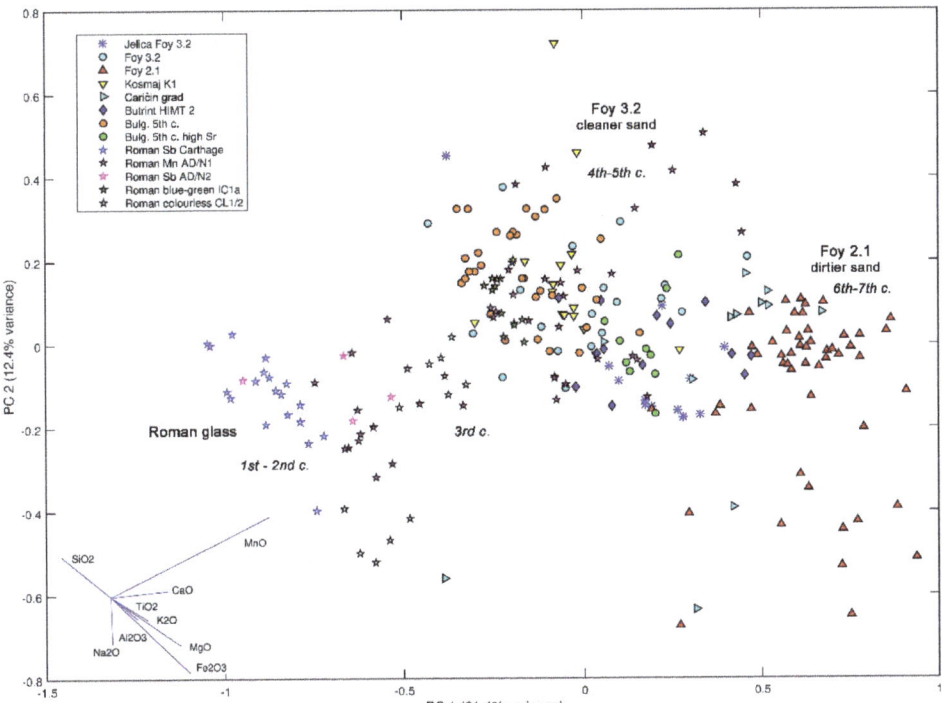

Figure 1. Principal component analysis of 256 glasses from 14 Roman and late antiquity natron glass groups. The diagram compares compositions of six glass groups from the Balkans, five characteristic Roman glass groups (with added manganese and with no added manganese, naturally colored and decolorized) and reference groups Foy 3.2 and 2.1. Vectors of oxides: Na_2O, MgO, Al_2O_3, SiO_2, K_2O, CaO, TiO_2, MnO, and Fe_2O_3 lower left. Data sources: (Jelica Foy 3.2—[1,2]; Foy 3.2, Foy 2.1—[3]; Kosmaj K1—[4]; Caričin grad—[5]; Bulg. 5th century—[7]; Butrint HIMT 2—[8]; Roman Sb Carthage—[10]; Roman AD/N1, AD/N2—[14]; Roman Ic1a—[15]; Roman CL1/2—[16]).

The PCA diagram can be interpreted chronologically. Vectors show a similar direction for an increase in oxides, reflecting both light and heavy minerals of the sand. Roughly congruent vector slopes reflect correlations between oxides reflecting light (alumina, potassium, and lime) and heavy minerals (titanium, iron, and magnesium). The change from antimony to manganese as a decolorizer, which started around the end of the third century [17], is manifested on the diagram by a shift in the direction of the manganese vector. Note that since manganese gradually replaced antimony as a decolorizer, the direction of the manganese vector on the PCA diagram roughly corresponds with the physical time for the period of the third–fifth century. Analogously, the direction of time in the period of the fifth–sixth century is roughly parallel to the vectors of oxides representing the heavy minerals of the sand.

There are three major compositional aggregations: Roman glass with no deliberately added manganese (left), Roman manganese-added glass and Foy 3.2 (middle), and Foy 2.1 glass (right). Oxide vectors, showing the directions of increase of corresponding oxides concentrations in the PCA diagram, demonstrate that the central aggregation differs from the left one mostly by the increased manganese. Analogously, the aggregation on the right side differs from the central one by a marked increase in heavy and light sand minerals.

How does Foy 3.2 differentiate from these glass groups? Compared to the "Roman glass" assemblage [18], Foy 3.2 has lower alumina and lime. Foy 3.2 has lower Al_2O_3 (1.92 ± 0.15% versus 2.46 ± 0.15%) and, to an extent, CaO (6.99 ± 0.74% versus 7.74 ± 0.53%) than Roman manganese-added group AD/N1, and a lower sum of light sand minerals (0.138% versus 0.158%), indicating the use of sand with less feldspars and lime (Table 1). Compared with the other major Roman glass types without added manganese, the antimony-added group AD/N2, Foy 3.2 differs by having higher lime (6.99% versus 5.14%) and alumina (1.92% versus 1.77%), as well as increased concentrations of oxides reflecting both heavy minerals (average of 2.1% against 1.4%) and light minerals (13.8% versus 11.2%), but it is also in lower zirconium (48 ppm for versus 73 ppm). This reflects different sand and lime sources, and different suits of heavy minerals. Regarding sand provenance, the TiO_2/Al_2O_3 ratio discriminates oxides reflecting heavy minerals rich Egyptian sands from alumina rich Levantine sands [10]. This ratio is around 2.5% for Levantine sand and almost double and more for the Egyptian sands. While manganese added Roman glass group AD/N1 is of a Levantine origin (TiO_2/Al_2O_3 ratio is 0.024), the antimony added group AD/N2 and série 3.2 originate in Egypt (0.047 and 0.049, respectively).

Table 1. Ratios of oxides reflecting heavy and light sand minerals and the provenance indicator TiO_2/Al_2O_3.

Group	Cent. AD	$(Fe_2O_3 + TiO_2 + MgO)/SiO_2$	$(Al_2O_3 + K_2O + CaO)/SiO_2$	Sum	TiO_2/Al_2O_3
Roman glass n = 227	1st–4th	0.019	0.156	0.175	0.05
AD/N1 (Mn added) n = 45	1st–4th	0.017 ± 0.004	0.158 ± 0.014	0.174 ± 0.016	0.024 ± 0.009
AD/N2 (Sb added) n = 4	2nd–3rd	0.014 ± 0.003	0.112 ± 0.007	0.127 ± 0.010	0.047 ± 0.015
série 3.2 (non-t.) n = 2	1st–2nd	0.023 ± 0.003	0.124 ± 0.007	0.147 ± 0.009	0.045 ± 0.011
série 3.2 n = 17	5th/6th	0.021 ± 0.004	0.138 ± 0.016	0.159 ± 0.020	0.049 ± 0.009
série 2.1 n = 51	6th–7th	0.043 ± 0.011	0.173 ± 0.012	0.216 ± 0.016	0.062 ± 0.007

Foy série 3.2 is compared to two common Roman glass types, with added manganese (AD/N1), and with no added manganese (AD/N2), an assemblage of "Roman" glass, and with Foy group 2.1. Data sources: [3,14,18].

Foy 2.1 is differentiated from série 3.2 by a significant increase in the sum of all oxides reflecting minerals in the sand (21.6% versus 15.9%) and in the TiO_2/Al_2O_3 ratio (0.062 versus 0.049). It has double the concentration of heavy sand minerals compared to Foy 3.2, and around 25% higher light minerals. Taking into account standard deviations, we propose the cutoff values for differentiating Foy 3.2 and 2.1 to be 0.03 for oxides reflecting heavy minerals in the sand, 0.16 for light minerals, and 0.19 for the sum of oxides reflecting both heavy and light minerals. There is a question regarding the alumina cutoff value between Foy 3.2 and 2.1. D. Foy reports alumina concentrations of 1.92 ± 0.15% for série 3.2, and 2.54 ± 0.15% for group 2.1 [3], allowing the cutoff might be set to, e.g., 2.3%. However, several papers have since reported Foy 3.2 type of glasses with higher alumina concentrations. The examples include alumina concentrations of up to 2.38%, 2.51%, and 2.59% [7,19,20], so higher alumina concentrations for Foy 3.2 should be allowed. What differentiates Foy 2.1 from 3.2 in such cases is the still higher heavy mineral concentrations and TiO_2/Al_2O_3 ratios.

Another yet unaddressed issue regarding série 3.2 is that it includes two non-tardifs samples from the first–second century AD, far earlier than the rest of the group. While these two are indeed very similar to the 17 samples dated to late fifth–early sixth century (being somewhat lower in light sand minerals), a question arises regarding the huge time-span between them. Indeed, D. Foy noted that série 3.2 is not specific for a single period, but that sand for its manufacturing was exploited for a longer period, and that other deposits were surely exploited at the same time [3]. This opens a question regarding the time duration of Foy 3.2 type. Whether the time gap between the two non-tardifs série 3.2 samples and fifth–sixth century série 3.2 is perhaps filled with assemblages dated continuously from the early second century to the sixth century AD? This time gap has been filled up, with several authors reporting Foy 3.2 assemblages from different periods and places. These include collections from the fifth century Bulgaria [7], fourth–fifth century Italy [19], late third–sixth century Italy [21], mid fourth–fifth century Germany [20], fourth century England [17], third–fourth century AD, outlier sample YAS-265 from Carthage [10], and second–fourth century AD Kosmaj in Serbia [4]. It seems that in the case of Foy 3.2 type, there is a need to overcome the traditional definition of a compositional group being tied to a single place and time of sand exploitation, and to introduce a definition that would encompass a greater area and longer time.

With the mentioned criteria, we searched the literature and carefully selected a total of 246 glasses that could be attributed to Foy 3.2-type (Supplementary Material Table S1, Foy 3.2 recycling). Many glasses are attributed as Foy 3.2 for the first time, and some are reattributed. Each new attribution and reattribution were performed very carefully, using all available compositional data and testing extensively through numerous diagrams. Indeed, some of these glasses seemed to further expand the timeframe of série 3.2 and fill the mentioned time gap, like AD-A-8 and AD-I-3 from Adria, dated from the second half first to the early second century and the third century AD, respectively. Their elevated TiO_2/Al_2O_3 and smaller Sr/Zr ratios fall well within série 3.2, with an Egyptian provenance of sand, and not within AD/N1 glasses, as is the case of manganese-added Roman glasses (Figure 2). In addition, elevated hafnium concentrations of AD-A-8 and AD-VC-1 (1.2 and 1.4 ppm, respectively) compared to the pertaining AD/N1 group (0.81 ± 0.18, excluding the outlier AD-A-11 with 6.6 ppm of hafnium) conform better to Roman antimony glass AD/N2 (1.13 ± 0.22 ppm), and MSG1c group from Padova (1.41 ± 0.81 ppm), classified as Foy 3.2, both with an Egyptian provenance. As shown by Barfod et al. [22], the hafnium concentration in sands decreases from the Nile delta towards the Levantine coast during the longshore transport of the Nile sediments. This indicates that AD-A-8 comes from a location that is likely closer to the Nile delta from the origin of the majority of AD/N1 glasses. In addition, this glass on the PCA diagram is close to the early Foy 3.2 glass VRR390 from Nabeul [3], dated also to the end of the first century AD, which further strengthens such an attribution. Likewise, Adria glass AD-VC-1 is dated to the second–fourth century, and AD-I-3 to the third century AD. This gives evidence to tentatively attribute these three glasses from Adria to Foy 3.2 type. Similarly, five seventh-century glasses from Crypta Balbi in Rome [23] and two seventh-century Merovingian glasses from Vicq in France [24] possibly extend the opposite end of the Foy 3.2 timespan.

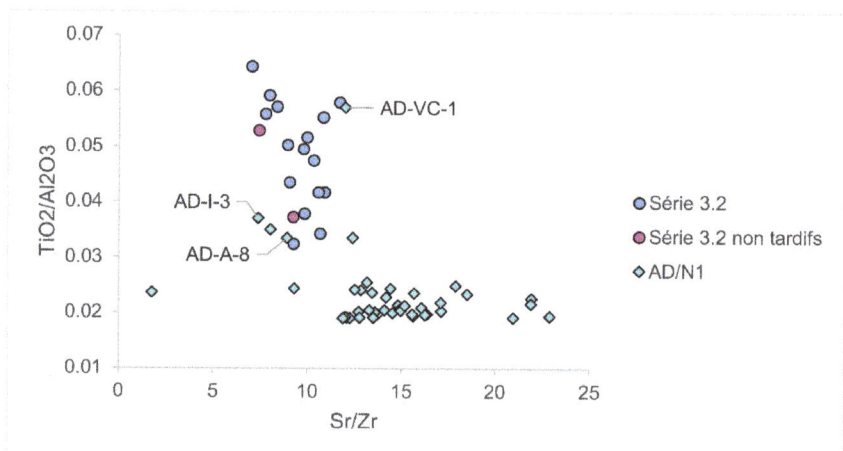

Figure 2. Provenance indicators. TiO$_2$/Al$_2$O$_3$ versus Sr/Zr, indicators of sand heavy minerals for Foy série 3.2 and Adria AD/N1 show that glasses AD-A-8, AD-I-3, and AD-VC-1 from AD/N1 (of Levantine origin) more likely utilized Egyptian than Levantine sand. Data sources: [3,14].

To try to gain some insight into the possible evolution of the composition, 246 Foy 3.2-type glasses were sorted by centuries according to their dating, and the averages and standard deviations of selected oxides and trace elements for each century were calculated and plotted (Figure 3). While some of the date attributions were indeed uncertain, many were reliable, and the observed timeframe was long enough that general trends could nevertheless be noticed with reasonable confidence. The relative stability of the mean concentrations of alumina and lime through the centuries (Figure 3a,b) indicate that sand -sources were located within a geologically similar area, poor in feldspars and lime. However, a gradual change also exists in the compositions with time. The overall tendency is a slow but steady increase of lime, strontium, zirconium (Figure 3c), iron, titanium, and magnesium, derived from heavy minerals of the sand. The sum of iron, titanium, and magnesium is 1.12% in the second century to 1.68% in the sixth century AD (Figure 3e). Antimony, derived from recycling, at first was relatively high (0.08% in group 2a of Foster and Jackson or 0.2% in VRR391), but decreased with time, as the antimony-decolorized cullet became less available, especially from the third century AD onwards (Figure 3d).

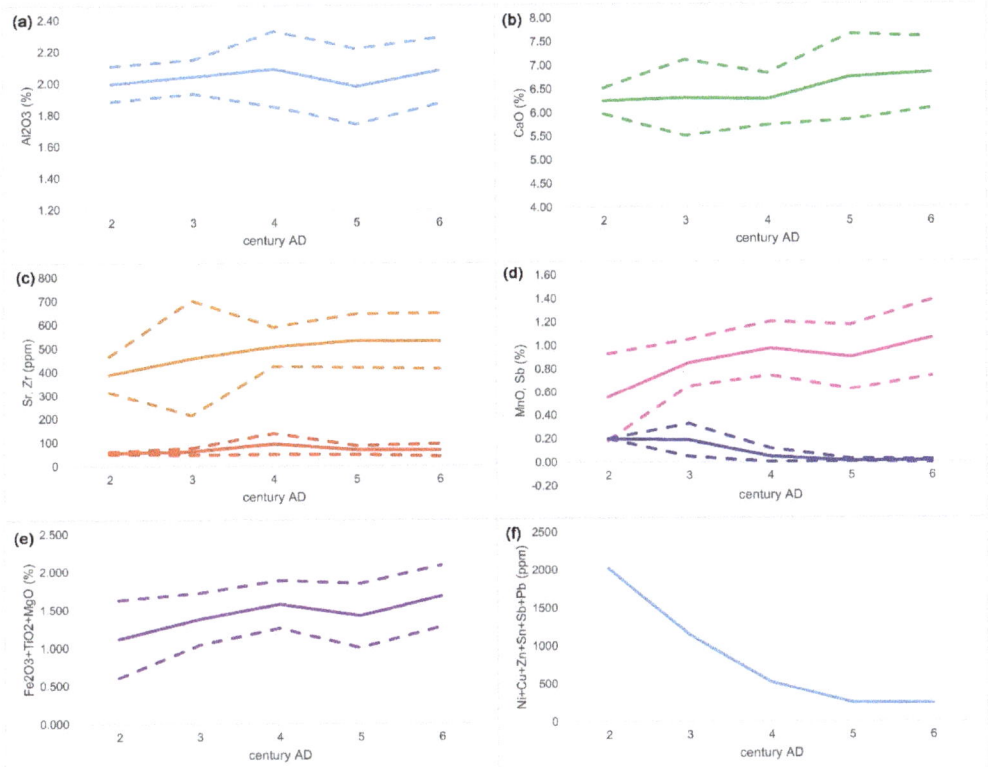

Figure 3. Evolution of the composition of série 3.2 in time. Selected oxides concentrations of 246 individual Foy 3.2 type glasses, belonging to 28 different glass groups across the Mediterranean and Western Europe, second–sixth century AD. Alumina (**a**), lime (**b**), strontium (orange) and zirconium (red) (**c**), manganese (magenta) and antimony (blue) (**d**), the sum of oxides reflecting heavy minerals of sand (**e**), and sum of trace elements reflecting recycling (**f**). Averages are straight lines, and standard deviations are dotted lines. Data sources: [1–3,5–10,12,14,19,23–30].

The decreasing degree of recycling is also notable, indicated by the constantly diminishing Ni + Cu + Zn + Sn + Sb + Pb sum, from over 2000 ppm in the second century to 237 ppm in the sixth century AD. A temporary exception to the trend of the increase of sand minerals is a slight change in the composition from the fourth to the fifth century AD, manifested by a decrease of alumina (2.09 to 1.98%), magnesium (0.79 to 0.65%), manganese (0.97 to 0.90%), and zirconium (93 to 67 ppm) and an increase of lime (6.29 to 6.75% on average). These changes likely reflect the use of different quarries over time [31]. It is important to stress that the time perspective or "evolution" of the composition gives quite different semantics to the meaning of a "compositional group"; while the classical compositional group reflects a single quarry and a few primary workshops in its vicinity, our approach implies a dynamic view: constant changing of sand quarries within the similar geochemical area and through a longer period. The concept tries to capture the dynamics of the process.

Série 3.2 glasses seem to form four distinct groups depending on their respective Nd/La and Ce/La values, namely: Nd/La < 1, Ce/La > 1 (group I); Nd/La > 1, Ce/La > 1 (group II); Nd/La > 1, Ce/La < 1 (group IIII); and Nd/La < 1, Ce/La < 1 (group IV; Figure 4a). MSG1c glass forms group I while AQ/3, FC/3, and CL3 glasses are distributed over groups II, III, and IV. Groups I and II are differentiated by Al_2O_3 and Ba concentrations (Figure 4d), showing differences in light sand minerals, while groups III and IV are differentiated by Fe_2O_3 and TiO_2 concentrations (Figure 4b), reflecting variations in heavy minerals. Groups III and IV are also differentiated by CaO, SrO, and Na_2O concentrations, possibly showing different sources of lime and different manufacturing recipes. This suggests the variability of sands used for manufacturing these groups. Different Al_2O_3/SiO_2 versus TiO_2/Al_2O_3 provenance indicators of the respective groups, albeit within a wider span of values characteristic for série 3.2 glasses, seem to confirm this. It should be noted that a relatively small number of samples with measured REEs (26) limits making more definitive conclusions, but it is still sufficient to gain insight into the compositional differences among série 3.2 glasses.

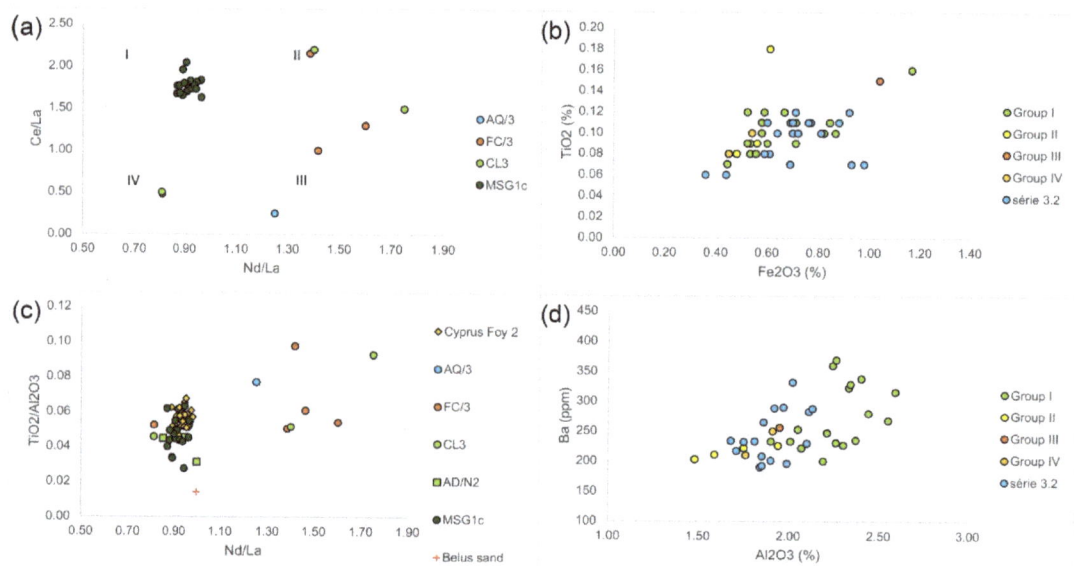

Figure 4. Groupings of Foy 3.2-type glasses according to trace elements. Nd/La versus Ce/La (**a**), Fe_2O_3 versus TiO_2 (**b**), Nd/La versus TiO_2/Al_2O_3 (**c**), and Al_2O_3 versus Ba (**d**). Data sources: [14,19,25,26].

Note that the AD/N2 group overlaps with the group I on Nd/La versus Ce/La diagram, further supporting the noted similarity between Roman antimony-decolorized glass and série 3.2 [10]. Note also that some groups classified as Foy 2 or Foy 2.1 share the Nd/La versus Ce/La space with group 1. This shows that group I of série 3.2 glasses has a similar origin to Foy série 2.1, while other types have not. Table 2 summarizes the Nd/La, Ce/La, and Zr/TiO2 ratios for various groups.

Table 2. Ratios and standard deviations of Nd/La, Ce/La, and Zr/TiO_2 were sorted by increasing Nd/La values.

Original Classification	Group/Location	n	Nd/La ppm/ppm	Ce/La ppm/ppm	Zr/TiO_2 ppm/pm
Foy 3.2	Group IV	2	0.81 ± 0.00	0.49 ± 0.02	56.88 ± 4.42
Foy 2.1 Fe-rich	Lower Danube	23	0.91 ± 0.03	1.44 ± 0.17	74.92 ± 4.78
Foy 3.2	Group I (MSG1c)	18	0.91 ± 0.03	1.77 ± 0.10	49.98 ± 6.58
Foy 2 High-Fe	Yeroskipou	2	0.92 ± 0.01	1.19 ± 0.22	52.58 ± 0.88
Foy 2.1	Lower Danube	43	0.92 ± 0.05	1.77 ± 0.12	74.71 ± 4.62
Ca-rich HIMT	Herdonia	4	0.93 ± 0.02	1.66 ± 0.10	72.80 ± 10.10

Table 2. Cont.

Original Classification	Group/Location	n	Nd/La ppm/ppm	Ce/La ppm/ppm	Zr/TiO$_2$ ppm/pm
HIMT 2	San Giusto	9	0.93 ± 0.03	1.69 ± 0.09	60.90 ± 2.95
Foy 2 High-Fe	Byz. glass weights	31	0.93 ± 0.04	1.19 ± 0.09	55.86 ± 2.80
Roman Sb	AD/N2	4	0.93 ± 0.08	1.81 ± 0.04	63.70 ± 6.12
Foy 2	Yeroskipou	34	0.94 ± 0.02	1.68 ± 0.07	53.44 ± 3.57
Levantine	San Giusto	21	0.94 ± 0.05	1.74 ± 0.08	62.30 ± 5.86
Foy 2	Byz. glass weights	85	0.95 ± 0.04	1.63 ± 0.10	56.78 ± 3.69
HIT	Cyprus	2	0.96 ± 0.00	1.90 ± 0.01	53.96 ± 9.95
High Mn Lev 1	Cyprus	6	0.96 ± 0.02	1.71 ± 0.07	54.58 ± 4.54
Roman Mn	AD/N1	43	0.96 ± 0.06	1.82 ± 0.12	69.12 ± 67.43
HIMTb	Cyprus	5	0.98 ± 0.02	1.28 ± 0.08	45.95 ± 0.77
HIMTa	Cyprus	14	1.00 ± 0.05	1.77 ± 0.13	48.05 ± 4.29
Foy 3.2	AQ/3	1	1.25 ± 0.00	0.25 ± 0.00	52.67 ± 0.00
Foy 3.2	FC/3	4	1.30 ± 0.34	1.23 ± 0.70	58.19 ± 4.50
Foy 3.2	CL3	3	1.32 ± 0.48	1.40 ± 0.85	54.40 ± 3.11
Foy 3.2	Group III	2	1.33 ± 0.12	0.63 ± 0.53	52.08 ± 0.82
Foy 3.2	Group II	4	1.53 ± 0.17	1.79 ± 0.46	57.67 ± 4.25

Some Foy 3.2 glasses have similar values of Nd/La as Foy 2.1, while some have higher values. Data sources: [6,12,14,19,21,25,26,32–34].

The different geology is manifested also in REE correlations. In Foy 3.2, La, Ce, and Pr are not correlated, whereas in Foy 2.1, all REE elements and yttrium are strongly correlated (typically $R^2 > 0.8$), while for La, Ce, and Pr are to a lesser degree with Th (>0.6), Hf (0.29–0.60), Zr (≥0.3), Nb (≥0.48), and Ti (0.22–0.47), indicating the presence of heavy minerals (e.g., allanite). Contrary to CL3, high correlations between titanium, niobium, and tantalum in Foy 2 from Cyprus indicate the presence of minerals like rutile [11].

As mentioned, Foy 3.2 has two distinct Nd/La ratio spans, Nd/La < 1 (groups I and IV) and Nd/La > 1 (groups II and III). Foy 2.1 glasses share this range, averaging from 0.92 to 0.95. For comparison, Belus river sand has Nd/La = 1 [35]. Group I is similar to Foy 2.1 in this regard, but can still be differentiated by a lower Zr/TiO$_2$ (around 50 versus 75). Group IV is distinguished from Foy 2.1 by a lower Nd/La and especially by a much lower Ce/La ratio (around 0.5 compared to above 1.5). Groups II and III have higher Nd/La than Foy 2.1 and HIMT glasses, above 1.2 compared to less than 1. It seems that Nd/La might be a good marker for differentiating Foy 3.2 from Foy 2.1 sand provenance, but caution should be kept because of the small number of available measurements of REEs in Foy 3.2 glasses. The ratio of Ce/La [11] for groups III and IV is on average lower than in Foy 2.1. The proxy for heavy minerals Zr/Ti, characterizing Egyptian sands, also differentiates these glasses from the HIMTa and HIMTb groups.

To determine the recycling degree of série 3.2-type glasses, we analyzed 20 raw glasses from the entire collection of 246 glasses (S1 Foy 3.2). We first eliminated glasses with values of Cu, Sb, and Pb > 100 ppm (criteria of Foy et al., 2003), then from the remaining glasses, we eliminated those with obviously elevated antimony (Sb > 10 ppm), obtaining 13 raw glasses that can be considered unrecycled. Cutoff values of the unrecycled raw glass were then determined as the mean values plus two standard deviations and were rounded up. The obtained cutoff values were 10 ppm for Sb, 40 ppm for Co, 50 ppm for Ni and Zn, 60 ppm for Pb, and 70 ppm for Cu. Above these levels, série 3.2-type glass can be considered surely recycled, and below it is modestly recycled or unrecycled.

Applying the obtained values to manufactured glasses, we obtained 54 glasses with means of 7 ppm, 26 ppm, 13 ppm, 1 ppm, and 17 ppm for nickel, copper, zinc, antimony, and lead, respectively. Eliminating the glasses with values of the recycling indicators above the mean plus two standard deviations, we obtained 35 glasses. The means plus two standard deviations of the recycling indicators can be considered as the cutoff values for pristine série 3.2-type glass. These values were 8 ppm, 37 ppm, 21 ppm, 2 ppm, and 29 ppm for nickel, copper, zinc, antimony, and lead, respectively.

Using these criteria, 14.3% of Foy 3.2 can be considered pristine, 47.1% is moderately recycled, and 38.6% surely recycled. The série 3.2 and the pristine Foy 3.2 are compositionally very similar, except for the pristine being somewhat lower in lime (6.7% versus 7%), iron (0.62% versus 0.7%), antimony (1 ppm versus 18 ppm), and lead (16 ppm versus 179 ppm). It is also mildly higher in zirconium (62 ppm versus 57 ppm) and titanium (0.1% versus 0.09%). These differences reflect a degree of recycling with other types of glasses, such as Roman Sb glass. Similar calculations were performed for pristine Levantine I glass from Cyprus, obtaining 3 ppm for cobalt and copper, and 10 ppm for zinc and lead [11]. This shows that glass of série 3.2-type was manufactured using sand somewhat richer in minerals than the Levantine, but it can nevertheless be considered quite clean.

3. Characteristics of Iron-Rich Foy 2.1 Glass and Source of Increased Iron

An extensive literature search yielded 125 glasses of Foy 2.1 type with elevated iron (Table 3, S2 Fe-rich Foy 2.1 correl). They are grouped in three groups regarding iron oxide concentrations: 47 high iron glasses (HI); 74 very high iron glasses (VHI); and 4 very-high iron, low lime glasses (VHILL). Apart from considerably higher average iron (2.29% versus 1.03%) iron-rich glass is also differentiated from série 2.1 by mildly higher alumina (2.75% versus 2.49%), and lower lime (7.42% versus 7.97%) and manganese (1.11% versus 1.69%). The VHILL was differentiated from VHI by lower lime (5.43% and 7.48%, respectively) and strontium (433 ppm versus 715 ppm), and lower magnesium (0.98% versus 1.32%). Taking into account two standard deviations and minimal/maximal values, we could conveniently draw the iron concentration limit between high and low iron Foy 2.1 to 1.3%, and between the HI and VHI to 2%. Similarly, we defined the lime limit between VHI and VHILL to 6%, and the upper titanium level of the entire high-iron glass to 0.2%.

Table 3. Foy 2.1 glass grouped according to iron concentrations.

	LI Série 2.1		HI		VHI		VHILL		Total Fe-Rich	
	n = 38		n = 47		n = 74		n = 4		n = 125	
wto%	aver.	std	aver.	std	aver.	std	aver.	std	aver.	std
Na_2O	18.46	1.26	17.67	2.76	17.63	1.19	17.43	0.62	17.78	1.20
MgO	1.21	0.15	1.28	0.29	1.32	0.21	0.98	0.11	1.28	0.22
Al_2O_3	2.49	0.13	4.02	9.06	2.78	0.26	2.67	0.18	2.75	0.26
SiO_2	64.50	1.08	63.81	9.55	64.61	1.22	67.88	0.79	64.93	1.48
K_2O	0.76	0.14	1.00	0.87	0.91	0.17	0.66	0.03	0.89	0.17
CaO	7.97	0.65	7.34	1.21	7.48	0.59	5.43	0.20	7.42	0.68
TiO_2	0.16	0.02	0.18	0.13	0.16	0.01	0.16	0.02	0.16	0.02
MnO	1.69	0.31	1.06	0.43	1.14	0.25	1.11	0.12	1.11	0.32
Fe_2O_3	1.03	0.11	1.64	0.21	2.69	0.41	2.57	0.15	2.29	0.61
ppm										
NiO			14	9	30	15	27	17	27	15
CuO	75	25	92	27	115	47	83	12	106	43
ZnO			60	110	48	18	46	14	46	16
SrO	798	86	701	147	715	99	433	222	705	120
ZrO_2	111	14	108	24	117	18	100	67	114	20
SnO_2			16	13	8	8	1	1	10	10
PbO	130	140	87	70	106	92	20	36	97	86
Sb_2O_3	188	289	155	75	134	119	12	13	135	107

Major and minor elements (wt%), and trace element concentrations (ppm) of Foy 2.1 glass. Groups: LI—low iron (Foy série 2.1 with $Fe_2O_3 \leq 1.3\%$); HI—high iron; VHI—very high iron; VHILL—very high iron low lime. Empty entries—data not available. Data sources: [1–3,5,6,12,13,23,24,28,32].

There are three glasses with TiO_2 contents higher than the rest of the Fe-rich glasses, averaging 0.16 and 0.23%, respectively (Figure 5a), but their high lime contents (7.8% on average) disqualified them from being classified as HIMT. The frequency of iron and titanium concentrations was also estimated by the kernel-density estimate (KDE) [36].

The optimal value of the bandwidth parameter (h) was sought by trial and error. In Figure 5b, we present distributions for three different values of h, for which we estimated the most representative for the density variation. The diagram shows three peaks in Fe_2O_3/TiO_2 distribution, centered at around values of 12, 14, and 18, respectively, indicating different mineral compositions. However, we need more glasses of this type to make firmer conclusions. Iron in Fe-rich glasses is not in correlation with trace elements indicative of recycling (NiO, CuO, ZnO, SnO_2, Sb_2O_3, and PbO), nor with pollutants (SO_3). It also does not have a negative correlation with Cl, which would indicate recycling, so its source needs to be found elsewhere.

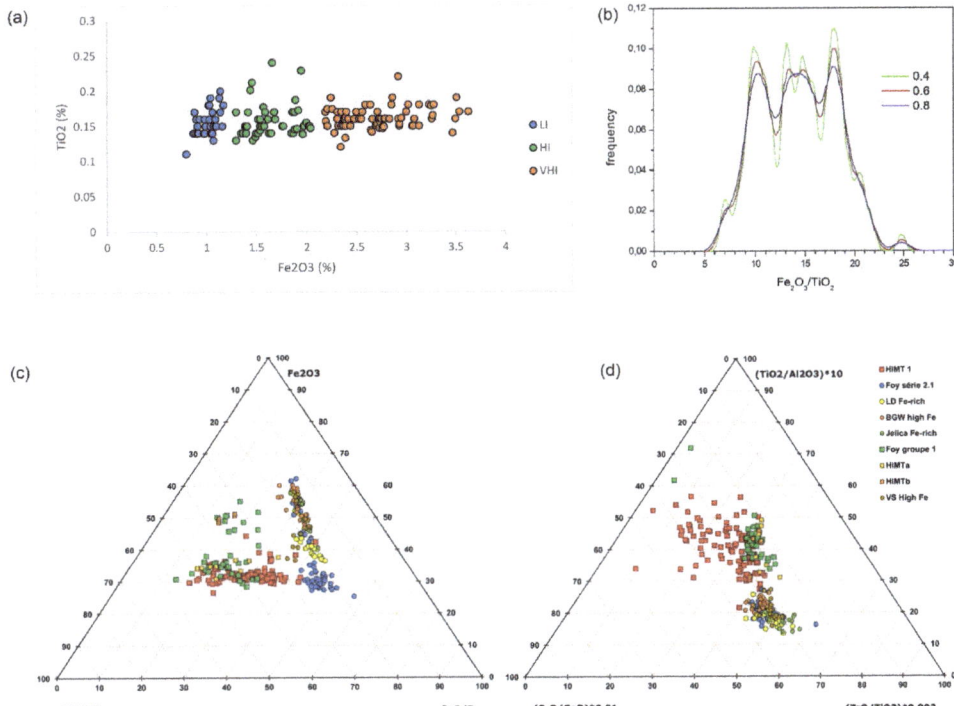

Figure 5. Iron and titanium distributions (up) and comparison between Fe-rich Foy 2.1 and similar late antiquity glass groups (down). Bi-plot of Fe_2O_3 and TiO_2 for Fe-rich Foy 2.1 glasses, including série 2.1 (**a**). Kernel density estimation for Fe-rich glasses, for three values of parameter (h) (**b**). Comparison between Fe-rich Foy 2.1 and similar late antiquity glass groups (lower). Compositional comparison of Fe-rich Foy 2.1 glasses with similar late antiquity glass groups (**c**). Provenance indicators based on sand mineralogy (**d**). Fe-rich Foy 2.1 glasses outlined in black. Values scaled to magnify and center the distributions.

Compared to other glass groups with elevated iron, Fe-rich Foy 2.1 is equal in iron to Foy 1 (2.29% versus 2.28%) [3]. However, its considerably lower titanium (0.16% versus 0.49%) and zirconium (114 ppm versus 216 ppm) differentiate the two. Its iron levels are intermediate between HIMT 1 (1.36%, [37]) and very high-iron HIMT groups like IIa Dichin 2.91% [38], AQ/1a 3.23% [21], FC/1a 3.69% [26], and CL1a, 3.38% [19]. HIMT glasses are analogously differentiated by iron levels (HIMTa and HIMTB, [32], also reported by [19,21]). While Foy 2.1 HI and HIMTa are on average comparable in iron (1.64% versus 1.81%), VHI is lower than HIMTb (2.69% versus 3.55%). In addition, in HI and VHI groups, iron and titanium are not correlated, contrary to HIMTa and HIMTb, indicating different mixtures of heavy minerals in the sand (like zircon, pyroxenes, and amphiboles).

Correlations of strontium with manganese (0.54) and of manganese with SrO/CaO ratio (0.6) in the HI group indicate that a part of its strontium comes from manganese ore. The SrO/CaO trendline extrapolates backwards at around 70, above the values characteristic of low-manganese Levantine glass (45–55). However, a part of Sr in Egyptian sands might also come from other Sr-bearing minerals or from the diagenetic alteration of aragonite into calcite [11]. The ratio of SrO/CaO in VHI is 88, intermediate between Egypt (around 70), and HI (98). VHILL is lower in Ba and Sr than VHI. As they have different Sr/Ca ratios (80 versus 96), this difference might come from sand with different amounts of Ba-bearing feldspar and Sr bearing minerals like witherite and strontianite.

The correlation of zirconium with titanium of the HI group is similar to the Egyptian glass (0.63 versus 0.56, [39]), but its Zr/Ti ratio is smaller (54 in HI and 59 in VHI versus 84 in Egypt), indicating different heavy minerals suites. Positive correlation Zr-Sr (0.58) is not congruent with regional geology, exhibiting zero or negative values. It is comparable to the Bulgarian Fe-rich (0.42), Cypriot Foy 2 (0.45), and Foy 3.2 (0.53). This indicates possibly different sand sources for these groups.

Figure 5c depicts some of these considerations. The iron-rich Foy 2.1 glasses are separated from HIMT groups by lower titanium and higher lime. Considering provenance, the TiO_2/Al_2O_3 ratio, differentiating heavy mineral-rich Egyptian sands from feldspar-rich Levantine sands [10], is compared with ratios of zirconium to titanium, characterizing regional sands, and of strontium to calcium, differentiating coastal from inland sands. Provenance regions for HIMT and Fe-rich Foy 2.1 are also clearly separated (Figure 5, right). This is further supported by their different Ce/La ratios of 1.51 and 1.2, respectively.

As mentioned by Ceglia et al. [32], the Ce/La ratio can be used for sand characterization. In this sense, it is noteworthy that in Fe-rich Foy 2.1 glasses, the Ce/La ratio decreases with the increase of iron, manifesting a strong negative correlation of 0.8 in Byzantine glass weights (BGW) and 0.85 in Cypriot Foy 2. The negative correlation of Ce/La with iron is also seen in HIMTa and HIMTb, although less pronounced (0.7) and with higher Ce/La ratios for the same concentration of iron, indicating different sand sources for iron-rich Foy 2.1 and HIMT glasses. Another possible provenance indicator is Ce/Gd, the ratio between the most abundant light and heavy REE element. Light and heavy REEs have a different hydrothermal mobility and reflect different hydrochemical and geochemical processes [40,41]. For VHI/VHILL, Ce/Gd is around 6.9, lower than for Levantine, Foy 2, and Egypt1 (all above 9), and comparable to HIMTb (6.5). This ratio is also negatively correlated with iron. In addition, Fe-rich Foy 2.1 glasses are differentiated from HIMTa and HIMTb by lower zirconium (around 100 ppm compared to above 200 ppm) and hafnium (2.1–2.4 ppm compared to 5.2 and 5.7 ppm), while the correlation of hafnium with zirconium varies more (0.66–0.98 compared to 0.98 and 0.96). The Byzantine glass weights (both Foy 2 and Fe-rich Foy 2) have strong correlations between REEs, while Foy 2 from Cyprus is mainly between the LREE. In addition, Ce and Gd are less correlated with other REEs in Fe-rich than in Foy 2 Byzantine glass, indicating different heavy minerals suits.

Regarding the origin of high iron, several assumptions can be made. Emphasized recycling signs allow for the hypothesis that this glass is perhaps manufactured by mixing low-iron Foy 2.1 with some contemporary available high-iron cullet. For example, mixing it with 60% glass similar to Dichin 2b HIMT or with 70% glass similar to Cyprus HIMTb yields adequate values of iron, alumina, and strontium, but higher titanium and zirconium and lower lime than expected. The composition of the hypothetical cullet needed to obtain Fe-rich Foy 2.1 composition, in the proportion of 70% of cullet and 30% of low-iron Foy 2.1 (similar percentages are reported by Silvestri 2008), comprises 3.6% iron, 0.17% titanium, 3% alumina, 7.3% lime, 128 ppm zirconium, and 727 ppm strontium. We are not aware of such contemporary glass. Other authors also exclude mixing with HIMT based on REE patterns [6].

Another hypothesis is the exploitation of specific, iron-rich sand quarries from the geologically related area, as evidenced by the similarity of their respective REE patterns (e.g., between Foy 2 and Fe-rich Foy 2 glasses from Cyprus and LI and HI glasses from Visighotic Spain). The case of different iron concentrations in Hambach glass factories [42,43], ranging from 1.4% to 1.8% to 2.2% on average, further makes this hypothesis plausible.

Technological interpretations for increased iron should also be considered, like contamination from oxidized scales from iron blow-pipes that forms during glassmaking [44], but the difference in trace elements ratios does not support this hypothesis.

Another candidate is the color-branding hypothesis [45]. This states that the HIMT glass was deliberately tinted by manganese ore to color-brand its superior working characteristics. Two different manganese ores were used according to the hypothesis, with different iron to manganese ratios, yielding two different types of glass, HIMTa and HIMTb, that are differentiated by Fe_2O_3/Al_2O_3 and Fe_2O_3/TiO_2 and Ce/La ratios [11,32].

Figure 6 shows a diagram of Fe_2O_3 versus MnO for Foy 2.1 assemblages from Visighotic Spain (upper) and Byzantine glass weights (middle). The low iron glasses form two distinct groups: lightly colored (mostly colorless, bluish and greenish) low iron glasses (LI cbg) and LI colorless glasses. The colorless glasses are decolored with manganese (MnO/Fe_2O_3 = 1.67). The very high iron glasses are darker colored, green and yellow (VHI yg). Lightly colored LI glasses lie on the same correlation line with darker colored VHI glasses (0.82). There is no such correlation for HI glasses. In the BGW collection, lightly colored LI glasses (different shades of aqua) fall on the manganese−iron correlation line with high iron olive and yellowish glass from the same collection (R^2 = 0.51). Considering only olive-green HI glasses, the correlation is more pronounced (0.59). Colorless low iron glasses from BGW assemblage are decolored with manganese (MnO/Fe_2O_3 = 1.7). Some low iron glasses in BGW assemblages are of an amber color (not shown for clarity). Their average manganese to iron ratio is similar to the decolorized glasses (1.65), but might derive their color from ferri-sulphide. The pronounced manganese−iron correlations between pale low iron glasses and darker high iron glasses in these two collections supports the color branding hypothesis.

A strong correlation between Fe_2O_3/Al_2O_3 and Fe_2O_3/TiO_2 ratios for HIMTa and HIMTb glasses from Cyprus, noticed by Ceglia and colleagues, also exists for the set of low iron and high iron Foy 2.1 glasses. (Figure 6, lower). The diagram shows this correlation between the low iron série 2.1 glasses of all colors and the most common Foy 2.1 high iron glasses, olive and yellowish. Different correlation lines of Foy 2.1 and HIMT glasses suggest that different manganese ore was used for color branding.

Figure 7 shows normalized REE patterns for glasses from Lower Danube (LD), Visigothic Spain (VS), and Byzantine glass weights (BGW), grouped by iron concentrations. With an increase in iron, the concentrations of REEs in all three compared datasets increased. An exception to this is HI glass compared to VHI glass from the Visighotic collection, where cerium and hafnium decrease with an increase of iron concentrations in high iron glasses, indicating a different type of mineral iron. Another discrepancy is the decrease of barium with an increase of iron in the same assemblage and the Lower Danube collection. The observed tendency is related to strontium. Correlations of barium with strontium are around 0.8 for HI glass from Lower Danube and Visigothic Spain, respectively.

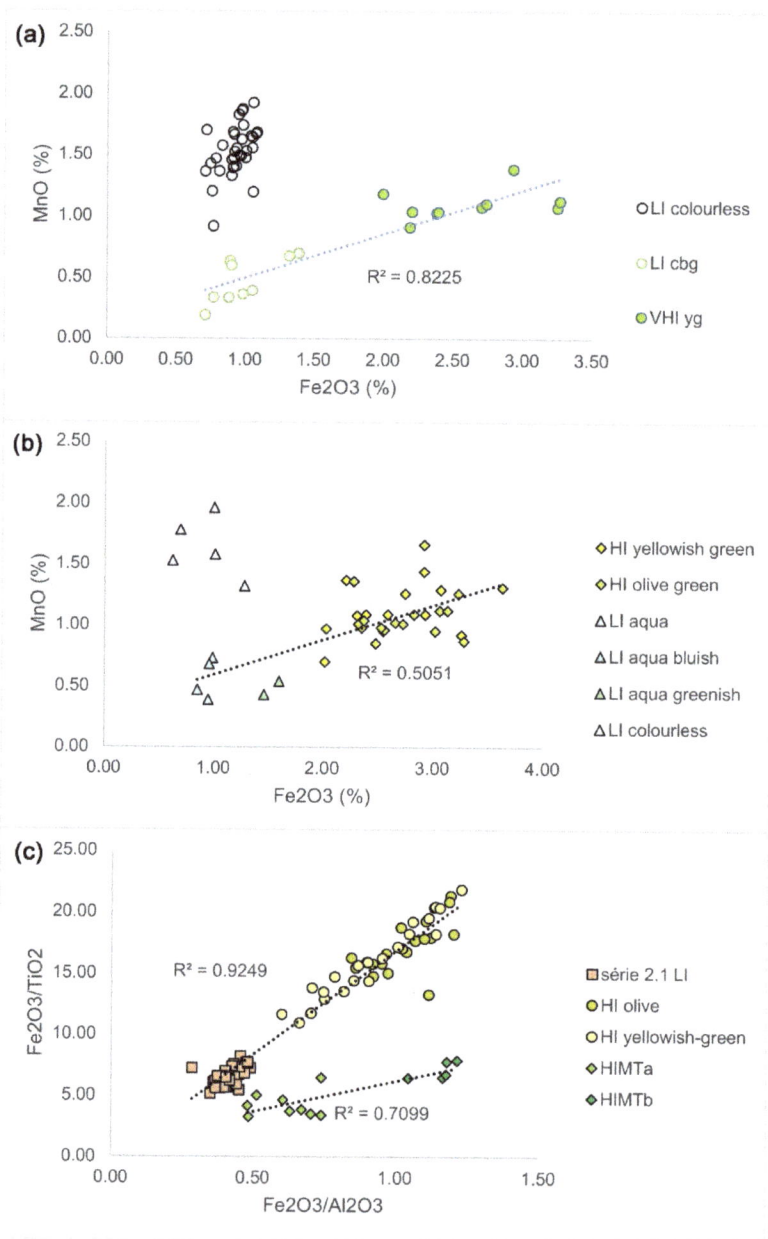

Figure 6. Color-branding hypothesis. Correlation between Fe_2O_3 and MnO for Foy 2.1 glasses from Visighotic Spain (VS) (**a**), Byzantine glass weights (BGW) (**b**); Fe_2O_3/Al_2O_3 versus Fe_2O_3/TiO_2 diagram of low iron Foy 2.1 and all high iron Foy 2.1-type of glasses, grouped according to color (**c**). HIMTa and HIMTb glasses from Cyprus are given for comparison. Data sources: [3,11–13].

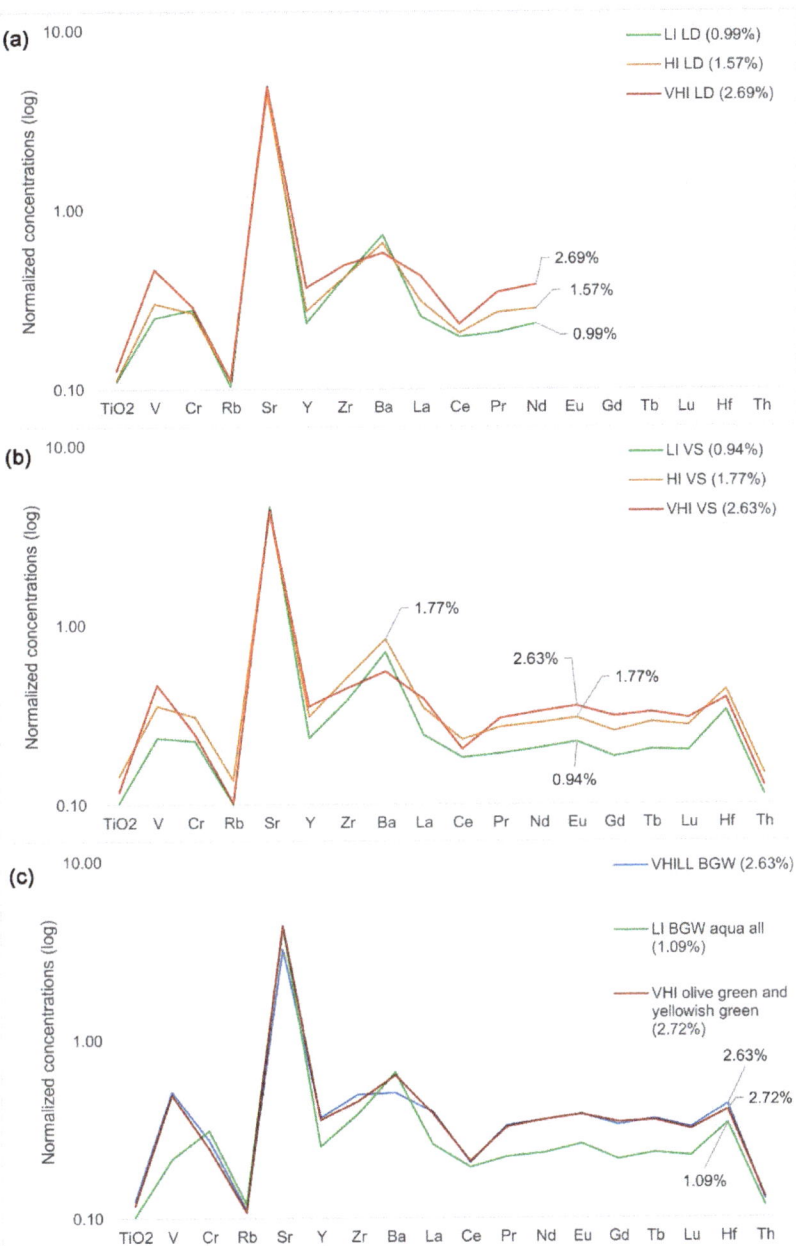

Figure 7. Trace element patterns of Fe-rich Foy 2.1 glasses, grouped by iron concentrations. Values normalized to the upper continental crust [46]. Groups from Lower Danube (**a**), Visighotic Spain (**b**), and Byzantine glass weights (**c**). Note that for the Lower Danube REE dataset, only La–Nd measurements are reported. Data sources: [6,12,13].

The REE patterns of low iron and high iron Foy 2.1 glasses from Visighotic Spain (Figure 7, middle) are very similar, which implies that differences in their heights are primarily related to iron concentrations (0.94% versus 1.77%). This implies a similar silica source but with higher concentrations of iron-bearing minerals, in other words, iron-based coloration. However, the REE pattern for VHI glass from the same collection is different from the HI REE patterns, which suggests different mineral compositions or the deliberate addition of some minerals to the glass-making mixture. Adding iron-bearing manganese ore, used for color branding, might account for this difference in REE concentrations. This would suggest that HI glasses from Visighotic Spain are colored naturally while VHI glasses are color branded. Indeed, almost all VHI glasses from this set are green-yellow, which is not so for the HI glasses from the same collection. Likewise, VHI Byzantine glass weights also seem color branded. Another sign of this is the average correlation of 0.47 between manganese and REEs for the set of olive and yellowish-green VHI and aqua LI BGW, quite similar to the correlation between manganese and iron for the same set (0.51). The relative decrease of cerium in these glasses with a higher iron can be accounted for by the fact that iron is not correlated with cerium, while it is well correlated with other REEs for the same collection (0.74).

This leaves us with two plausible hypotheses for the origin of high iron in Foy 2.1 glass, namely: variability of sand source and the color-branding. Both might even be correct, depending on the particular glass group. The candidates for the former are HI and LD VHI groups, with LD VHI having strong correlations of LREEs with iron and magnesium, and none with manganese, indicating that REEs are related to the heavy mineral fractions of the sand. The candidate for the latter is Fe-rich Byzantine glass weights, where iron and magnesium are not correlated with REEs. They are almost all olive green or yellowish-green (contrary to Foy-2 glass weights that are of many different colors), further supporting the color-branding hypothesis.

4. Discussion

4.1. Changes in Distribution of Glass Types from Fourth to Sixth Century AD in the Balkans

Changes in the proportion of glass types from the fourth to the sixth century in the central and eastern Balkans are depicted in Figure 8 (upper). The first change, from the fourth to the fifth century, is characterized by the disappearance of the Roman glass, an increase in the proportion of HIMT type, and the appearance of some Levantine type of glass in the record (12.2%). At the same time, série 3.2 remains an almost constant and dominating type throughout this period (changing from 50% to 51.2%).

Another change, from the fifth–sixth century, is characterized by the disappearance of HIMT and Levantine types and the appearance and almost total domination of a new type, série 2.1 (almost 90%), with the rest being série 3.2. It is plausible to explore if this change might be correlated with turbulent events of the fifth century in the Balkans, brought about by Hunnish plunder. The sixth century brought different kinds of change compared to the fifth. HIMT and Levantine types virtually disappeared, and a new type appeared that would dominate (85.7%). The older type, série 3.2, strongly diminished (9.2%). It would be plausible to explore if these changes happened gradually or abruptly.

The sixth century for northern Italy shows quite a different pattern from the contemporary central and eastern Balkans (Figure 8, middle). Contrary to the Balkans, with the strong domination of one type, there is a more balanced proportion of several types in Italy: Roman (37.9%), HIMT (29.9%), equal amounts of série 3.2 and série 2.1 (10.3%), and Levantine (5.7%). This might indicate that the Italian glass markets were more diversified than the contemporary Balkan ones, or that Balkan glass import was perhaps more centralized than in Italy. The apparent scarcity of the Levantine in the sixth century inner Balkans (2.6%) is notable and yet to be explained.

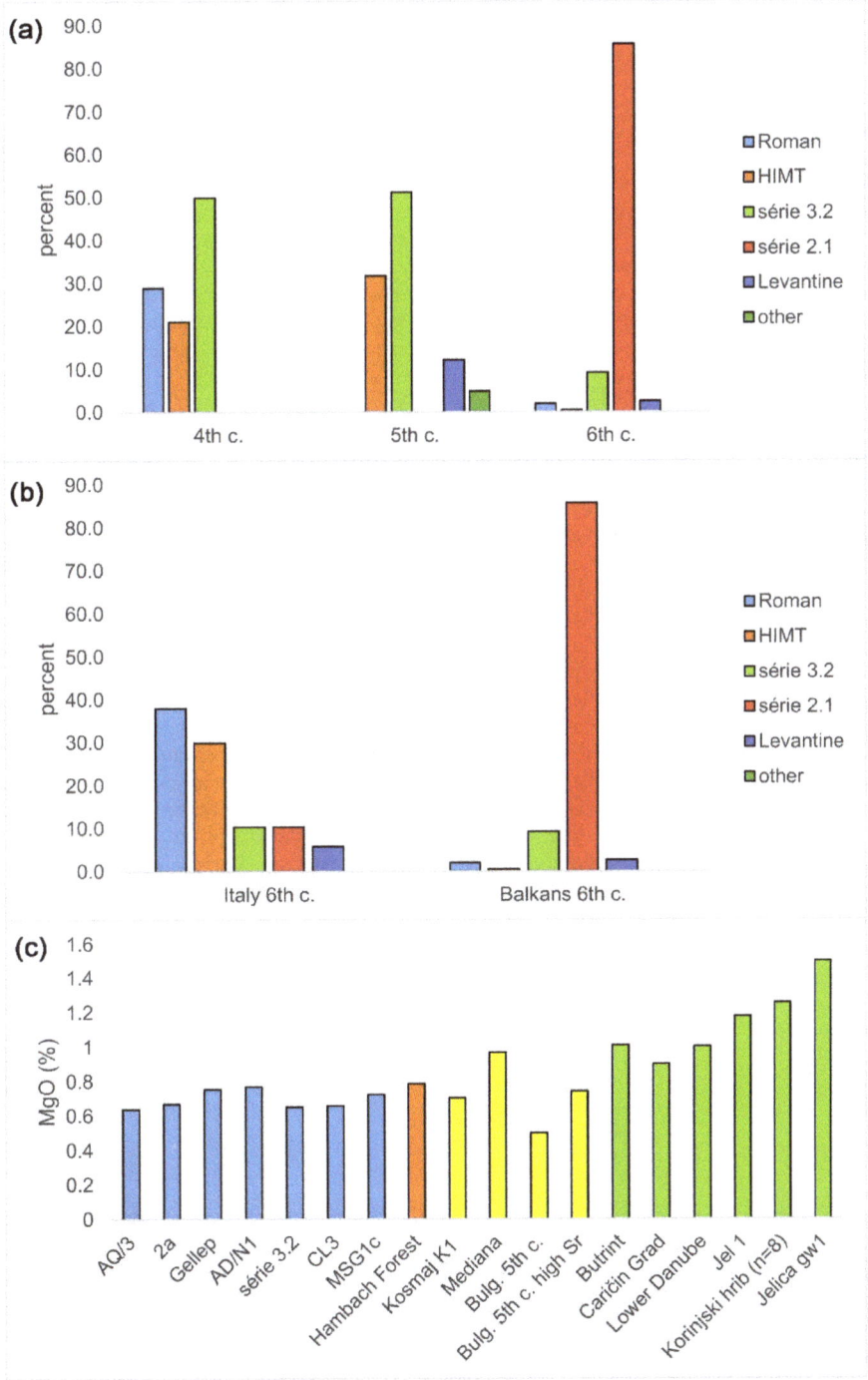

Figure 8. Relative frequencies of glass types by time and region and MgO concentrations by location. Relative frequencies of glass types from fourth century (38 glasses), fifth century (82 glasses), and sixth

century contexts (196 glasses) from central and eastern Balkans (**a**). Relative frequencies of compositional types among glasses from sixth century contexts from the central and eastern Balkans (196 glasses from three locations) and northern Italy (87 glasses from two locations, (**b**). MgO contents in série 3.2-type glasses (**c**)–from the fourth to eight century western Mediterranean and Europe (blue), second to fifth century Balkans (yellow), and sixth century Balkans, in all 213 glasses from 18 different groups (green). Data sources: [1–10,14,17,19,23,25,26,28–30,33,34,47–49].

The continuous presence and the widespread findings of several major glass types during the fourth and the fifth centuries suggests that the Balkans imported glass artefacts and raw glass regularly from various Eastern Mediterranean production centers, implying regular economic activity. This further suggests that the traditional view of the dramatic economic decline during the late antiquity turmoil in peripheral parts of the Balkans might not be completely accurate. The analogous situation is described in the late antiquity Carthage after the Vandals conquest [49].

4.2. Evolving Chemical Composition of Série 3.2

More than 200 glasses of Foy 3.2 type that are now recognized (compared to the 19 originally described) give strong evidence to the hypothesis that série 3.2 was not limited to the turn of the fifth–sixth centuries and France and Tunisia [3]. It also weakens the arguments to include them under the umbrella name Foy 2 [10]. This type spread across the Mediterranean and beyond, to Germany, continental France, and Britain. It lasted from the second to the beginning of the seventh century, peaking during the fifth and the sixth centuries AD (Figure 9, left, right). It is distributed mostly around the Adriatic (Figure 9—middle), in Italy and the Balkans, highlighting the importance of the Adriatic trade route for the import of this type of glass and reflecting analogous findings regarding the distribution of Ca-rich HIMT, HIMT, and Foy 3.2 types [19,30]. This, together with the wide presence of Foy 3.2 glass across central parts of the Balkans, suggests that at least some Foy 3.2 raw glass was imported to the central Balkans over the Adriatic. However, during the sixth century, Foy 3.2 glasses in the Balkans showed a significant increase in the average MgO concentrations (1–1.25%) compared to the fifth century (Figure 8, lower), and to Foy 3.2 glasses outside the Balkans (0.6–0.7%), implying different sand quarries and, possibly, different trade routes for its import. This might reflect a marked shift in the Byzantine trade routes during the sixth century. The question of Foy 2.1 raw glass import is not clear. It could have been imported over the Adriatic, also over the Aegean and the Black Sea ports such as Odessus.

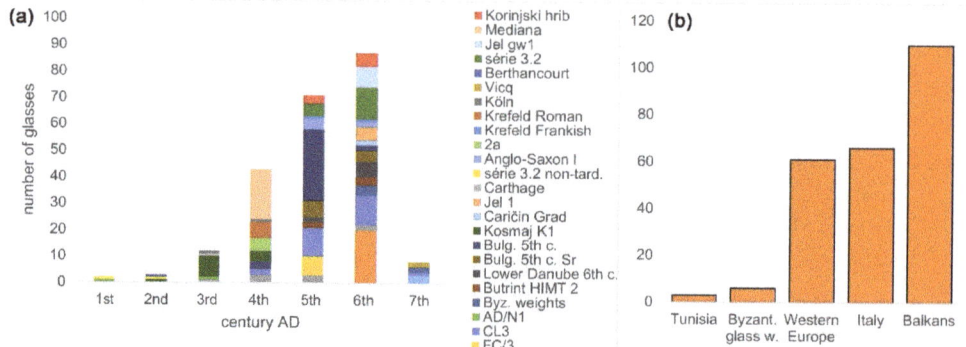

Figure 9. Distribution of série 3.2 in time and region. Histogram showing number of glasses belonging to the Foy série 3.2 against time, divided to particular groups (**a**). The absolute number of glasses of série 3.2 by the region (**b**). Data sources: [1,3–10,12,14,17,19,24,28–30].

Foy 3.2 type was manufactured from the second century on, from geologically similar sand as Roman-Sb glass. Its production gradually increased through the period when manganese was replacing antimony as a decolorant and peaked during the fifth and the first half of the sixth century. Its production decreased rapidly during the second half of the sixth century, simultaneously to a rapid increase in Foy 2.1 production, and ceased altogether during the seventh century.

One explanation might be that this change took place when the search for new sand quarries finally led to the geologically different area, characterized by higher alumina, lime, iron, and magnesium. This hypothesis is built upon the observation that primary glass workshops were located in the vicinity of sand and wood resources, and that workshops regularly moved from place to place once resources were exhausted [31]. While it is well understood that the glass composition predominantly reflects the primary glassmaking source, rather than the secondary workshop, this model does not account for the probable and frequent local migration of workshops in search of raw materials. In Bet Eli'ezer (Hadera), 17 furnaces were in use for a year or two, and then moved to another location in the same area, in search of wood and sand [50]. It is thus reasonable to suppose that the primary glass composition would reflect these frequent location changes in somewhat increased compositional spread compared to a single workshop—single sand quarry production model. Therefore, as long as sands derived from the same geological process are exploited, this would result in the same primary glass type. The changes of glass composition in time reflecting such activities are termed compositional "evolution". However, if at a particular moment in time, the quest for resources leads to exploiting sand with geologically markedly different characteristics (reflecting the different geological processes), we would all of a sudden have another basic type of primary glass, even from the same furnace. In such a case, a small geographical step might have led to a quite different type of primary glass.

To try to terminologically capture this complex picture of ever-varying exploitation conditions, we propose the term "generic composition/type" or "(geochemical) class" to glass manufactured from all quarries possessing a similar geological composition. Reflecting further the Linnaean approach, the term "family" denotes glass manufactured from all batches produced from the same sand quarry, and "species" denotes glass manufactured from a single batch.

4.3. Distribution of Fe-Rich 2.1

Percentages of iron-rich glasses among Foy 2.1 type vary considerably from site to site, with a 25% overall average (Table 4). Taking into account only more numerous collections, the values span from 5.6% in Cyprus to 47.7% in Odartsi. Summing and comparing by regions, the values range from 5.6% in Cyprus to 70% in Lower Rhine (Figure 10, upper). Note, however, that only the Balkans and Visighotic Spain collections were more numerous, and in these two regions, the percentages were around 30%, which is close to the overall average of 25%. There is no notable correlation between the percentage of high-iron glass among Foy 2.1 type at some particular locations and its geographical distance from Egypt, where it was probably produced. This would indicate that this type of glass was freely exported and widely popular across the Mediterranean.

Table 4. Percentages of iron-rich among Foy 2.1 glasses, by archaeological site or collection.

Site/Collection	Type of Object	Fe-Rich Foy 2.1	Total Foy 2.1	Percent Fe-Rich
Jelica	windowpane, glassware	16	63	25
Caričin grad	raw glass	2	24	8
Serdica	glassware, raw glass	4	17	23
Odartsi	glassware	21	44	48
Dichin	glassware, windowpane	5	19	26
Cyprus	not specified	2	36	6
Krefeld-Gellep	glassware, bead	5	8	62

Table 4. *Cont.*

Site/Collection	Type of Object	Fe-Rich Foy 2.1	Total Foy 2.1	Percent Fe-Rich
Köln	glassware	2	2	100
Crypta Balbi	glassware, windowpane	4	8	50
Bordeaux	raw glass	3	5	60
Marseille	glassware, raw glass, debris	1	6	17
Gémenos	glassware	4	4	100
Nabeul	glassware	4	6	67
Sidi Jdidi	glassware	1	2	50
Maguelone	glassware, raw glass, debris	1	20	5
Visigothic Spain	glassware	19	62	31
Byzantine glass weights	glass weights	31	174	18
Total		125	500	25.0

Data sources: [1–3,5,6,11–13,23,24,28].

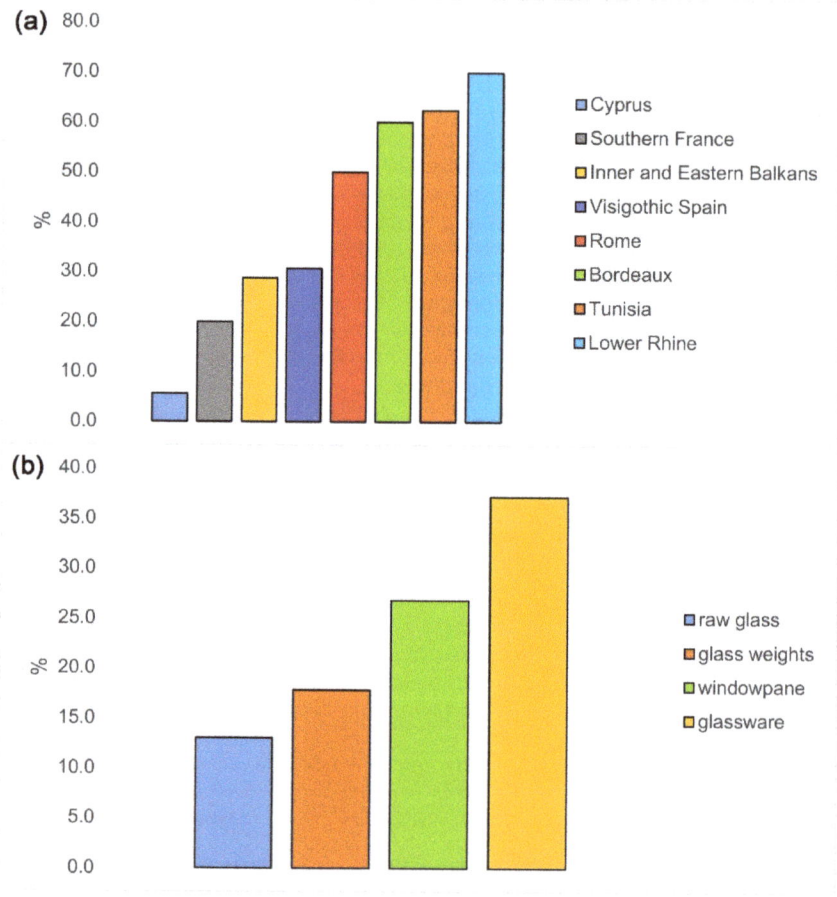

Figure 10. Percentages of Fe-rich glasses by region and type. Percentages of Fe-rich glasses among Foy 2.1 glasses by the region (**a**); percentages of Fe-rich glasses among Foy 2.1 glasses by object type (**b**). Data sources: [1–3,5,6,11–13,23,24,28].

The percentages of high-iron samples are lower in the raw glass than in the manufactured glass (13% compared to 26.4%). This might suggest that iron was being added predominately, albeit not exclusively, in the secondary workshops, i.e., for color branding. Among the manufactured glass, the percentage is greater in glassware than in windowpane glass, perhaps also for marketing purposes (Figure 10, lower).

Glass products with elevated iron, color-branded or sand-derived, covered virtually the entire Mediterranean, from Cyprus, over the Balkans, Tunisia, Italy, Spain, France, and beyond, to inland France and Germany. The regional distribution might have been influenced by economic factors such as local purchasing power, as illustrated by the case of the dominance of more expensive Levantine glass in Carthage and Cyprus [10,32] compared to the inner Balkans. The production of this type of glass peaked during the sixth and seventh century AD, but lasted perhaps for some time before and after this period, showing the lasting popularity and competitiveness of this glass, apparently more economical in comparison to the more expansive and luxurious Levantine.

Supplementary Materials: The following supporting information can be downloaded at: https://www.mdpi.com/article/10.3390/ma15031086/s1; Table S1: Foy 3.2 recycling.

Author Contributions: Conceptualization, R.V.B. and Ž.Š.; methodology, R.V.B.; software, Ž.Š.; validation, R.V.B. and Ž.Š.; formal analysis, R.V.B.; R.V.B.; writing—review and editing, R.V.B.; visualization, R.V.B. and Ž.Š. All authors have read and agreed to the published version of the manuscript.

Funding: This research received no external funding.

Acknowledgments: Roman Balvanović acknowledges the support by the Ministry of Education, Science, and Technological Development of Serbia through the project "Physics and Chemistry with Ion Beams", no. III45006. The work of Žiga Šmit was partly supported by the Slovenian Research Agency (research core funding no. P6-0283). Archaeological and Archaeometric Research of Portable Archaeological Heritage). The authors kindly thank Harlavan Yehudit of the Geological Survey of Israel, Division of Geochemistry and Environmental Geology, for her valuable advice regarding geological issues.

Conflicts of Interest: The authors declare no conflict of interest.

References

1. Balvanović, R.; Stojanović, M.M.; Šmit, Ž. Exploring the unknown Balkans: Early Byzantine glass from Jelica Mt in Serbia and its contemporary neighbours. *J. Radioanal. Nucl. Chem.* **2018**, *317*, 1175–1189. [CrossRef]
2. Balvanović, R.; Šmit, Ž. Sixth-century AD glassware from Jelica, Serbia—an increasingly complex picture of late antiquity glass composition. *Archaeol. Anthr. Sci.* **2020**, *12*, 1–17. [CrossRef]
3. Foy, D.; Picon, M.; Vichy, M.; Thirion-Merle, V. Caractérisation des verres de la fin de l'Antiquité en Méditerranée occidentale: L'émergence de nouveaux courants commerciaux. In *Échanges et Commerce du Verre Dans le Monde Antique: Actes du Colloque International de l'Association Française pour l'Archéologie du Verre*; Montagnac, Ed.; Mergoil: Drémil-Lafage, France, 2003; pp. 41–86.
4. Stojanović, M.M.; Šmit, Ž.; Glumac, M.; Mutić, J. PIXE–PIGE investigation of Roman Imperial vessels and window glass from Mt. Kosmaj, Serbia (Moesia Superior). *J. Archaeol. Sci. Rep.* **2015**, *1*, 53–63. [CrossRef]
5. Drauschke, J.; Greiff, S. Early Byzantine glass from Caričin Grad/Iustiniana Prima (Serbia): First results concerning the composition of raw glass chunks. In *Glass along the Silk Road from 2000 BC to AD 1000*; Zorn, B., Hilgner, A., Eds.; Verlag des Römisch-Germanisches Zentralmuseums: Mainz, Germany, 2010; pp. 53–67.
6. Cholakova, A.; Rehren, T.; Freestone, I.C. Compositional identification of 6th c AD glass from the Lower Danube. *J. Archaeol. Sci. Rep.* **2016**, *7*, 625–632. [CrossRef]
7. Cholakova, A.; Rehren, T. A Late Antiquity manganese-decolourised glass composition: Interpreting patterns and mechanisms of distribution. In *Things that Travelled: Mediterranean Glass in the First Millennium ce*; Rosenow, D., Phelps, M., Meek, A., Freestone, I., Eds.; UCL Press: London, UK, 2018. [CrossRef]
8. Conte, S.; Chinni, T.R.; Arletti, R.; Vandini, M. Butrint (Albania) between eastern and western Mediterranean glass production: EMPA and LA-ICP-MS of late antique and early medieval finds. *J. Archaeol. Sci.* **2014**, *49*, 6–20. [CrossRef]
9. Milavec, T.; Šmit, Ž. Analyses of glass from late antique hilltop site Korinjski hrib above Veliki Korinj (Slovenia). *Arheol. Vestn.* **2020**, *71*, 271–282. [CrossRef]
10. Schibille, N.; Sterrett-Krause, A.; Freestone, I.C. Glass groups, glass supply and recycling in late Roman Carthage. *Archeol. Antrhropol. Sci.* **2016**, *9*, 1223–1241. [CrossRef]

11. Ceglia, A.; Cosyns, P.; Schibille, N.; Meulebroeck, W. Unravelling provenance and recycling of late antique glass from Cyprus with trace elements. *Archaeol. Anthropol. Sci.* **2017**, *11*, 279–291. [CrossRef]
12. Schibille, N.; Meek, A.; Tobias, B.; Entwistle, C.; Avisseau-Broustet, M.; Da Mota, H.; Gratuze, B. Comprehensive chemical characterisation of Byzantine glass weights. *PLoS ONE* **2016**, *11*, e0168289. [CrossRef]
13. Ares, J.J.; Guirado, V.E.; Gutiérrez, Y.C.; Schibille, N. Changes in the supply of eastern Mediterranean glasses to Visigothic Spain. *J. Archaeol. Sci.* **2019**, *107*, 23–31. [CrossRef]
14. Gallo, F.; Silvestri, A.; Molin, G. Glass from the Archaeological Museum of Adria (North-East Italy): New insights into Early Roman production technologies. *J. Archaeol. Sci.* **2013**, *40*, 2589–2605. [CrossRef]
15. Silvestri, A. The coloured glass of Iulia Felix. *J. Archeol. Sci.* **2008**, *35*, 1489–1501. [CrossRef]
16. Silvestri, A.; Molin, G.; Salviulo, G. The colourless glass of Iulia Felix. *J. Archeol. Sci.* **2008**, *35*, 331–341. [CrossRef]
17. Jackson, C.M. Making colourless glass in the Roman period. *Archaeometry* **2005**, *47*, 763–780. [CrossRef]
18. Nenna, M.D.; Vichy, M.; Picon, M. L'atelier de verrier de Lyon du 1er siècle après J.-C, et l'origine des verre "romains". *Rev. d'Archaéométrie* **1997**, *21*, 81–87. [CrossRef]
19. Maltoni, S.; Chinni, T.; Vandini, M.; Cirelli, E.; Silvestri, A.; Molin, G. Archaeological and archaeometric study of the glass finds from the ancient harbour of Classe (Ravenna- Italy): New evidence. *Herit. Sci.* **2015**, *3*, 13. [CrossRef]
20. Rehren, T.; Brüggler, M. The Late Antique glass furnaces in the Hambach Forest were working glass-not making it. *J. Archaeol. Sci. Rep.* **2020**, *29*, 102072. [CrossRef]
21. Gallo, F.; Marcante, A.; Silvestri, A.; Molin, G. The glass of the "Casa delle Bestie Ferite": A first systematic archaeometric study on Late Roman vessels from Aquileia. *J. Archaeol. Sci.* **2014**, *41*, 7–20. [CrossRef]
22. Barfod, G.H.; Freestone, I.C.; Lesher, C.E.; Lichtenberger, A.; Raja, R. 'Alexandrian' glass confirmed by hafnium isotopes. *Sci. Rep.* **2020**, *10*, 11322. [CrossRef]
23. Mirti, P.; Lepora, A.; Saguì, L. Scientific analysis of fragments from the seventh-century glass Crypta Balbi in Rome. *Archaeometry* **2000**, *42*, 359–374. [CrossRef]
24. Velde, B. Aluminum and calcium oxide content of glass found in western and northern Europe, first to ninth centuries. *Oxf. J. Archaeol.* **1990**, *9*, 105–117. [CrossRef]
25. Silvestri, A.; Tonietto, S.; Molin, G. The palaeo-Christian glass mosaic of St. Prosdocimus (Padova, Italy): Archaeometric characterization of 'gold' tesserae. *J. Archaeol. Sci.* **2011**, *38*, 3402–3414. [CrossRef]
26. Maltoni, S.; Silvestri, A.; Marcante, A.; Molin, G. The transition from Roman to Late Antiquity glass: New insights from the *Domus of Tito Macro* in Aquilea (Italy). *J. Archaeol. Sci.* **2016**, *73*, 1–16. [CrossRef]
27. Foster, H.E.; Jackson, C.M. The composition of late Romano-British colourless vessel glass: Glass production and consumption. *J. Archaeol. Sci.* **2010**, *37*, 3068–3080. [CrossRef]
28. Wedepohl, K.H.; Pirling, R.; Hartmann, G. Römische und fränkische Gläaser aus dem Gräberfeld von Krefeld-Gellep. *Bonn. Jahrbücher* **1997**, *197*, 177–189.
29. Stamenković, S.Z. Glass Production Technology and Production Centres in Dacia Mediterranea. Ph.D. Thesis, University of Belgrade, Faculty of Philosophy, Department of Archaeology, Belgrade, Sebia, 2015. Available online: https://nardus.mpn.gov.rs/handle/123456789/4259?locale-attribute=en (accessed on 20 July 2021). (In Serbian).
30. Freestone, I.C.; Hughes, M.J.; Stapleton, C.P. The composition and production of Anglo-Saxon glass. In *Catalogue of Anglo-Saxon Glass in the British Museum*; Evison, V.I., Ed.; BMP: London, UK, 2008; pp. 29–46.
31. Gorin-Rosen, Y. The Ancient Glass Industry in Israel: Summary of the Finds and New Discoveries. In *La Route du Verre: Ateliers Primaires et Secondaires du Second Millénaire av. J.-C. au Moyen Âge*; Nenna, M.-D., Ed.; Maison de l'Orient et de la Méditerranée Jean Pouilloux: Lyon, France, 2000; pp. 49–63.
32. Ceglia, A.; Cosyns, P.; Nys, K.; Terryn, H.; Thienpont, H.; Meulebroeck, W. Late antique glass distribution and consumption in Cyprus: A chemical Study. *J. Archaeol. Sci.* **2015**, *61*, 213–222. [CrossRef]
33. Gliozzo, E.; Turchiano, M.; Giannetti, F.; Santagostino, B.A. Late Antique glass vessels and production indicators from the town of Herdonia (Foggia, Italy): New data on CaO-rich/weak HIMT glass. *Archaeometry* **2015**, *58*, 81–112. [CrossRef]
34. Gliozzo, E.; Braschi, E.; Giannetti, F.; Langone, A.; Turchiano, M. New geochemical and isotopic insights into the Late Antique Apulian glass and the HIMT1 and HIMT2 glass productions—The glass vessels from San Giusto (Foggia, Italy) and the diagrams for provenance studies. *Archaeol. Anthropol. Sci.* **2019**, *11*, 141–170. [CrossRef]
35. Brill, R.H. Scientific investigations of Jalame glass and related finds. In *Excavations in Jalame: Site of a Glass Factory in Late Roman Palestine*; Weinberg, G.D., Ed.; University of Missouri: Columbia, MO, USA, 1988; pp. 257–294.
36. Baxter, M.J.; Buck, C.E. Data handling and statistical analysis. In *Modern Analytical Methods in Art and Archaeology*; Ciliberto, E., Spoto, G., Eds.; John Wiley & Sons: New York, NY, USA, 2000; pp. 681–746.
37. Foster, H.; Jackson, C. The composition of 'naturally coloured' late Roman vessel glass from Britain and the implications for models of glass production and supply. *J. Archeol. Sci.* **2009**, *36*, 189–204. [CrossRef]
38. Rehren, T.; Cholakova, A. The early Byzantine HIMT glass from Dichin, Northern Bulgaria, UCL. *Interdiscip. Stud.* **2010**, *22*, 81–96.
39. Shortland, A.; Rogers, N.; Eremin, K. Trace element discriminants between Egyptian and Mesopotamian Late Bronze Age glasses. *J. Archaeol. Sci.* **2007**, *34*, 781–789. [CrossRef]
40. Åström, M. Abundance and fractionation patterns of rare earth elements in streams affected by acid sulphate soils. *Chem. Geol.* **2001**, *175*, 249–258. [CrossRef]

41. Åström, M.; Corin, N. Distribution of rare earth elements in anionic, cationic and particulate fractions in boreal humus-rich streams affected by acid sulphate soils. *Water Res.* **2003**, *37*, 273–280. [CrossRef]
42. Wedepohl, K.H.; Baumann, A. The use of marine molluskan shells for Roman glass and local raw glass production in the Eifel area (western Germany). *Naturwissenschaften* **2000**, *87*, 129–132. [CrossRef] [PubMed]
43. Wedepohl, K.H.; Gaitzsch, W.; Follmann-Shultz, A.-B. Glassmaking and Glassworking in Six Roman Factories in the Hambach Forest, Germany. *Ann. AIHV* **2003**, *15*, 53–55.
44. Schibille, N.; Freestone, I.C. Composition, production and procurement of glass at San Vincenzo Al Volturno: An early Medieval monastic complex in Southern Italy. *PLoS ONE* **2013**, *8*, e76479. [CrossRef]
45. Freestone, I.C.; Degryse, P.; Lankton, J.; Gratuze, B.; Schneider, J. HIMT, glass composition and the commodity branding in the primary glass industry. In *Things That Travelled, Mediterranean Glass in the First Millennium CE*; Rosenow, D., Phelps, M., Meek, A., Freestone, I., Eds.; UCL Press, University College London: London, UK, 2018; pp. 159–190. [CrossRef]
46. Kamber, B.S.; Greig, A.; Collerson, K.D. A new estimate for the composition of weathered young upper continental crust from alluvial sediments, Queensland, Australia. *Geochim. Cosmochim. Acta* **2005**, *69*, 1041–1058. [CrossRef]
47. Arletti, R.; Vezzalini, G.; Benati, S.; Mazzeo Saracino, L.; Gamberini, A. Roman Window Glass: A Comparison of Findings from Three Different Italian Sites. *Archaeometry* **2010**, *52*, 252–271. [CrossRef]
48. Smith, T.; Henderson, J.; Faber, E.W. Early Byzantine glass supply and consumption: The case of Dichin, Bulgaria. In *Recent Advances in the Scientific Research On Ancient Glass And Glaze*; Gan, F., Li, Q., Henderson, J., Eds.; World Scientific: Singapore, 2016; pp. 207–231.
49. Siu, I.; Henderson, J.; Faber, E. The production and circulation of Carthaginian glass under the rule of the romans and the vandals (fourth to sixth century ad): A chemical investigation. *Archaeometry* **2017**, *59*, 255–273. [CrossRef]
50. Freestone, I.C. Primary glass sources in the mid first millennium AD. In Proceedings of the Annales du 15e Congrès de l'Association Internationale pour l'Histoire du Verre, Corning, NY, USA, 15–20 October 2003; Nottingham AIHV and Authors, UK. pp. 111–115.

Article

Shedding Light on Roman Glass Consumption on the Western Coast of the Black Sea

Roxana Bugoi [1,*], Alexandra Țârlea [2], Veronika Szilágyi [3], Ildikó Harsányi [3], Laurențiu Cliante [4], Irina Achim [5] and Zsolt Kasztovszky [3]

1. Horia Hulubei National Institute for Nuclear Physics and Engineering, 30 Reactorului Street, 077125 Măgurele, Romania
2. Faculty of History, University of Bucharest, 030018 Bucharest, Romania; alexandra.tarlea@istorie.unibuc.ro
3. Centre for Energy Research, 29-33 Konkoly-Thege Street, H-1121 Budapest, Hungary; szilagyi.veronika@ek-cer.mta.hu (V.S.); harsanyi.ildiko@ek-cer.mta.hu (I.H.); kasztovszky.zsolt@ek-cer.mta.hu (Z.K.)
4. Museum of National History and Archaeology, 12 Piața Ovidiu, 900745 Constanța, Romania; cliante@gmail.com
5. Vasile Pârvan Institute of Archaeology, 11 Henri Coandă Street, 010667 Bucharest, Romania; irina.adriana.achim@gmail.com
* Correspondence: bugoi@nipne.ro

Abstract: The chemical composition of 48 glass finds from Histria and Tomis, Romania, chiefly dated to the 1st–4th c. AD, was determined using prompt gamma activation analysis (PGAA) at the Budapest Neutron Centre (BNC). Most fragments have composition typical for the Roman naturally colored blue-green-yellow (RNCBGY) glass; Mn-colorless, Sb-colorless, and Sb–Mn colorless glass finds were evidenced, too. Several Foy *Série 2.1* and Foy *Série 3.2* glass fragments, as well as an HIMT and a plant ash glass sample, were identified in the studied assemblage. The archaeological evidence, the glass working waste items, and the samples with compositional patterns suggestive of recycling are proofs of the secondary glass working activities at Tomis during the Early Roman Empire period.

Keywords: Roman glass; PGAA; Histria; Tomis; chemical composition; Early Roman Empire

Citation: Bugoi, R.; Țârlea, A.; Szilágyi, V.; Harsányi, I.; Cliante, L.; Achim, I.; Kasztovszky, Z. Shedding Light on Roman Glass Consumption on the Western Coast of the Black Sea. *Materials* **2022**, *15*, 403. https://doi.org/10.3390/ma15020403

Academic Editors: Žiga Šmit and Eva Menart

Received: 10 November 2021
Accepted: 30 December 2021
Published: 6 January 2022

Publisher's Note: MDPI stays neutral with regard to jurisdictional claims in published maps and institutional affiliations.

Copyright: © 2022 by the authors. Licensee MDPI, Basel, Switzerland. This article is an open access article distributed under the terms and conditions of the Creative Commons Attribution (CC BY) license (https://creativecommons.org/licenses/by/4.0/).

1. Introduction

Compositional analyses of archaeological glass finds can provide insights into raw materials and working techniques, demonstrating the skills of our predecessors [1]. Compositional data can also shed light on the glass provenance or manufacturing practices [2].

The archaeometric research on Roman glass finds evidence remarkably similar compositional patterns, especially during the Early Roman Empire period [3–5].

Roman glass is mostly identified as soda–lime–silica glass, typically containing 63–75 wt% silica (SiO_2), 11–22 wt% soda (Na_2O), and 4–12 wt% lime (CaO). Flux (the soda component) was introduced in the recipe to allow the melting of the mixture at temperatures attainable in ancient furnaces, while the stabilizer (lime) was needed to make glass stable in the presence of water, as pure soda–silica glass would be soluble in aqueous solutions.

In ancient times, two major sources of soda were available to the glass workers from Europe, North Africa, and the Near East: the ashes of halophytic plants and the mineral natron. "Natron" is a term used in archaeology to describe the deposits of salts containing sodium carbonate minerals, along with chlorides and sulfates that formed through the evaporation of soda lakes, mostly in Wadi Natrun in Egypt. However, east of the Euphrates, in the Persian Empire, plant ash was used as a soda source in glass making without any interruption during the 1st millennium AD [2].

During the Roman period, glass was fused in a restricted number of primary workshops located on the Egyptian or Syro-Palestine shores of Mediterranean Sea. It is worth mentioning here the remarkable tank furnaces discovered in the Levant and in Egypt [6–8], in which huge blocks of raw glass were made from carefully selected calcareous sands (i.e., shell-containing sands from particular beaches) and the mineral flux natron originating from Egypt [2].

From these primary workshops, glass was shipped via maritime routes, preferably as small pieces and just occasionally as finite products, toward the so-called consumption places spread all over the Roman Empire.

Of special importance for the history of glass was the discovery of several Roman shipwrecks with glass chunks as cargo, such as the *Embiez* (2nd c. AD), which contained lumps of fresh glass, as well as finished vessels and windows [9,10], and the *Iulia Felix* (3rd c. AD), which carried a barrel with fragments of glass bottles, plates, and cups collected for recycling [11].

A large number of small workshops scattered all over the empire and also in the regions under the influence of Rome dealt with the manufacturing of finished objects: vessels, window panes, and adornments. The basic raw material for these secondary workshops consisted of glass chunks dispatched from the primary workshops. As in many other historical periods, recycling was often encountered in secondary glass making, involving the collection, melting, and mixing of broken vessels or discarded adornments. When glass was recycled and reused, the color criteria were also taken into account [12].

During the Roman period, the manipulation of glass color was quite sophisticated. In particular, colorless glass was highly appreciated. However, items of intentionally colored glass were also occasionally encountered in archaeological records: beads, bangles, colored vessels, and decorative features (stripes, blobs, and handles) applied on naturally colored vessels.

In principle, Roman glass contains ≥0.3 wt% Fe_2O_3, the iron component being mainly inherited from the glassmaking sands. This accidentally incorporated iron provides glass a distinctive blue or green-blue tint. The typical aqua blue color of Roman glass is due to iron in reduced state (Fe^{2+} cation), while the fully oxidized state iron (Fe^{3+} cation) leads to a pale-yellow color. The typical green hue of Roman glass is a consequence of a relatively well-balanced mixture of the two cations of iron [2].

Antimony minerals are powerful decolorizers that were preferentially employed in Roman glass making until the 4th c. AD. Thus, the addition of antimony compounds to the glass batch led to the removal of the faintest tints of blue, green, or yellow and upon analyzing it, the resulting glass was classified as Sb-colorless. The extensive use of this particular additive coincided with the spread of glass blowing across the Roman Empire.

The competitor decolorizers were the manganese compounds; however, in order to produce a similar effect on glass appearance, manganese compounds had to be added in higher amounts (~1 wt% MnO). This decolorizer prevailed especially during the Late Roman Empire period; however, its occasional presence in earlier artefacts is also attested in the archaeometric literature [13].

The most-often employed glass chromophores in ancient times were the ions of particular metals: copper, cobalt, tin, antimony, lead, manganese, and iron. One of the most remarkable is cobalt, an element that even if present in extremely low amounts (some hundreds of parts per million) would induce a typical deep-blue hue to the glass. Copper in the upper oxidation state would impart glass a completely different turquoise blue or green hue, depending on the presence of other elements, such as iron or lead, while copper in a reducing state would lead to the production of the highly prized opaque red glass. Other chromophores used in Roman glass making were iron (for black) and manganese (brown or violet). To summarize, intentionally colored glass resulted from the interplay of the oxidation states of particular metallic ions and the concentrations at which various chromophore compounds were encountered in the glass batch [2].

Roman glass artefacts discovered at archaeological sites on the western coast of the Black Sea and in the Lower Danube region have been rarely investigated from compositional point of view [14–18]. The chemistry of ancient glass finds excavated in Romania started to be systematically studied only during recent years [19–23].

This paper reports the chemical composition of 48 glass fragments excavated at Histria (33 samples) and Tomis (15 samples), Romania, mostly dated to the Early Roman Empire period (1st–3rd c. AD). A small number of samples (six fragments discovered at Histria) turned out to be actually dated to the Late Roman Empire period (5th–6th c. AD). The measurements were performed at the Budapest Neutron Centre (BNC), using a bulk non-destructive analytical technique, namely prompt gamma activation analysis (PGAA).

The main aim of this study was to assign the analyzed fragments to well-established chemical groups of Roman glass from the archaeometric literature [3,5,13,24–26] to ease the comparison with data on coeval finds discovered in other regions of the Roman Empire. This classification was subsequently used to draw some conclusions about the provenance of glass from which these vessels were shaped, as well as on the manufacturing techniques.

This study is part of a larger project targeting to unravel the story of the circulation and consumption of glass in the Black Sea zone during the Roman period, in a trial to integrate this area with the impressive corpus of chemical data on Roman glass existing in the archaeometric literature.

2. Materials and Methods

Archaeological Background and Analyzed Samples

Histria (Istros) is one of the most significant archaeological sites of Romania. Founded on the western coast of the Black Sea by the Ionian Greek city of Miletus, Histria played an important political and economic role from the end of the 7th c. BC until the beginning of the 7th c. AD (see Figure 1). After its integration into the Roman Empire during the last half of the 1st c. BC, Histria became a part of the Roman production and trade network [27,28].

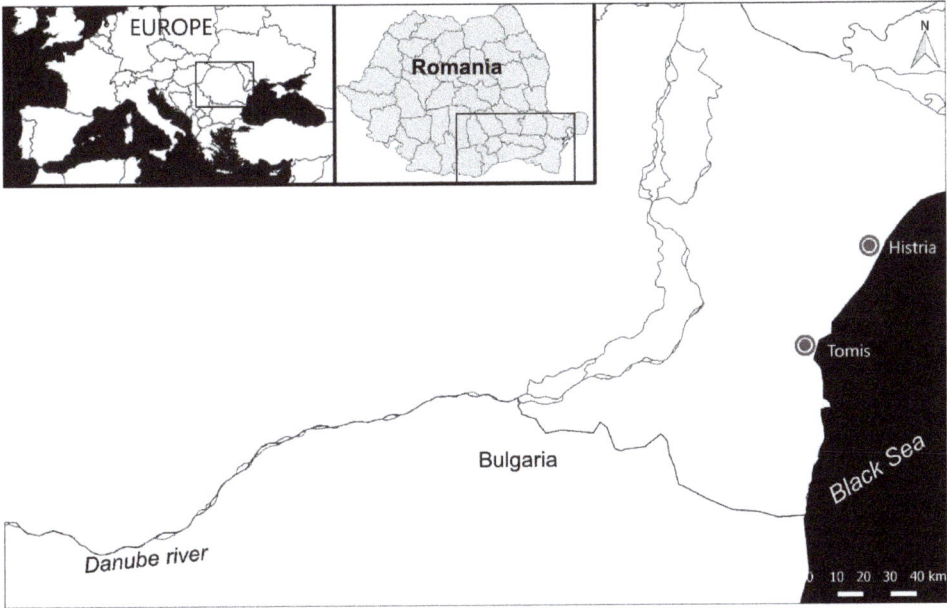

Figure 1. Maps showing the location of the archaeological sites where the glass fragments reported in this paper were discovered.

Different types of glassware are well represented in the archaeological record from Histria, as intact vessels, but more often as fragments [29–33].

So far, no glass working kilns or other indicators of glass manufacturing activities, such as crucibles or glass working waste fragments, have been discovered at Histria. However, taking into account the abundance of glass finds—and one should mention here the 1600 lamps and goblets (mostly Isings 111 forms) discovered in the basilica from Histria and published in [29,30]—the local production of glass vessels at Histria during the Late Roman Empire was postulated.

A project supported by the University of Bucharest (2013–2016 and 2017–2021) led to the opening of a new excavation sector at Histria, named Sector Acropola Centru-Sud/Acropolis Center-South (ACS) Sector, which revealed a 6th c. AD building [34,35].

The Crypt Basilica Sector, informally called the Basilica Florescu (BFL) Sector, situated in the central-northern part of the Late Roman/Early Byzantine city, has been excavated, with some interruptions, since 2002. The research is conducted by the Vasile Pârvan Institute of Archaeology, Bucharest, and has as the main objective the restoration of the Christian monument, based on a better understanding of the topographic situation and of the archaeological contexts from the entire area [34,36,37].

Tomis, another significant archaeological site of Romania, is also situated on the western coast of the Black Sea, approximately 50 km south of Histria. Its remains are practically located under the city of Constanța (see Figure 1). It was a colony of the Greek city of Miletus, too, being founded during the second half of 6th c. BC. Tomis came under Roman control at the same time as Histria and enjoyed a flourishing economic life during the Early Roman Empire period (1st–3rd c. AD), being the largest city of Moesia Inferior province [38].

The excavations conducted at Tomis during the 20th and 21st c. led to the discovery of a large number of Early Roman Empire glass finds (jugs, cups, decanters, and *unguentaria*) in well-dated contexts, especially in graves [39–43], as well as several kilns for glass working [32,39,44].

Because the finds related to glass production were the result of rescue excavations in specific conditions, i.e., incomplete excavations or archaeological contexts affected by the subsequent interventions, as the modern city of Constanța superposes upon the ancient city of Tomis, most of them were either not published or just mentioned in the literature, without providing too many details [44]. Consequently, for the time being, the knowledge on glass working activities at Tomis during the Early Roman Empire period is rather sketchy.

During the Roman period, the import of glass vessels represented an important part of the economic exchanges between the western coast of the Black Sea and the Mediterranean zone [29,30,32].

So far, six (possibly seven) glass kilns and remains related to secondary glass working have been identified in Tomis [32,39,44]. These finds have been loosely dated to the 1st–4th c. AD. All these glass kilns are of small size and circular shape, consistent with the secondary production of glass items relying on the use of imported raw glass and recycled cullet; most likely, they manufactured vitreous items for the local and regional market. These discoveries confirm the hypothesis of local glass working at Tomis during the Early Roman Empire period. Most likely, Tomis was a supplier of glass objects for the neighboring cities, too—e.g., Histria.

The large number of glass vessels of various types, differing greatly in shape, size, and quality, that were discovered in funerary contexts at Tomis, mostly dated to the 1st–3rd c. AD, brought further support to the idea of local glass working.

From this perspective, three vessels with globular bodies and long necks, identified as grave goods, are particularly important as they had the moil undetached at the moment of their deposition in the graves [32,39,40,44].

To get some hints about glass consumption during the Early Roman Empire period in these two cities on the western coast of the Black Sea, an archaeometric study was initiated, targeting the compositional analysis of 48 glass fragments. This study intended to provide chemical data that would allow speculating about the raw materials and the provenance (i.e., the place where glass was fused from raw materials) and also about the way these glass items were made, i.e., manufacturing techniques, for example, whether there was any recycling involved in their making.

Twenty-nine glass fragments from Histria were analyzed using the PGAA technique at BNC in April 2019. They included the following: 14 vessel fragments, out of which 5 dated to the 1st–3rd c. AD and 9 dated to the 4th–6th c. AD, discovered in the layer of the Late Roman building in the ACS Sector, and 15 vessel fragments from the 1st–3rd c. AD, unearthed in the BFL Sector.

This selection of fragments aimed the determination of the chemical composition of coeval glass fragments excavated at Histria in slightly different contexts. Thus, the BFL Sector offered the opportunity of recovering glass fragments from a context dated to the 1st–3rd c. AD, while in the ACS Sector, the Early Roman Empire glass fragments were excavated from the 6th c. AD layer, the last inhabitation phase of the city; in particular, the items analyzed in this study were discovered in secondary positions.

For comparison, 19 glass samples from Tomis (14 vessel fragments and 5 glass working waste items discovered in two kilns), dated to the 1st–4th c. AD, were measured as well. Fragments from two glass vessels excavated from a dated funerary context (3rd–4th c. AD), namely Tomis-2 and Tomis-3, are of particular interest, as they are unfinished products, i.e., the moil is still undetached from the rim, suggesting that they were locally manufactured with the intention to be deposited as grave goods [44].

Given the fact that some finds belong to the same type, the archaeologists also wanted to check for eventual similarities in their chemical composition; this particular point will be tackled in the Results and Discussion section.

All these glass fragments offer the advantage of having been discovered in well-defined archaeological contexts. Moreover, they can be considered, up to a certain point, representative of the repertoire of Early Roman Empire vessels unearthed at Histria and Tomis (thousands of such items: some intact glass vessels, but mostly fragments).

The detailed typological and chronological classification of the analyzed samples, made according to stylistic and stratigraphic criteria, is given in Table S1 and illustrated by Figure 2.

Tomis: 1, 2, 9
Histria: 16, 17, 21, 27, 31, 37, 38

Figure 2. Photos of selected glass samples reported in this study.

This study aimed to identify the finds made of fresh glass and to distinguish, if possible, the fragments whose composition would suggest their manufacture using recycled glass. In particular, the compositional data were used to provide additional, circumstantial evidence for the hypothesis that local glass working activities took place in this region during the Roman period.

3. Experimental

The bulk elemental composition of the glass samples was determined at the Budapest PGAA facility. The present setup of the BNC facility is described in detail in [45]. The concentrations of different chemical elements are determined using the detection of gamma rays emitted during (n, γ) reactions.

The method is applicable for the quantitative determination of the bulk elemental composition, comprising the major components and some minor and trace elements, in particular, to measure the concentrations of those chemical elements with high-neutron-absorption cross sections.

In PGAA, quantitative results are determined using the prompt-k_0 method, which is an internal standardization method [46,47]. The standardization of PGAA at BNC has been published in [48,49].

Because of the high penetration of neutrons, the technique provides the average composition of the irradiated volume (usually on the order of a few centimeter cubes). Thanks to the relatively low intensity of the neutron beam, the induced radioactivity decays quickly after the irradiation and the artefacts can be safely returned to the owners. PGAA is well suited for the study of cultural heritage objects since this method does not require any sample preparation or extraction of small samples from the artefacts, allowing the completely non-destructive characterization of unique archaeological finds.

The PGAA facility from BNC has been extensively used to analyze various kinds of cultural heritage artefacts, including ancient glass items [15,17,50–53].

The accuracy and precision of the PGAA setup from BNC has been checked by repeated measurements of several certified glasses; for details on this issue, refer to [52,53]. In particular, the PGAA results on Corning Museum of Glass archaeological reference glasses Brill B and Brill C are provided in Table S2, along with some published values [54].

According to [55], when applying the k_0 method in quantitative PGAA, the most significant source of uncertainty for the concentration values is related to the peak area determination, i.e., counting statistics. Therefore, it was not necessary to perform repeated measurements on each sample.

A total of 48 samples, namely 42 glass fragments originating from different types of vessels, 1 bead, and 5 glass waste chunks, were analyzed during the PGAA experiment from BNC.

The glass samples were irradiated with an external cold neutron beam of 9.6×10^7 cm^{-2} s^{-1} thermal equivalent intensity for 1700–20,000 s. The acquisition times were optimized to collect gamma spectra with statistically significant peaks for the elements of interest. The cross section of the external neutron beam varied between 24 mm^2 and 400 mm^2; thus, in most cases, the entire glass fragment was exposed to the neutron beam.

The prompt-gamma spectra were collected using a 64 k MultiChannel Analyzer, and the Hypermet-PC software was used for spectrum evaluation. The quantitative analysis is based on the k_0 principle using a PGAA library developed at the BNC [49].

For radiation safety reasons, in particular because of the activation of Na (the half-life ($T_{1/2}$) of ^{23}Na is 14.96 h), it was mandatory to keep the glass samples in the laboratory for 2–3 more days after the experiment to allow the decay of the ^{23}Na nuclei.

Before the PGAA experiment, all glass fragments were washed with water and gently brushed to remove the superficial dirt and soil deposits; however, on several samples, despite the cleaning procedures, visible thick layers of weathering products remained firmly attached to the glass surface. In principle, the chemical composition of an ancient glass fragment that has been buried for centuries is altered to a depth of some tens of micrometers compared to the bulk composition. In these thin layers, the alkali, alkali earth elements, and silica concentrations are modified by leaching phenomena [56]. However, due to the high penetration of cold neutrons in tandem with the detection of gamma rays, the reported PGAA concentrations can be considered as representative of the chemistry of the pristine glass of all samples.

Using PGAA, it was possible to quantify H, B, Na, Al, Cl, Si, Ti, K, Ca, Mn, Fe, Sm, and Gd in all glass fragments and Mg, S, Co, and Sb only in certain samples, where these elements are present above the detection limits. The data expressed in wt%, as well as the detection limits of the PGAA setup, are given in Table S3. The concentrations of all elements are expressed as oxides, except for Cl, which is given in elemental form. The overall uncertainties of the reported concentrations were estimated to be less than 10% (relative values) from the reported numbers.

4. Results and Discussions

All analyzed glass fragments were identified as soda–lime–silica glass fragments, as indicated by the average values of soda (17.2 ± 1.7 wt% Na_2O), lime (6.9 ± 0.8 wt% CaO), and silica (68.4 ± 2.3 wt% SiO_2).

The magnesia and potash content for most samples is less than 1.5 wt%, a consequence of the fact that natron was the mineral flux used in making these glass items [57]. While potash was quantitatively determined in all samples, in some of them, magnesia is under the detection limits of the PGAA setup, estimated to be 1.0 wt% MgO.

However, several samples (Tomis-15, Tomis-48, Histria-17, Histria-31, Histria-38, and Histria-39) feature magnesia concentrations slightly higher than the 1.5 wt% value that in the literature separates the natron glass from that made of plant ashes [57].

In particular, sample Histria-39 has both magnesia and potash content higher than 1.5 wt%. This finding was interpreted as indicative of a plant ash component in the composition of this glass fragment. Recycling procedures involving the addition of plant ash glass cullet to natron glass chunks might be a possible explanation for this particular compositional pattern. This idea is also supported by the chlorine content, which is the lowest in this sample set (0.66 wt% Cl), a further argument for recycling, taking into account that chlorine is a volatile element [58].

High hydrogen content (H_2O concentrations varying from 0.9 up to 3.5 wt%) was detected in several fragments (Histria-23, Histria-36, Histria-39, Tomis-2, Tomis-9, Tomis-13, Tomis-14, and Tomis-15). Thick layers of soil and weathering products are visible on these particular samples, despite the cleaning with water performed before the PGAA experiment. The explanation might be the hydrated mineral compounds that still cover the surfaces of these glass fragments.

A summary of the glass groups identified in the studied assemblage is given in Table S3, including the mean values for larger groups, as well as some comparison terms from the archaeometric literature [3,5,13,24,26,59].

The glass fragments reported in this paper are either colorless or naturally colored; the detailed sample description is given in Table S1. Therefore, the first criterion taken into account when performing the classification of the samples into compositional groups was their content of chemical elements indicating the intentional addition of decolorizing compounds, i.e., Sb and Mn oxides. Further, the concentrations of major and some minor elements of the vitreous matrix (Si, Na, Ca, K, Mg, Ti, Fe, and Al) were considered as discriminating factors.

The fragment of a flask Histria-41 is a sample whose lack of color can be explained by its relatively high content of antimony (7500 ppm Sb_2O_3), accompanied by a small manganese concentration (320 ppm MnO), close to the limits given in the literature that would suggest the recycling and mixing of Sb-colorless with Mn-colorless glass, i.e., 250 ppm MnO [13]. The compositional pattern of this sample, with little alumina (1.68 wt% Al_2O_3), titanium (0.073 wt% TiO_2), and lime (5.03 wt% CaO), indicates that mature sand was used to manufacture this glass, most likely of Egyptian origin [13]. Sb-decolorizing procedures were widespread during the Early Roman Empire period, being frequently used until the 4th c. AD, when a steep decline began [13].

Two other colorless samples from this assemblage might be also attributed to the Sb-colorless group: Tomis-11 and Tomis-15, containing relatively high amounts of antimony (8900 and 7800 ppm Sb_2O_3, respectively) and low amounts of manganese (440 and 460 ppm MnO, respectively), accompanied by low concentrations of lime (5.14 and 6.03 wt% CaO, respectively) and low concentrations of alumina (2.08 and 1.97 wt% Al_2O_3, respectively); these compositional patterns are characteristic for Sb-colorless glass finds. One should mention though that the values of the MnO concentrations are a bit higher those from the archaeometric literature, which indicates the exclusive use of antimony as a decolorizer, i.e., above the background level of MnO naturally contained in the sands used for making glass [13,60].

Nine colorless samples from the studied sample set owe their lack of color to a relatively high manganese content, or 1.62 wt% MnO on average (see Table S3), suggesting the prevalence in the studied sample set of colorless glass made using this a decolorizer; compare this with the three fragments discussed before, representative of the alternative technological solution, namely the use of antimony decolorizing compounds. Mn-colorless glass is often encountered in artefacts dated to the Late Roman Empire period, but this type of glass also existed during the Early Roman Empire period, though less frequently reported in the literature [13]. In all these samples, antimony, even if present, is under the detection limits, estimated to be approximately 1000 ppm Sb_2O_3 for the PGAA setup.

The samples from this group have different typological and chronological attributes, varying from the Early Roman Empire to the Late Roman Empire period. For example, samples Histria-33, Histria-43, and Histria-44 were dated to the 6th c. AD.

The overall inhomogeneity of this group, chronological, typological, and compositional, suggests that the raw glass from which these vessels were shaped might had originated in Levantine and Egyptian workshops [13].

Nine samples from the studied sample set fall into the group of the Sb–Mn colorless glass, being characterized by intermediate amounts of both decolorizers (~4000 ppm MnO and 4900 ppm Sb_2O_3 on average). Their compositional pattern is in good agreement with the values for the Roman Sb–Mn colorless glass from the literature (1st–4th c. AD) [5,13]; see Table S3. These samples are clearly the result of recycling procedures involving both types of colorless glass. It is worth mentioning that most of these fragments were found in excavations in Tomis (six out of nine samples).

The graph shown in Figure 3 is a good illustration of the division of the colorless samples into these three groups; in particular, it supports the attribution of samples Tomis-11 and Tomis-15 to the Sb-colorless group, despite their slightly high manganese content.

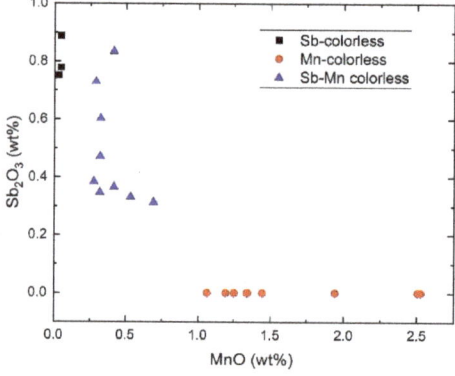

Figure 3. Plot of Sb_2O_3 versus MnO concentrations in the colorless glass samples from Histria and Tomis discussed in this publication.

All these classifications were made considering the groups of colorless glass defined in [13].

Most of the naturally colored and colorless glass fragments (15 items) were attributed to the well-known chemical group of Roman glass from the literature, the so-called RNCBGY2 group defined by Gliozzo and colleagues (2016), where RNCBGY2 stands for Roman Naturally Colored Blue Green Yellow 2, thought to originate in some Levantine primary workshops [26]. Similar compositions, i.e., data for naturally colored Early Roman Empire glass assemblages, were also reported in [3,5].

Two samples dated to the 4th c. AD, namely Histria-20 and Histria-39, were initially identified as HIMT (high iron manganese titanium) glass, according to their high iron and manganese content. In particular, sample Histria-20 can be assigned to the HIMTb sub-group; for details on this refined classification, see [61–63].

The term HIMT was coined by Freestone [59], though this type of glass was evidenced for the first time by Mirti and colleagues [64]. It is the equivalent of Foy *Groupe 1* [24]. This is an example of a Roman glass chemical type that appeared sometimes during the 4th c. AD, made using particular Egyptian sand, rich in iron-bearing minerals [63].

However, as stated above, according to its high magnesia (1.67 wt% MgO) and potash (1.77 wt% K_2O), sample Histria-39 suggests a plant ash glass component in its recipe. Plant ash glass items were not frequently encountered in the zones under the Roman influence, where natron glass was the prevailing type of glass. An interesting overview about the occurrence of plant ash glass finds in the Roman world is given in [65]. The water content of this sample is also high (3.494 wt% H_2O), possibly a consequence of the thick deposits of hydrated minerals still present on the surface of this fragment.

Several samples from the ACS Sector in Histria dated from the 4th c. onwards, namely Histria-34, Histria-35, Histria-36, and Histria-38, were identified as Foy *Série 2.1* [24]. The last two samples have relatively high amounts of H_2O, which can be correlated with the thick layers of soil/weathering compound deposits covering their surfaces.

Sample Histria-37 is a special one. According to its high manganese and iron content, this pale blue handle of lamp dated to the 6th c. AD was included in the Foy *Série 2.1* group. However, its composition resembles that of RNCBGY2 glass, too, particularly considering its low Al_2O_3 and TiO_2 concentrations, at least compared to the other Foy *Série 2.1* samples reported in this paper. Taking into account the 110 ppm CoO from its composition, we can infer that this fragment was produced by recycling, the manufacturing process involving cullet belonging to Foy *Série 2.1* and RNCBGY2 glass groups. Decorative elements made of intentionally colored blue glass might have accidentally entered the batch, thus the explanation for the presence of cobalt in minute amounts.

The presence of cobalt traces (100 ppm CoO) in the colorless glass fragment Tomis-2, a Mn-colorless sample, is also suggestive of recycling procedures. Co-containing glass might have been carelessly introduced into the batch (for example, fragments of vessels with blue decoration might have been included in the cullet), which would have contaminated the entire melt.

From the entire sample set, only fragments Histria-37 and Tomis-2 show unambiguous evidence of recycling, provided by the presence of small amounts of a chromophore trace-element (namely cobalt) in naturally colored and colorless glass, respectively. It is worth mentioning here the detection limits for CoO of the PGAA setup, estimated to be 100 ppm.

Glass waste Tomis-1, a fragment found in a glass kiln, is another example of recycled sample, as this green piece of glass has a compositional pattern resembling that of Foy *Série 2.1*, RNCBGY2, and colorless glass samples.

Compositionally speaking, glass working waste fragments Tomis-45, Tomis-46, and Tomis-48, fragments of molten glass with tiny pieces of wattle from the kiln walls attached on their surfaces, are clear outliers. A possible interpretation of these usual compositional patterns is that these chunks resulted from recycling many types of glass. In particular, Tomis-45 and Tomis-48 are somehow compositionally similar, featuring remarkably high alumina content (6.23 and 7.46 wt% Al_2O_3, respectively). Considering the composite structure of these samples (photos of these particular items were published in [44], one might interpret these results as stemming from the alumino-silicates compounds (wattle accretions) that are still present on their surfaces, resulting in the relatively high concentrations of alumina. Further analyses using optical microscopy and scanning electron microscopy (SEM) might be performed in the future on these particular samples to clarify these results, taking into account the fact that the PGAA technique provides the bulk chemical composition of the analyzed object.

Histria-31, a fragment from a colorless beaker, is also an interesting sample. Its color and its overall compositional pattern did not allow its assignation to any of the colorless/intentionally decolorized glass groups, as it contains antimony and manganese in small amounts. Moreover, its chemical composition does not present any similarity with RNCBGY2 glass. Dated on archaeological grounds to 4th–5th c. AD, it was identified as a Foy *Série 3.2* glass [24].

Tomis-2 and Tomis-3 are both fragments from vessels with spherical bodies, long necks, and flaring mouths, similar not only in shape but also in dimensions; both items were discovered in graves. Tomis-2 is colorless, while Tomis-3 has a yellow tinge. According to archaeological arguments, these containers were considered as local products; see the Introduction. Chemically speaking, Tomis-2 turned out to be Mn-colorless glass, while Tomis-3 was assigned to the RNCBGY2 group, i.e., it has a composition typical for the Early Roman glass.

Tomis-4 and Tomis-5 belong to the same type of colorless flask; they originated from the same context, i.e., the same grave, being dated to the 2nd–3rd c. AD. Both items were made of Sb–Mn colorless glass involving the recycling and mixing of two different chemical types of intentionally decolorized glass.

Tomis-8 and Tomis-9 are fragments from the same type of bulbous *unguentarium* (Isings 105 form) dated to the 4th c. AD. Both fragments are colorless, but Tomis-8 has olive tinges and Tomis-9 has green tinges. Both of them were identified as Mn-colorless glass.

Histria-16 and Histria-32 are fragments of square bottles, a well-known type of container from the 1st–2nd c. AD (Isings 50 form). Histria-16 has a pale-blue color, and Histria-32 has green tinges. Histria-16 was found in the BFL Sector and Histria-32 in the ACS Sector. Histria-16 was identified as RNCBGY2 glass/typical Early Roman glass, while Histria-32 is Sb–Mn colorless glass.

Two fragments of plates or shallow bowls were found in the BFL Sector, Histria-21 and Histria-27, both with blue-green tinges. They were made of RNCBGY2 glass, i.e., typical Early Roman glass of Levantine origin.

Histria-35 and Histria-36 are fragments of Late Roman Empire goblets (Isings 111) that were discovered in the ACS Sector. Both samples belong to Foy *Série 2.1*. A relatively large number of similar finds were also identified at Tropaeum Traiani, a site relatively close to both Histria and Tomis [16].

Histria-43 and Histria-44 are fragments of some Late Roman Empire goblets, found in the ACS Sector; both are Mn-colorless glass.

We tried to compare the composition of the glass samples unearthed in the BFL and ACS Sectors, the two particular places where the Histria fragments analyzed in this study were found. Thus, most of the samples from the BFL Sector were assigned to the RNCBGY2 group. The following exceptions occurred: Histria-23 bead, dated to the 2nd–3rd c. AD, assigned to Mn-colorless glass group; the fragment of a honeycomb cup Histria-20, dated to the 4th c. AD, attributed to the HIMT group; and the flask fragment Histria-30, pertaining to the Sb–Mn colorless glass group. However, all samples originating from the ACS Sector belong to several compositional groups discussed in the text: Foy *Série 2.1* and *Série 3.2*, Sb-colorless, Mn-colorless, and Sb–Mn colorless. Remarkably, a single RNCBGY2 sample was identified among those discovered at the ACS Sector, namely Histria-40. This might be explained by the fact that many of the ACS fragments chosen for this study turned out to be dated slightly later than those from the BFL Sector. Thus, the BFL Sector provided most of the Early Roman Empire samples *stricto sensu* (i.e., dated until the 4th c. AD), and, in turn, most of the BFL samples were included in the RNCBGY2 group.

Regarding the Tomis samples, most of the Sb–Mn colorless samples are from this site and none of them was identified as HIMT, Foy *Série 2.1*, or Foy *Série 3.2*, i.e., none of them pertains to any Late Roman Empire glass groups. In accordance with their archaeological and stylistic dating, the glass fragments discovered at Tomis proved to have compositional patterns typical for the Early Roman Empire period.

The Roman glass samples (15 fragments from the RNCBGY2 group) were shaped from glass fused in primary workshops located on the Levantine coast of the Mediterranean Sea [26]. The relatively small concentrations of titanium and iron oxides (0.07 wt% TiO_2 and 0.35 wt% Fe_2O_3 on average) suggest the use of mature sand in their making.

According to the literature data [13,63], the primary glass of the Sb-colorless glass, of some of the Mn-colorless fragments, as well as of the Late Roman Empire items, those belonging to HIMT, Foy *Série 2.1*, and Foy *Série 3.2* groups, was most likely produced in Egyptian primary workshops.

Of clear interest is the composition of the glass working waste fragments discovered in the excavations from Tomis reported in this paper. It turned out that they are clear outliers compared to the rest of the glass samples published here, which are mostly vessel fragments. New questions were triggered by the bulk PGAA results that might be only answered by further analyses using some microscopy techniques involving minimally invasive sampling of these artefacts. In any case, the composition of these items suggests recycling procedures, having as input various types of glass, certainly a feature of the glass manufacturing in the secondary workshops from Tomis during the Early Roman Empire period.

The recycling of many types of decolorized glass is also evidenced in the nine Sb–Mn colorless glass samples from this assemblage.

Other samples resulting from glass recycling are two naturally colored fragments, Tomis-2 and Histria-37 that contain cobalt in minute amounts, as well one in Histria-39 sample, whose composition indicates the use of plant ash cullet in its making.

Based on the PGAA data reported in this paper, we can speculate that glass recycling procedures were used in the manufacturing of the vitreous finds discovered at Histria and Tomis.

This study indicates that both vessels made of fresh glass imported from the Mediterranean region as well as containers made of recycled glass, the latter ones probably locally manufactured in workshops active at Tomis, were used by the inhabitants of Histria and Tomis during the Roman period.

It is worth stressing a methodological conclusion, namely that all glass samples reported in this paper, exclusively analyzed using PGAA technique, were relatively easy assigned to well-known chemical groups from the archaeometric literature, allowing straightforward comparisons with coeval finds from other sites from all over the Roman Empire and further speculations on provenance and manufacturing practices.

5. Conclusions

This publication reports the composition of 48 glass fragments discovered during archaeological excavations at Histria and Tomis, Romania, mainly dated to the Early Roman Empire period. PGAA data allowed the attribution of the analyzed glass fragments to several well-established glass types from the archaeometric literature: Sb-colorless, Mn-colorless, Sb–Mn colorless, and RNCBGY2. Several finds from the Late Roman Empire period were identified as being made of HIMT, Foy *Série 2.1*, and Foy *Série 3.2* glass.

Using the PGAA data, clues for glass recycling were evidenced in several samples. Archaeological and compositional arguments support the idea that during the Early Roman Empire period, recycling practices were in use in the secondary glass workshops from Tomis.

The reported glass finds, discovered in archaeological excavation at Histria and Tomis, mostly dated according to typological and stratigraphic criteria to the Early Roman Empire period, were manufactured from raw glass fused in primary workshops in both Egypt and on the Levantine coast of the Mediterranean Sea. Moreover, PGAA data bring additional circumstantial evidence for the hypothesis of secondary local glass working in Tomis, glass manufacturing of different types of vessels involving the use of various types of cullet.

This study contributes to the understanding on life during the Roman time in two cities on the western shore of the Black Sea: Histria and Tomis. An impressive number of different glass vessels with various chemical compositions were used in everyday life, as well as in the burial rituals, offering a better perception on the glass consumption mechanisms during the Roman period in this zone. This publication brings details on the circulation of this fascinating material in the Black Sea area, harmoniously integrating this region within the landscape of archaeometric publications on Roman glass.

Supplementary Materials: The following supporting information can be downloaded at: https://www.mdpi.com/article/10.3390/ma15020403/s1. Table S1 Archaeological description of the investigated glass samples [29,30,39,44,66–82]. Table S2 PGAA data for Brill B and Brill C archaeological reference glasses from Corning Museum of Glass [52] and published values [54]. Table S3 PGAA data for Histria and Tomis glass samples expressed in oxide wt% (except for Cl that is given in elemental form). Blank cells mean below the detection limits of the PGAA setup that are given in the second line of the same table. Samples were grouped according to their chemical composition; the mean values of the larger groups are provided, too. The corresponding comparison terms from the literature are also shown.

Author Contributions: Conceptualization: R.B. and A.Ț.; Methodology: Z.K., V.S. and I.H.; Software: Z.K., V.S., I.H. and R.B.; Validation: R.B., A.Ț., I.A., L.C., Z.K., V.S. and I.H.; Formal analysis: R.B., L.C., A.Ț. and I.A.; Investigation: R.B., A.Ț., L.C., Z.K., V.S. and I.A.; Resources: Z.K. and V.S.; Data curation: R.B., Z.K., V.S., A.Ț., L.C. and I.A.; Writing—original draft and editing: R.B., A.Ț., I.A., L.C., Z.K. and V.S.; Writing—review and editing: R.B., A.Ț., I.A., L.C., Z.K. and V.S.; Supervision: R.B., A.Ț. and Z.K.; Project administration: R.B. and Z.K.; Funding acquisition: R.B. and Z.K. All authors have read and agreed to the published version of the manuscript.

Funding: The present work has received funding from the Access to Research Infrastructures activity of the EU HORIZON 2020 IPERION CH programme (grant agreement No. 654028).

Data Availability Statement: All the data were published in this paper; the tables are in the supplementary material available at the link indicated above.

Acknowledgments: Vasile Opriș from Muzeul de Istorie al Municipiului București is gratefully thanked for his kind help in drawing the maps shown in Figure 1.

Conflicts of Interest: The authors declare no conflict of interest.

References

1. Shortland, A.J.; Rehren, T. Glass. In *Archaeological Science—An Introduction*; Richards, M.P., Britton, K., Eds.; Cambridge University Press: Cambridge, UK, 2020; pp. 347–364.
2. Freestone, I.C. Glass production in the first millennium CE: A compositional perspective. In *Ancient Glass and Glass Production, Edition TOPOI*; Klimscha, F., Ed.; Berlin Studies of the Ancient World 67: Berlin/Heidelberg, Germany, 2021; pp. 243–262.
3. Nenna, M.D.; Vichy, M.; Picon, M. L'atelier de verrier de Lyon du 1er siècle après J.-C. et l'origine des verres "romains". *Revue d'Archéométrie* **1997**, *21*, 81–87. [CrossRef]
4. Sayre, E.V.; Smith, R.W. Compositional categories of ancient glass. *Science* **1961**, *133*, 1824–1826. [CrossRef] [PubMed]
5. Silvestri, A.; Gallo, F.; Maltoni, F.; Degryse, P.; Ganio, M.; Longinelli, A.; Molin, G. *Things that Travelled: A Review of the Roman Glass from Northern Adriatic Italy in Things That Travelled. Mediterranean Glass in the First Millennium AD*; Rosenow, D., Phelps, M., Meek, A., Freestone, I.C., Eds.; UCL Press: London, UK, 2018; pp. 346–367.
6. Gorin-Rosen, Y. The ancient glass industry in Israel: Summary of the finds and new discoveries. In *La route du Verre: Ateliers Primaires et Secondaires du Second Millénaire Av J-C au Moyen Âge*; Nenna, M.D., Ed.; Maison de l'Orient Méditerranéen: Lyon, France, 2000; Volume 33, pp. 49–63.

7. Nenna, M.D. Primary glass workshops in Graeco-Roman Egypt: Preliminary report on the excavations of the site of Beni Salama, Wadi Natrun (2003, 2005–2009). In *Glass of the Roman World*; Bayley, J., Freestone, I.C., Jackson, C.M., Eds.; Oxbow Books: Oxford, UK, 2015; pp. 1–22.
8. Tal, O.; Jackson-Tal, R.E.; Freestone, I.C. New evidence of the production of raw glass at late Byzantine Apollonia-Arsuf, Israel. *J. Glass Stud.* **2004**, *46*, 51–66.
9. Fontaine, S.D.; Foy, D. L'épave Ouest-Embiez 1, Var: Le commerce maritime du verre brut et manufacturé en Méditerranée occidentale dans l'Antiquité. *Rev. Archéol. Narbonn.* **2007**, *40*, 235–265. [CrossRef]
10. Thirion-Merle, V.; Nenna, M.D.; Picon, M.; Vichy, M. *Un Nouvel Atelier Primaire Dans le Wadi Natrun (Egypte) et les Compositions des Verres Produits Dans Cette Région*; Bulletin de l'AFAV: Dijon, France, 2002–2003; pp. 21–24.
11. Silvestri, A.; Molin, G.; Salviulo, G. The colourless glass of *Iulia Felix*. *J. Archaeol. Sci.* **2008**, *35*, 331–341. [CrossRef]
12. Freestone, I.C. The recycling and reuse of Roman glass: Analytical approaches. *J. Glass Stud.* **2015**, *57*, 29–40.
13. Gliozzo, E. The composition of colourless glass: A review. *Archaeol. Anthropol. Sci.* **2017**, *9*, 455–483. [CrossRef]
14. Bugoi, R.; Alexandrescu, C.G.; Panaite, A. Chemical composition characterization of ancient glass finds from *Troesmis*—Turcoaia, Romania. *Archaeol. Anthropol. Sci.* **2018**, *10*, 571–586.
15. Bugoi, R.; Talmațchi, G.; Szilágyi, V.; Harsányi, I.; Cristea-Stan, D.; Boțan, S.; Kasztovszky, Z. PGAA analyses on Roman glass finds from Tomis. *Rom. J. Phys.* **2021**, *66*, 906.
16. Bugoi, R.; Panaite, A.; Alexandrescu, C.G. Chemical analyses on Roman and Late Roman Empire glass finds from the Lower Danube: The case of Tropaeum Traiani. *Archaeol. Anthropol. Sci.* **2021**, *13*, 148.
17. Bugoi, R.; Țârlea, A.; Szilágyi, V.; Harsányi, I.; Cliante, L.; Kasztovszky, Z. Colour and beauty at the Black Sea coast: Archaeometric analyses of selected small finds from Histria. *Rom. Rep. Phys.* **2022**, *74*.
18. Stawiarska, T. *Roman and Early Byzantine Glass from Romania and Northern Bulgaria. Archaeological and Technological Study*; Biblioteca Antiqua Volume XXIV, Academia Scientiarum Polona: Warsaw, Poland, 2014.
19. Bugoi, R.; Poll, I.; Mănucu-Adameșteanu, G.; Neelmeijer, C.; Eder, F. Investigations of Byzantine glass bracelets from Nufăru, Romania using external PIXE-PIGE methods. *J. Arch. Sci.* **2013**, *40*, 2881–2891. [CrossRef]
20. Bugoi, R.; Poll, I.; Manucu-Adamesteanu, G.; Calligaro, T.; Pichon, L.; Pacheco, C. PIXE-PIGE analyses of Byzantine glass bracelets (10th–13th centuries AD) from Isaccea, Romania. *J. Radioanal. Nucl. Chem.* **2016**, *307*, 1021–1036. [CrossRef]
21. Bugoi, R.; Poll, I.; Mănucu-Adameșteanu, G.; Pacheco, C.; Lehuédé, P. Compositional study of Byzantine glass bracelets discovered at the Lower Danube. *Microchem. J.* **2018**, *137*, 223–230. [CrossRef]
22. Bugoi, R.; Măgureanu, A.; Măgureanu, D.; Lemasson, Q. IBA analyses on glass beads from the Migration Period. *Nucl. Instrum. Method B* **2020**, *478*, 150–157. [CrossRef]
23. Bugoi, R.; Mureșan, O. A brief study on the chemistry of some Roman glass finds from Apulum. *Rom. Rep. Phys.* **2021**, *73*, 803.
24. Foy, D.; Picon, M.; Vichy, M.; Thirion-Merle, V. Caractérisation des verres de la fin de l'Antiquité en Mediterranée occidentale: l'émergence de nouveaux courants commerciaux. In *Échanges et Commerce du Verre dans le Monde Antique. Actes du Colloque de l'AIHV, Aix-en-Provence et Marseille, Juin 2001. Monographies Instrumentum 24*; Foy, D., Nenna, M.D., Eds.; Monique Mergoil: Montagnac, France, 2003; pp. 41–85.
25. Foy, D.; Thirion-Merle, V.; Vichy, M. Contribution à l'étude des verres antiques décolorés à l'antimoine. *Revue d'Archéométrie* **2004**, *28*, 169–177.
26. Gliozzo, E.; Turchiano, M.; Giannetti, F.; Santagostino Barbone, A. Late Roman Empire glass vessels and production indicators from the town of *Herdonia* (Foggia, Italy): New data on CaO-rich/weak HIMT glass. *Archaeometry* **2016**, *58*, 81–112. [CrossRef]
27. Condurachi, E. (Ed.) *Histria. Monografie Arheologică Vol. I*; Editura Academiei RPR: București, Romania, 1954.
28. Condurachi, E. (Ed.) *Histria. Monografie Arheologică Vol. II*; Editura Academiei RSR: București, Romania, 1966.
29. Băjenaru, C.; Bâltâc, A. Depozitul de candele de sticlă descoperit la basilica episcopală de la Histria. *Pontica* **2000–2001**, *33–34*, 469–513.
30. Băjenaru, C.; Bâltâc, A. Histria—Basilica Episcopală. Catalogul descoperirilor de sticlă (1984–2000). *Pontica* **2006**, *39*, 219–247.
31. Boțan, S.P. *Vase de Sticlă în Spațiul dintre Carpați și Prut (Secolele II a.Chr.–II p.Chr.)*; Ed. Mega: Cluj-Napoca, Romania, 2015.
32. Chiriac, C.; Boțan, S.P. Sticlăria elenistică și romană din Pontul Euxin. Între producție și import. In *Poleis în Marea Neagră. Relații Interpontice și Producții Locale, Pontica et Mediterranea I*; Panait Bîrzescu, F., Birzescu, I., Matei-Popescu, F., Robu, A., Eds.; Ed. Humanitas: București, Romania, 2013.
33. Țârlea, A.; Cliante, L. 'Put the lights on': Early Byzantine stemmed goblets and lamps from the Acropolis Centre-South Sector in Histria (I). *Peuce SN* **2020**, *18*, 301–332.
34. Achim, I.; Bottez, V.; Angelescu, M.; Cliante, L.; Țârlea, A.; Lițu, A. A city reconfigured: Old and new research concerning Late Roman urbanism in Istros. In *The Greeks and Romans in the Black Sea and the Importance of the Pontic Region for the Graeco–Roman World (7th Century BC–5th Century AD): 20 Years On (1997–2017); Tsetskhladze, G.R., Avram, A., Hargrave, J., Eds.; In Proceedings of the Sixth International Congress on Black Sea Antiquities, Constanța, Romania, 18–22 September 2017, Dedicated to Prof. Sir John Boardman to Celebrate His Exceptional Achievement and his 90th Birthday*; Archaeopress Archaeology: Oxford, UK, 2021; pp. 477–487.
35. Bottez, V.; Lițu, A.; Țârlea, A. Preliminary results of the excavations at Histria. The Acropolis Centre-South Sector (2013–2014). *MCA SN* **2015**, *11*, 157–192. [CrossRef]
36. Achim, I. La basilique à crypte d'Istros: Dix campagnes de fouilles (2002–2013). *MCA* **2014**, *10*, 265–287. [CrossRef]

37. Achim, I.; Dima, M.; Beldiman, C.; Surdu, V.; Băcăran, M.; Munteanu, F. Comuna Istria, jud. Constanța. "Basilica cu criptă". In *Cronica Cercetărilor Arheologice din România, Campania 2013*; MInisterul Culturii, Institutul National al Patrimoniului: Oradea, Romania, 2014; pp. 64–66.
38. Bărbulescu, M.; Buzoianu, L. *Tomis. Comentariu Istoric și Arheologic*; Ed. Ex Ponto: Constanța, Romania, 2012.
39. Bucovală, M. *Vase Antice de Sticlă la Tomis*; Muzeul de Arheologie: Constanța, Romania, 1968.
40. Bucovală, M. Tradiții elenistice în materialele funerare de epocă romană timpurie la Tomis. *Pontice* **1969**, *2*, 297–332.
41. Bucovală, M. Roman glass vessels discovered in Dobrudja. *J. Glass Stud.* **1984**, *26*, 59–63.
42. Drăghici, C. Glassware from Tomis. Chronological and typological aspects. In *Annales AIHV Thessaloniki 2009*; Drăghici, C., Drăghici, C., Eds.; Ziti Publications: Thessaloniki, Greece, 2012; pp. 211–216.
43. Lungu, V.; Chera, C. Importuri de vase de sticlă suflate în tipar descoperite în necropolele Tomisului. *Pontica* **1992**, *25*, 273–280.
44. Cliante, L.; Țârlea, A. Secondary glass kilns and local glass production in Tomis during the Roman times. *CICSA J.* **2020**, *6*, 117–133.
45. Szentmiklósi, L.; Belgya, T.; Révay, Z.; Kis, Z. Upgrade of the prompt gamma activation analysis and the neutron-induced prompt gamma spectroscopy facilities at the Budapest research reactor. *J. Radioanal. Nucl. Chem.* **2010**, *286*, 501–505. [CrossRef]
46. Révay, Z.; Molnár, G.L. Standardisation of the prompt gamma activation analysis method. *Radiochim. Acta* **2003**, *91*, 361–369. [CrossRef]
47. Yonezawa, C. Quantitative analysis. In *Handbook of Prompt Gamma Activation Analysis with Neutron Beams*; Molnár, G.L., Ed.; Kluwer Academic Publishers: Dordrecht, The Netherlands; Boston, MA, USA; New York, NY, USA, 2004; pp. 113–135.
48. Révay, Z.; Belgya, T.; Kasztovszky, Z.; Weil, J.L.; Molnar, G.L. Cold neutron PGAA facility at Budapest. *Nucl. Instrum. Method B* **2004**, *213*, 385–388. [CrossRef]
49. Révay, Z. Determining elemental composition using Prompt γ Activation Analysis. *Anal. Chem.* **2009**, *81*, 6851–6859. [CrossRef]
50. Constantinescu, B.; Cristea-Stan, D.; Szőkefalvi-Nagy, Z.; Kovács, I.; Harsányi, I.; Kasztovszky, Z. PIXE and PGAA—Complementary methods for studies on ancient glass artefacts from Byzantine, Late Medieval to modern Murano glass. *Nucl. Instrum. Method B* **2018**, *417*, 105–109. [CrossRef]
51. Kasztovszky, Z.; Kunicki-Goldfinger, J.J.; Dzierżanowski, P.; Nawrolska, G.; Wawrzyniak, P. (2005) PGAA and EPMA as complimentary nondestructive methods for analysis of boron content in historical glass. In Proceedings of the Art'05–8th International Conference on Non Destructive Investigations and Microanalysis for the Diagnostics and Conservation of the Cultural and Environmental Heritage, Lecce, Italy, 15–19 May 2005.
52. Moropoulou, A.; Zacharias, N.; Delegou, E.T.; Maróti, B.; Kasztovszky, Z. Analytical and technological examination of glass tesserae from Hagia Sophia. *Microchem. J.* **2016**, *125*, 170–184. [CrossRef]
53. Zacharias, N.; Kaparou, M.; Oikonomou, A.; Kasztovszky, Z. Mycenaean glass from the Argolid, Peloponnese, Greece: A technological and provenance study. *Microchem. J.* **2018**, *141*, 404–417. [CrossRef]
54. Adlington, L.W. The Corning archaeological reference glasses: New values for "old" compositions. *Pap. Inst. Archaeol.* **2017**, *27*, 1–8.
55. Révay, Z. Calculation of uncertainties in Prompt Gamma Activation Analysis. *Nucl. Instrum. Method A* **2006**, *564*, 688–697. [CrossRef]
56. Schreiner, M.; Woisetchlaeger, G.; Schmitz, I.; Wadsak, M. Characterisation of surface layers formed under natural environmental conditions on medieval stained glass and ancient copper alloys using SEM, SIMS and atomic force microscopy. *J. Anal. At. Spectrom.* **1999**, *14*, 395–403. [CrossRef]
57. Shortland, A.J.; Schachner, L.; Freestone, I.; Tite, M. Natron as a flux in the early vitreous materials industry: Sources, beginnings and reasons for decline. *J. Archaeol. Sci.* **2006**, *33*, 521–530. [CrossRef]
58. Al-Bashaireh, K.; Al-Mustafa, S.; Freestone, I.C.; Al-Housan, A.Q. Composition of Byzantine glasses from Umm el-Jimal, northeast Jordan: Insights into glass origins and recycling. *J. Cultural Herit.* **2016**, *21*, 809–818. [CrossRef]
59. Freestone, I.C. Appendix: Chemical analysis of "raw" glass fragments. In *Excavations at Carthage II, 1. The Circular Harbor, the Site and Finds Other Than Pottery*; Hurst, H.R., Ed.; Oxford University Press for British Academy: Oxford, UK, 1994.
60. Brill, R.H. Scientific investigations of the Jalame glass and related finds. In *Excavations at Jalame: Site of a Glass Factory in Late Roman Palestine*; Weinberg, G.D., Ed.; University of Missouri Press: Columbia, MO, USA, 1988; pp. 257–294.
61. Ceglia, A.; Cosyns, P.; Nys, K.; Terryn, H.; Thienpont, H.; Meulebroeck, W. Late Roman Empire glass distribution and consumption in Cyprus: A chemical study. *J. Archaeol. Sci.* **2015**, *61*, 213–222. [CrossRef]
62. De Juan Ares, J.; Schibille, N.; Molina Vidal, J.; Sanchez de Prado, M.D. The supply of glass at *Portus Ilicitanus* (Alicante, Spain): A meta-analysis of HIMT glasses. *Archaeometry* **2019**, *61*, 647–662. [CrossRef]
63. Freestone, I.C.; Degryse, P.; Lankton, J.; Gratuze, B.; Schneider, J. HIMT, glass composition and commodity branding in the primary glass industry. In *Things That Travelled, Mediterranean Glass in the First Millenium CE*; Rosenow, D., Phelps, M., Meek, A., Freestone, I., Eds.; UCL Press: London, UK, 2018; pp. 159–190.
64. Mirti, P.; Casoli, A.; Appolonia, L. Scientific analysis of Roman glass from Augusta Praetoria. *Archaeometry* **1993**, *35*, 225–240. [CrossRef]
65. Drauschke, J.; Greiff, S. (Eds.) Chemical aspects of Byzantine glass from Caričin Grad/Iustiniana Prima (Serbia). In *Glass in Byzantium: Production, Usage, Analyses: International Workshop Organised by the Byzantine Archaeology Mainz, 17–18 of January 2008*; Verlag des Römisch-Germanischen Zentralmuseums: Mainz, Germany, 2010; pp. 25–46.

66. Adam-Veleni, P. (Ed.) *Glass Cosmos*; Archaeological Museum of Thessaloniki: Thessaloniki, Greece, 2010.
67. Antonaras, A. *Fire and Sand. Ancient Glass in the Princeton University Art Museum*; Princeton University Art Museum: Princeton, NJ, USA, 2012.
68. Arveiller-Dulong, V.; Arveiller, J. *La Verre D'epoque Romaine au Musée Archéologique de Strasbourg*; Éditions de la Réunion des Musées Nationaux: Strasbourg, France, 1985.
69. Arveiller-Dulong, V.; Nenna, M.D. *Les Verres Antiques du Musée du Louvre, Vol. II: Vaisselle et Contenants du Ier Siècle au Début du VIIe Siècle Après J.-C., Musée du Louvre*; Editions D'Art Somogy: Paris, France, 2005.
70. Atila, C.; Gürler, B.; Özerler, M.; Ünsalan, D. *Glass Objects from Bergama Museum/Bergama Muzesi Cam Eserleri*; Zero Books: Izmir, Turkey, 2009.
71. Buljević, Z. Imprints on the bottoms of glass bottles from Dalmatia held in the Archaeological Museum in Split. In *Corpus des Signatures et Marques sur Verres Antiques*; Foy, D., Nenna, M.D., Eds.; Montagnac: Aix-en-Provence–Lyon, France, 2011; Volume 3, pp. 179–195.
72. Foy, D. *Les Verres Antiques d'Arles*; La collection du Musée Départamental Arles Antique: Paris, France, 2010.
73. Gorin-Rosen, Y. The glass finds from Horbat Zefat 'Adi (east). *Hadashot Arkheologiyot* **2015**, *127*, 1–11.
74. Gorin-Rosen, Y. Glass finds and remains of a glass industry from Miska. *Atiqot* **2020**, *99*, 135–168.
75. Gorin-Rosen, Y.; Jackson-Tal, R. Chapter 9: Area F: The glass finds. In *Paneas I, IAA Reports 37*; Tzaferis, V., Israeli, S., Eds.; Israel Antiquities Authority Publications Department: Jerusalem, Israel, 2008; pp. 141–154.
76. Hayes, J.W. *Roman and Pre-Roman Glass in the Royal Ontario Museum: A Catalogue*; Royal Ontario Museum: Toronto, ON, Canada, 1975.
77. Israeli, Y. *Ancient Glass in the Israel Museum*; The Eliahu Dobkin Collection and Other Gifts, with Contributions by Dan Barag and Na'ama Brosh; The Israel Museum: Jerusalem, Israel, 2003.
78. Lightfoot, C.S. *Ancient Glass in National Museums Scotland*; NMSE Publishing Ltd.: Cambridge, UK, 2007.
79. Lightfoot, C.S. *Cesnola Collection of Cypriot Art*; Ancient Glass; The Metropolitan Museum of Art: New York, NY, USA, 2017.
80. Lightfoot, C.S.; Arslan, M. *Ancient Glass of Asia Minor*; The Yüksel Erimtan Collection: Ankara, Turkey, 1992.
81. Whitehouse, D. *Roman Glass in the Corning Museum of Glass*; Corning: New York, NY, USA, 1997; Volume I.
82. Whitehouse, D. *Roman Glass in the Corning Museum of Glass*; Corning: New York, NY, USA, 2001; Volume II.

Article

Architectural Glazed Tiles Used in Ancient Chinese Screen Walls (15th–18th Century AD): Ceramic Technology, Decay Process and Conservation

Jingyi Shen [1], Li Li [2,*], Ji-Peng Wang [3,*], Xiaoxi Li [4], Dandan Zhang [1], Juan Ji [2] and Ji-Yuan Luan [3]

- [1] School of History and Culture, Shandong University, Jinan 250100, China; jingyi.shen@sdu.edu.cn (J.S.); echo_zhangdandan@126.com (D.Z.)
- [2] Shaanxi Institute for the Preservation of Cultural Heritage, Xi'an 710075, China; jijuan107@163.com
- [3] School of Civil Engineering, Shandong University, Jinan 250012, China; Jasperluan@outlook.com
- [4] Department of Conservation, Emperor Qinshihuang's Mausoleum Site Museum, Xi'an 710600, China; riverlx2@126.com
- * Correspondence: lili@sxwby.org.cn (L.L.); ji-peng.wang@sdu.edu.cn (J.-P.W.)

Citation: Shen, J.; Li, L.; Wang, J.-P.; Li, X.; Zhang, D.; Ji, J.; Luan, J.-Y. Architectural Glazed Tiles Used in Ancient Chinese Screen Walls (15th–18th Century AD): Ceramic Technology, Decay Process and Conservation. *Materials* **2021**, *14*, 7146. https://doi.org/10.3390/ma14237146

Academic Editors: Ziga Smit and Eva Menart

Received: 8 October 2021
Accepted: 20 November 2021
Published: 24 November 2021

Publisher's Note: MDPI stays neutral with regard to jurisdictional claims in published maps and institutional affiliations.

Copyright: © 2021 by the authors. Licensee MDPI, Basel, Switzerland. This article is an open access article distributed under the terms and conditions of the Creative Commons Attribution (CC BY) license (https://creativecommons.org/licenses/by/4.0/).

Abstract: The glazed tile is an important building material used throughout the history of traditional Chinese architecture. Architectural glazed tiles used to decorate the screen walls of ancient China are studied scientifically for the first time. More than 30 glazed tile samples from the screen walls of the 15th to 18th century AD of the Hancheng Confucian Temple and Town God's Temple in Shaanxi Province were carefully investigated using SEM–EDS and XRD. Microstructure and chemistry indicated the raw materials, the recipes and the technological choices used to produce the paste and glaze of the glazed tile samples studied. The causes for the key degradation processes of these glazed tiles used as building materials in the screen walls have also been discussed. This work has clear implications for the restoration and conservation treatments on these kinds of ancient Chinese building materials.

Keywords: glazed tile; screen wall; ceramic technology; decay; conservation; ancient China

1. Introduction

1.1. Chinese Glazed Tiles

The glazed tile is an important building material used throughout the history of traditional Chinese architecture due to its artistic appearance and waterproof qualities. Additionally, glazed tiles were widely used in ancient top-tier Chinese architectural structures such as royal palaces, mausoleums, imperial gardens and temples as a significant part of ancient architectural etiquette. Architectural glazed tiles were used to decorate imperial palaces in China as far back in history as the Northern Wei Dynasty (386–534 AD) [1]. Until the Ming (1368–1644 AD) and Qing (1644–1911 AD) dynasties, the use of architectural glazed tiles grew to the peak period. According to the different positions and functions used in ancient architecture, glazed tiles can be divided into two categories. The first is the glazed tiles used as building materials on the roofs, such as roofing tile-ends (known as Wadang in Chinese), pan-roofing tiles and roll-roofing tiles (Figure 1a). The other kind is glazed tiles decorated on the walls, such as the walls of pagodas and screen walls (known as Yingbi in Chinese) (Figure 1b).

Previous scientific research on Chinese glazed tiles has primarily focused on the following areas: the manufacturing technique of glazed tiles production [2–4]; the composition and provenance of their raw materials [5,6] and the main deterioration mechanism and conservation including glaze fading and the shedding of the glaze layer [7–9]. However, the glazed tiles that have been studied are mainly from the Beijing city and its surrounding areas, while architectural ceramic tiles from other areas have rarely been studied scientifically. In addition, the majority of previous studies mainly assessed the glazed tiles used on the roofs, whereas studies on the glazed tiles used to decorate the walls of ancient architecture are scarce. The only scientific research conducted to date is a study of the manufacturing technique of glazed tiles used to decorate the walls of the Ming dynasty Baoensi Pagoda located in Nanjing City in South China [10].

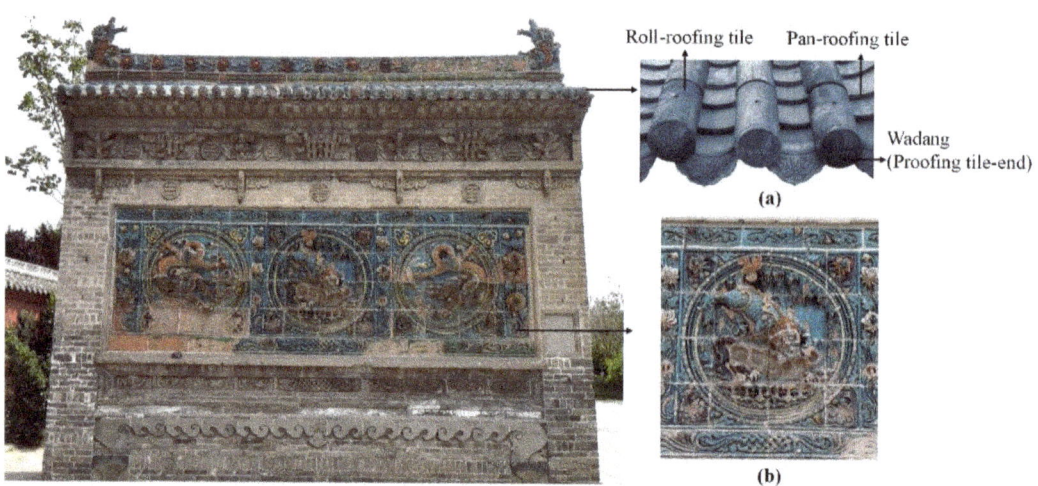

Figure 1. Glazed tiles decorated in different positions of ancient Chinese architecture. (This picture is the South screen wall in the Hancheng Confucian Temple) (**a**) Glazed tiles decorated on the roof. (**b**) Glazed tiles decorated on a screen wall.

Glazed tiles decorated on the roofs and walls are used in different positions of architectures with different shapes. This means that they may have distinct ceramic technology, suffer from different decay processes and have variation in how well they can be preserved. Moreover, these differences may also exist in the glazed tiles produced and preserved in different areas of China. Therefore, it is of great significance to especially study the representative glazed tiles used as ancient building materials to decorate walls, which is crucial for taking effective protection measures for such building materials.

The present paper scientifically studies the Chinese glazed tiles used to decorate ancient screen walls, which is rarely considered in previous studies. Raw materials used, the manufacturing techniques applied and the degradation processes of this kind of glazed tiles in the Hancheng Confucian Temple and Town God's Temple are discussed in detail for the first time. This research enables us to propose correspondingly helpful suggestions on the preservation and restoration of these tiles.

1.2. The Glazed Tiles on the Screen Walls of the Hancheng Confucian Temple and Town God's Temple Building Complexes

The screen wall is an isolated standing wall normally built inside the courtyard of a traditional Chinese building complex, or located on each side of the gate, to shield the rest of the buildings from view. This paper focuses on the exquisite glazed tiles which decorate the screen walls of the Hancheng Confucian Temple and the Hancheng Town God's Temple (known as Chenghuang Miao in Chinese) building complexes in Hancheng City, Shaanxi province. The locations of these buildings are shown in Figure 2.

The Confucian Temple is built in the memory of Confucius, who is the originator of Confucius culture, while the Town God's Temple is the Taoist temple of the town god. The Hancheng Town God's Temple is the most integrated Town God's Temple building complex in Northwest China, and the Hancheng Confucian Temple is the third-largest Confucian temple building complex. Both the Hancheng Confucian Temple and Town God's Temple building complex have a prestigious reputation because of their exquisite glazed tiles decorated on the screen walls. All these enable them to be concerned as key historical and cultural sites under state protection in China.

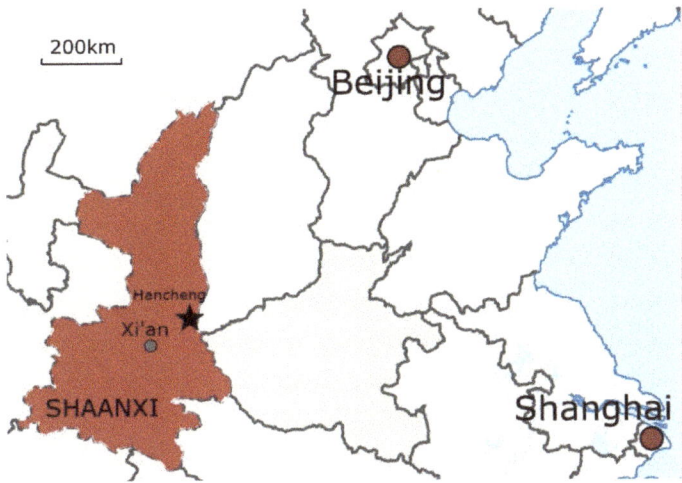

Figure 2. The location of the Hancheng city where the screen walls studied are located.

In the Hancheng Confucian Temple building complex, the glazed tile decoration is mainly concentrated on five screen walls comprising a total area of approximately 139.34 m². The main decorative patterns of these glazed tiles are dragons and phoenixes, as well as some flower patterns. All these dragon and phoenix patterns are vivid in colors and shapes, and the total number of dragon patterns is exactly nine, representing the highest level of architectural etiquette. Several screen walls with glazed tiles can be seen in Figure 3.

(a) Five-Dragon screen wall

(b) East screen wall of Lingxing Gate

(c) Glazed tiles on west screen wall of Lingxing Gate

Figure 3. Screen walls with glazed tile decoration in the Hancheng Confucian Temple building complex.

In the Hancheng Town God's Temple building complex, the glazed tile decoration is mainly centered on five screen wall buildings with a total area of around 225.34 m^2. Almost all the screen walls are mainly decorated with glazed tiles of turquoise color, and supplemented by green, yellow and brown glazed tiles which are used to form patterns. The main decorative patterns of these glazed tiles are dragons, phoenixes and tigers, as well as scenery with hills, water, pavilions, flowers and little animals. Several screen walls with glazed tiles can be seen in Figure 4.

(a) Screen wall of Biping Gate and glazed tiles

(b) Screen walls of Xizhi Gate and glazed tiles

Figure 4. Screen walls with glazed tile decoration in the Hancheng Town God's Temple building complex.

2. Materials and Methods

2.1. Sample Information

Thirty-three fragments of glazed tiles traced back to several different periods of the Ming and Qing dynasties (ca. 1482–1722 AD) were collected in this study. They include 17 pieces from four different screen walls of the Hancheng Confucian Temple and 16 samples from 4 different screen walls of the Town God's Temple. The glazed tile samples cover all glaze colors, namely green, yellow, brown and turquoise. Details are shown in Table 1. The appearances of some typical samples are shown in Figure 5.

Table 1. Sample information.

Sample No.	Location	Production Year	Glaze Color	Sample No.	Location	Production Year	Glaze Color
Glazed tile samples from the Hancheng Confucian Temple				Glazed tile samples from the Hancheng Town God's Temple			
C-1	Five-Dragon screen wall	within the Wanli Period of the Ming Dynasty (1573–1620 AD)	green	T-1	South screen wall	the 44th year of the Wanli Period of the Ming Dynasty (1616 AD)	turquoise
C-2			green and yellow	T-2			turquoise
C-3			turquoise	T-3			yellow
C-4			brown	T-4			yellow
C-5	East screen wall of Lingxing Gate	the 18th year of the Chenghua Period of the Ming Dynasty (1482 AD)	green	T-5	West screen wall of Biping Gate	the 44th year of the Wanli Period of the Ming Dynasty (1616 AD)	turquoise
C-6			green and yellow	T-6			turquoise
C-7			yellow	T-7			yellow
C-8	East screen wall of Halberd Gate	the 18th year of the Chenghua Period of the Ming Dynasty (1482 AD)	green	T-8	East screen wall of Biping Gate	the 44th year of the Wanli Period of the Ming Dynasty (1616 AD)	turquoise
C-9			turquoise	T-9			yellow
C-10			turquoise	T-10			yellow
C-11			brown	T-11			turquoise
C-12	West screen wall of Halberd Gate	the 18th year of the Chenghua Period of the Ming Dynasty (1482 AD)	yellow	T-12	Screen walls of Xizhi Gate	within the Kangxi Period of the Qing Dynasty (1662–1722 AD)	yellow
C-13			green and yellow	T-13			green and brown
C-14			green and yellow	T-14			green
C-15			yellow and turquoise	T-15			green
C-16			brown	T-16	East screen wall of Biping Gate	the 44th year of the Wanli Period of the Ming Dynasty (1616 AD)	turquoise
C-17			turquoise				

Figure 5. Photographs of the samples representing the four glaze colors in this study. (**a**) Green; (**b**) Turquoise; (**c**). Yellow; (**d**). Brown.

2.2. Analytical Methods

2.2.1. Scanning Electron Microscopy (SEM) with Energy Dispersive Spectroscopy (EDS)

A ZEISS instrument, model EVO-250 SEM (Zeiss, the instrument is located in Xi'an, China), coupled with an Oxford Instruments X-MAX20 EDS system, was used to study the sample's microstructure and the body and glaze compositions. A cross-section of each representative sample was taken off and mounted as a polished block. The acceleration voltage used for observation and analysis was 20 keV with a working distance of 8 mm. SEM backscattered electron (BSE) images of the microstructure of the body paste and glaze were recorded. For each sample, the chemical compositions of the tile body and glaze were analyzed in three micro areas and the average values were calculated. The size analyzed by EDS both for the body and the glaze is around 200 µm × 300 µm. The test time of EDS was 90 s. Oxford Instruments standards were used to quantify 10 elements (Si, Al, Na, K, Mg, Ca and Fe for the tile body, while Pb, Si, Al, Na, K, Mg, Ca, Fe, Cu and Ti for the tile glaze). The results were converted into oxide percentages and normalized to 100%. The detection limit of the instrument is 0.1%. The glaze chemical compositions of all 33 samples have been analyzed in this study. For the chemical compositions of paste, only 23 samples (C1–10 and T1–13) have been analyzed. The reason is that the pastes of samples C11–17 and samples T14–16 have already decayed heavily, so the pastes of these samples look loose and fragile (having suffered from salt efflorescence); therefore, if these samples were tested for their chemical compositions, inaccurate results would be obtained.

2.2.2. X-ray Diffraction (XRD)

Eighteen tile bodies of the samples and two samples of solute salt powder were tested by XRD (RIGAKU SmartLab) for their phase analysis. Before the test, a small amount of the tile body of each sample was taken off and ground into fine powder. A 2θ range of 5–55° was used, with the detector type being a D/teX Ultra 250 silicon strip detector device (SmartLab, the instrument is located in Xi'an, China). A tube voltage of 30 kV with a current of 300 mA was applied.

2.2.3. Analytical Methods for Degradation Process

To discuss body salt efflorescence, a glazed tile sample (T-15) with intact glaze but weathered-looking paste was scanned by a micro-CT scanner (Nanovoxel 3000, the instrument is located in Tianjing, China) with 190 keV voltage, and a 25 W power was used on the X-ray source. Sequences of grayscale images were obtained, which can be reconstructed into a 3D image. The image resolution is 0.5 µm per voxel length.

The moisture contents at different heights (25 to 250 cm from the ground with 25 cm as the measurement interval) of five screen walls have been tested with the HF SENSOR MOIST 210 non-destructive moisture measurer (HF SENSOR, the instrument is located in Xi'an, China). Due to the fact that variances in the moisture content at different depths of the screen wall also need to be compared, the areas for the test were measured at the gray brick zone to ensure consistency.

3. Results and Discussion

3.1. Manufacturing Techniques Reflected by Composition and Textural Morphology

3.1.1. Body

The SEM–EDS analysis revealed that all the glazed tile samples are made by siliceous paste (60.2–65.7 wt.% SiO_2) and contain 15.0–18.7 wt.% of Al_2O_3. The pastes have relatively high components of CaO (7.9 wt.% on average), Na_2O (2.3 wt.% on average), K_2O (2.8 wt.% on average) and MgO (2.9 wt.% on average) as a fluxing agent. Additionally, 4.7 wt.% FeO content on average was observed, which explains the samples' red hue fired in an oxide atmosphere. No significant differences in the major chemical compositions of glaze tile pastes could be observed between those used in the Confucian Temple screen walls and the Town God's Temple screen walls. The same applied to the glazed tile pastes made in different periods. The chemical compositions of loess-based north Chinese ceramics and Chinese loess were also listed in Table 2. The paste composition characteristic of these glazed tiles is similar to that of loess-based glazed tiles of the Palace Museum and the loess of North China, which has the characteristics of having low SiO_2 and Al_2O_3 content with a high content of fluxing oxides such as CaO and Na_2O. This implies that the glazed tiles in this study are possibly also made by loess. Small differences in chemical composition could be found between the loess and tile pastes because the loess used for the tile pastes was prepared by sorting and elutriation. It also needs to be pointed out that the chemical composition of some early period loess-based north Chinese ceramics is different from that of the glazed tiles in this study, especially for their lower content of CaO. This demonstrates that the kind of loess used to make Hancheng glazed tiles is characterized by its Ca-rich minerals.

Table 2. Major chemical compositions of the body of Hancheng glazed tile samples and loess from North China (wt.%). The data of loess from North China is taken from Wu et al. (1996) [11]; loess-based north Chinese ceramics is from Wood (2011, 197) [12]; glazed tiles of the Palace Museum is from Kang et al. (2018) [13] and the N means number of samples.

Sample No.	SiO_2	Al_2O_3	Na_2O	K_2O	MgO	CaO	FeO
Glazed tile samples from the Hancheng Confucian Temple							
C-1	60.2	18.7	1.8	4.1	3.1	5.9	6.1
C-2	61.7	18.1	1.9	3.3	3.4	5.2	6.3
C-3	61.9	18.5	2.5	2.8	3.0	5.8	5.4
C-4	62.5	17.2	2.7	3.1	3.0	8.1	3.5
C-5	62.4	16.5	2.1	2.9	2.6	8.3	5.2
C-6	64.5	16.4	3.2	2.4	2.5	6.8	4.2
C-7	61.3	15.0	2.9	4.1	2.8	8.1	6.0
C-8	64.2	17.4	1.4	2.4	2.5	8.8	3.3
C-9	62.5	17.0	2.3	2.4	3.0	8.7	4.0
C-10	60.9	15.9	2.1	2.2	2.9	10.5	5.3
Glazed tile samples from the Hancheng Town God's Temple							
T-1	61.8	16.6	1.4	2.8	2.2	9.3	5.9
T-2	60.9	17.3	2.6	2.7	3.6	7.5	5.5
T-3	61.8	16.9	2.4	2.2	3.5	10.2	3.2
T-4	63.4	17.7	1.9	1.9	2.6	9.8	2.9
T-5	63.5	17.8	1.7	2.5	2.9	7.1	4.6
T-6	61.8	16.7	2.9	3.1	2.9	8.3	4.0
T-7	62.6	14.6	3.2	3.5	2.7	9.6	3.7
T-8	62.9	15.8	2.8	3.4	2.3	7.4	5.3
T-9	61.7	16.0	2.7	2.7	4.0	6.8	6.0
T-10	65.7	17.7	2.5	2.3	2.5	5.4	4.0
T-11	64.5	16.1	2.6	1.8	3.0	7.7	4.3
T-12	65.7	16.5	2.3	2.1	2.4	7.3	3.7
T-13	62.1	16.9	1.7	4.0	3.0	8.0	4.4
Loess of North China (chemical compositions on average) [11]							
Sandy loess (N = 23)	69.3	12.9	2.2	2.5	2.6	7.8	2.8
Loess (N = 97)	66.3	13.7	1.9	2.6	2.7	9.3	3.3
Clay loess (N = 27)	65.6	13.9	1.8	2.6	2.7	10.1	3.4
Analysis of loess-based North Chinese ceramics [12]							
Qin terracotta warrior, Lishan	66.3	16.6	2.0	3.3	2.3	2.0	6.1
Qin terracotta horse, Lishan	63.2	15.9	1.6	2.9	2.1	2.6	6.1
Han dynasty funerary jar	65.8	15.8	1.6	3.3	2.1	2.1	5.2
Shang dynasty earthenware pipe	66.5	16.9	1.3	3.0	2.0	2.8	6.5
Yuan dynasty glazed tiles (N = 8), the Palace Museum [13]	62.1	16.8	2.2	2.6	2.6	8.2	4.6

Some inclusions of quartz, K-feldspar, calcite and hematite could be found in sample paste by SEM–SDS analysis, and sometimes illite, biotite and zircon also could be observed (the SEM images of sample C-1, C-8, T-1 and T-2 can be seen in Figure 6). The mineral phases of the paste samples identified by XRD have been summarized in Table 3; quartz (SiO_2), K-feldspar ($K_2O \cdot Al_2O_3 \cdot 6SiO_2$), plagioclase ($Na(AlSi_3O_8)$–$Ca(Al_2Si_2O_8)$) and hematite

(Fe_2O_3) were observed in all the samples studied, and K-feldspar, plagioclase and hematite should be the main sources of Na_2O, K_2O and Fe_2O_3 in these pastes, respectively. This mineralogy, consistent with the SEM–EDS analyses, is indicative of common loess, which is mainly composed of non-clay minerals including quartz, K-felspar, plagioclase and calcite as well as clay minerals [14].

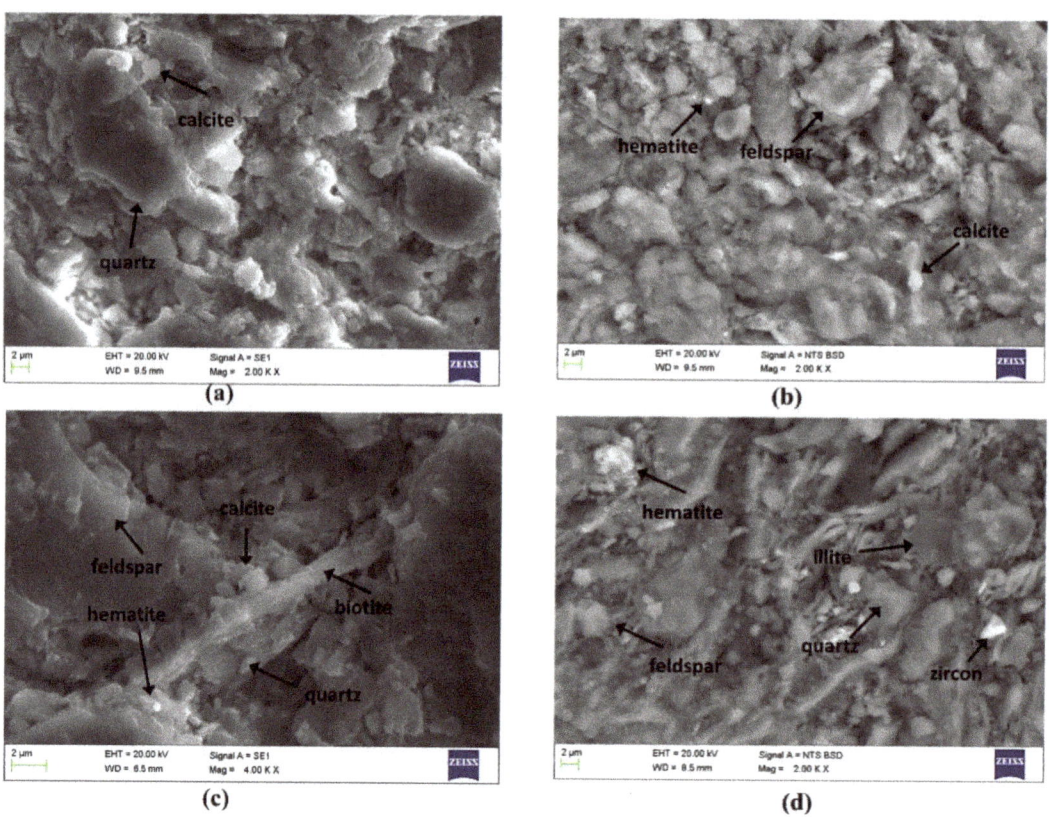

Figure 6. SEM backscattered electron mode images representative of inclusions of sample paste. (**a**) C-1; (**b**) C-8; (**c**) T-1 and (**d**) T-2.

One significant difference between the paste samples was identified. Mullite was detected in most samples, while illite was found in five samples with no mullite detected. The presence of mullite in the paste samples indicates a relatively high firing temperature of at least 1000 °C, since this phase starts nucleation at a range of 1000–1100 °C from the calcium silicates in the matrix [15,16]. However, the paste samples that have mullite also contain dolomite and/or calcite, which usually do not exist above 900 °C [17]. A possible reason for this contradiction might be understood by the following explanation. Due to the influence of many factors, such as the poor control of firing atmosphere in ancient kilns, ancient glazed tiles might be heated unevenly in the firing process. The existence of mullite in tile paste at least implies that the local firing temperature for these paste samples may reach more than ca. 1000 °C. Calcite and dolomite in clay might decompose and produce CaO and MgO during high-temperature firing (above ca. 900 °C). Then CaO and MgO in tile paste might react with H_2O and CO_2 when the glazed tiles are exposed to outer conditions long-term. Finally, small amounts of $CaCO_3$ (calcite) and $CaMg(CO_3)_2$ (dolomite) are formed. In contrast, the paste samples C5, T1, T2, T3 and T4, which have the mineral illite, might have been fired at a relatively low temperature, below 1000 °C [16]. This indicates that these architectural glazed tiles might have been produced at the firing temperature of ca. 900–1000 °C (this is only a tentative inference, and further analysis of sintering temperature is needed). They were produced in batches and at different times, or made in a different workshop, and the firing temperature of the different batches fluctuated. However, the difference in firing temperature is not reflected in the chronological sequence because the tiles decorated on the South screen wall and the screen wall of Biping Gate, both made in 1616 AD, seem to have been produced by different firing temperatures.

Table 3. Mineral composition of the body of Hancheng glazed tile samples.

Sample No.	Location	Mineral Composition
C-1	Five-Dragon screen wall (1573–1620AD)	quartz, K-feldspar, plagioclase, mullite, calcite, dolomite, hematite
C-2		quartz, K-feldspar, plagioclase, gypsum, mullite, hematite
C-3		quartz, K-feldspar, plagioclase, mullite, dolomite, hematite
C-5	East screen wall of Lingxing Gate (1482AD)	quartz, K-feldspar, plagioclase, calcite, illite, hematite
C-6		quartz, K-feldspar, plagioclase, dolomite, mullite, hematite
C-8	East screen wall of Halberd Gate (1482AD)	quartz, K-feldspar, plagioclase, mullite, calcite, dolomite, hematite
C-9		quartz, K-feldspar, plagioclase, mullite, gypsum, calcite, dolomite, hematite
C-10		quartz, K-feldspar, plagioclase, gypsum, mullite, hematite
T-1	South screen wall (1616AD)	quartz, K-feldspar, plagioclase, calcite, illite, hematite
T-2		quartz, K-feldspar, plagioclase, calcite, illite, dolomite, hematite
T-3		quartz, K-feldspar, plagioclase, gypsum, illite, hematite
T-4		quartz, K-feldspar, plagioclase, calcite, illite, hematite
T-5	West screen wall of Biping Gate (1616AD)	quartz, K-feldspar, plagioclase, mullite, dolomite, hematite
T-8		quartz, K-feldspar, plagioclase, dolomite, mullite, hematite
T-9	East screen wall of Biping Gate (1616AD)	quartz, K-feldspar, plagioclase, dolomite, mullite, hematite
T-10		quartz, K-feldspar, plagioclase, mullite, hematite
T-12	Screen walls of Xizhi Gate (1662–1722AD)	quartz, K-feldspar, plagioclase, mullite, calcite, hematite
T-13		quartz, K-feldspar, plagioclase, mullite, calcite, hematite

Both the chemical composition and mineral phase results of the tile samples demonstrate that all the tile pastes in this study were highly possibly made by local loess. However, according to published studies on the Chinese architectural glazed tiles used as building materials in the roof, the majority of the tile pastes were made from local clay or porcelain stone [18]. So far, it has been solely found that the paste of turquoise glazed tiles from the Beijing Palace Museum was produced from loess [13]. Loess has been selected as a raw material for ceramic production in North China for a significantly long period due to its good plasticity. For example, it was used in the Neolithic pottery of the Banpo Site and the terracotta sculptures depicting the armies of Qin Shi Huang, both located in the Shaanxi Province [12].

3.1.2. Glaze

Fifteen samples were observed by SEM backscattered electron mode images. The thickness of the glaze ranges from around 200 to 600 μm. The majority of the samples (13 of 15 samples) were found to contain a small number of quartz inclusions in the glaze as shown in Figure 7a–c, while two samples contain a large number of quartz particles in the glaze, as shown in Figure 7d. The quartz inclusions show variable sizes, between ca. 5 and 150 μm, and variable shapes: irregular, rectangular and oval. Additionally, bubbles of different sizes were also observed in the glaze. The sizes of bubbles are between ca. 5 and 50 µm in diameter. The quartz inclusions and bubbles, especially those of a large size, increase the opacity of the glaze and also contribute to the cracking and shedding of the glaze.

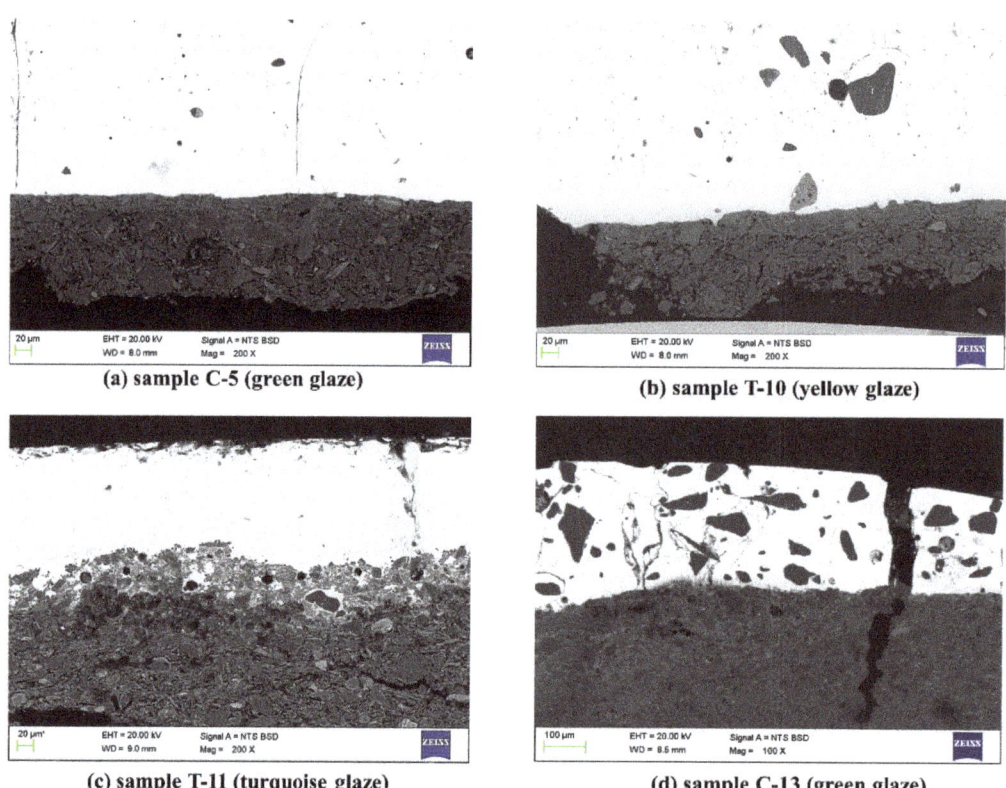

Figure 7. SEM backscattered electron mode images representative of glazed tile samples. (**a**–**c**) Glaze with a small number of quartz inclusions; (**d**) glaze with a large number of quartz particles.

Images obtained by SEM also show that most of the samples in this study have almost no body-glaze interface (similar to the samples in Figure 7a,b,d) or a very clear and thin interface (similar to the sample in Figure 7c). This indicates that the glazed tiles were likely produced by a double-firing process. In specific, during the first time firing of the tile paste itself, most of the paste raw materials-loess form a relatively stable phase at high temperature. Then when the glaze suspension is applied on the already biscuit-fired body for the second time firing, a fewer concentration of diffused elements between glaze and paste is formed, and then leads to a less obvious or thinner glaze-body interface [19].

As highlighted in Table 4, the glaze of the tile samples is mainly comprised of SiO_2, Al_2O_3, PbO, K_2O, Na_2O, MgO, CaO, CuO and FeO. Among them, SiO_2 and Al_2O_3 are indispensable components in all types of ceramic glaze. They react with flux oxides at a high temperature to form the glaze. PbO, K_2O, Na_2O, MgO and CaO are common flux oxides and are used in different types of glazes to reduce the firing temperature of glaze production. CuO and FeO are the main colorants for glazes.

Table 4. Major chemical compositions of the glaze of Hancheng glazed tile samples (wt.%, b.d. means below the detection limits).

Sample No.	Glaze Colour	PbO	SiO$_2$	Al$_2$O$_3$	FeO	CuO	Na$_2$O	K$_2$O	CaO	MgO	TiO$_2$
C-1G	Green	62.7	28.3	1.9	0.8	2.6	1.1	0.6	1.2	0.8	b.d.
C-2G		59.0	31.6	1.8	0.5	3.7	1.0	0.3	1.3	0.3	0.2
C-5G		61.3	30.1	2.3	0.5	2.9	0.4	0.3	0.9	0.9	0.1
C-6G		62.8	30.6	1.8	0.6	2.3	0.4	0.4	0.8	0.2	0.1
C-8G		56.4	33.2	2.7	0.7	4.3	0.5	0.9	0.5	0.8	b.d.
C-13G		58.3	32.3	2.0	0.5	4.6	0.4	0.9	0.5	0.5	0.2
C-14G		57.6	31.7	2.1	0.7	3.7	0.7	0.7	1.3	0.9	0.2
T-13G		54.5	33.8	1.3	0.9	5.5	1.1	0.8	1.3	0.6	0.1
T-14G		55.2	33.7	2.0	0.9	4.5	1.3	0.8	0.9	0.6	0.2
T-15G		60.1	30.6	2.2	0.6	3.1	0.9	0.8	1.0	0.4	b.d.
C-2Y	Yellow	58.5	32.1	2.2	3.1	0.2	1.1	0.7	1.1	0.5	0.2
C-6Y		57.0	33.2	2.5	3.0	b.d.	1.1	0.8	1.6	0.9	b.d.
C-7Y		55.9	33.6	2.7	2.7	b.d.	1.2	0.9	1.6	0.7	0.2
C-12Y		59.4	32.2	1.7	3.3	b.d.	1.0	0.9	0.9	0.4	0.2
C-13Y		60.6	30.7	1.5	4.0	0.1	1.0	0.8	0.8	0.5	b.d.
C-14Y		57.3	34.9	0.9	3.7	b.d.	0.9	1.1	0.8	0.4	b.d.
C-15Y		61.8	31.1	0.9	3.3	b.d.	0.8	0.9	0.7	0.6	b.d.
T-3Y		58.5	31.8	1.7	4.6	b.d.	0.7	0.8	0.8	0.7	0.1
T-4Y		59.9	31.2	2.0	4.5	0.1	0.7	0.5	0.9	0.2	b.d.
T-7Y		59.7	30.6	2.0	4.6	b.d.	0.7	0.7	1.1	0.4	0.2
T-9Y		57.2	32.6	1.8	5.0	b.d.	0.9	0.7	1.3	0.5	0.2
T-10Y		60.3	29.6	1.8	4.0	b.d.	0.9	1.0	1.8	0.6	b.d.
T-12Y		59.4	32.1	2.4	2.6	0.1	1.2	0.7	0.6	0.7	b.d.
C-4B	Brown	58.9	32.2	2.2	4.2	b.d.	0.9	0.8	0.6	0.3	b.d.
C-11B		60.8	31.2	1.5	4.3	b.d.	0.6	0.6	0.6	0.5	b.d.
C-16B		58.3	35.5	1.6	2.4	b.d.	0.7	0.7	0.5	0.3	b.d.
T-13B		58.8	35.4	1.6	2.2	b.d.	0.6	0.5	0.7	0.2	b.d.
C-3T	Turquoise	18.6	62.6	1.5	0.6	2.0	4.7	7.7	0.9	0.9	b.d.
C-9T		16.7	62.1	2.0	0.3	2.3	4.5	9.5	1.4	0.9	b.d.
C-10T		17.7	62.3	1.1	0.7	3.4	4.8	8.5	0.8	0.4	b.d.
C-15T		18.8	60.3	2.6	0.7	2.6	4.1	8.2	0.9	1.4	0.3
C-17T		21.2	59.3	1.2	0.4	3.8	4.2	7.7	0.9	0.7	0.2
T-1T		22.0	56.3	1.5	0.9	4.7	3.2	7.7	1.5	1.8	0.1
T-2T		24.0	56.7	1.8	0.6	4.4	3.9	6.9	0.9	0.6	0.2
T-5T		19.9	59.0	2.1	0.9	4.1	3.6	8.2	0.9	1.0	b.d.
T-6T		18.4	60.6	2.0	0.3	4.3	4.1	7.9	1.1	1.1	b.d.
T-8T		17.4	60.1	1.6	1.0	4.9	4.9	7.9	0.8	1.0	0.1
T-11T		19.8	60.2	1.3	0.7	3.4	4.6	7.8	1.0	0.5	0.2
T-16T		19.4	60.4	2.3	0.8	3.2	5.0	6.5	0.9	1.1	0.2

Samples of the glazed tiles considered in this study can be classified into two subgroups, according to their color and chemical composition.

Green, Yellow and Brown Glazes

Table 4 highlighted that the glaze samples which have a green, yellow and brown colors contain similar levels of SiO_2 (ranging from 28.3 to 35.5 wt.% and 32.1 wt.% on average), Al_2O_3 (ranging from 0.9 to 2.7 wt.% and 1.9 wt.% on average), PbO (ranging from 54.5 to 62.8 wt.% and 58.9 wt.% on average), K_2O+Na_2O (ranging from 0.7 to 2.1 wt.% and 1.5 wt.% on average) and MgO + CaO (ranging from 0.8 to 2.5 wt.% and 1.5 wt.% on average). This indicates that all these glazes are a SiO_2-Al_2O_3-PbO system, with lead oxide as the primary flux in the glaze. This corresponds to the results of previous studies which demonstrated that architectural tiles are normally decorated with a typically high-lead glaze, such as the glazed tiles which decorate the palaces and halls in the Beijing Palace Museum [20].

The analysis revealed that the green glaze samples have the largest copper oxide (CuO) component, ranging from 2.3 to 5.5 wt.%. This reveals that that CuO is the coloring agent. The yellow and brown glaze samples have a relatively high iron oxide (FeO) content ranging from 2.2 to 5.0 wt.%. This infers that Fe_2O_3 was used as the coloring agent for the yellow and brown glaze. In Chinese low-fired lead glaze, both the brown and yellow glaze are colored by iron oxide and fired in oxide atmospheres. In some cases, brown glaze is named as dark yellow glaze, and sometimes brown glaze and yellow glaze are called as dark amber and light amber, respectively. The color tones of the glazes are affected by the colorant content, firing atmosphere, ratios of Fe^{3+}/Fe^{2+} and other comprehensive factors.

There are two historical records that mentioned the glaze recipe of architectural glazed tiles, and the details have been organized in Table 5. The chemical compositions of tile glazes in this study are basically consistent with the glaze formula recorded in ancient literature. The lithargite mentioned in the ancient glaze recipe provides the content of PbO in glaze. The Luohe stone and Maya Stone are both quartzite collected from different areas, which provide the SiO_2 content in the glaze. They contain a high SiO_2 content, generally higher than 90% and up to more than 98%. Copper powder and ochre are the coloring materials of the green glaze and the yellow glaze, respectively. However, if we ignore the impurities in the raw materials and only calculate the contents of SiO_2, Al_2O_3, CuO and FeO (the results are shown in Table 5), it can be found that compared to the chemical compositions of green and yellow glazes analyzed in this study, the glaze recipe in historical records had a higher PbO content combined with a lower SiO_2 content. This infers that the glaze recipe mentioned in historical records has higher PbO/SiO_2 ratio (3:1) than that of glazes in this study (ranging from 1.6 to 2.2, and 1.8: 1 on average).

The lower PbO/SiO_2 ratios in tile glazes studied might be understood by the following three explanations: (A). In the actual glazed tile production, the glaze-making formula may be slightly different in different periods and regions; (B). In the actual firing process of the glaze, lead oxide might volatilize to some extent; (C). Following hundreds of years of erosion, the lead oxide in the glaze may have been dissolved by precipitation. This could have then led to the reduction of the lead content.

Turquoise Glaze

As Figure 8 highlights, in relation to the composition of the turquoise glazes, they are significantly different to the glaze samples of green, yellow and brown. Specifically, the turquoise glaze samples have lower contents of PbO (ranging from 16.7 to 24.0 wt.% and 19.5 wt.% on average) with higher contents of SiO_2 (ranging from 56.3 to 62.6 wt.% and 60.0 wt.% on average), K_2O (ranging from 6.5 to 9.5 wt.% and 7.9 wt.% on average) and Na_2O (ranging from 3.2 to 5.0 wt.% and 4.3 wt.% on average). This indicates that the turquoise glaze is a lead-alkaline glaze, mainly fluxed by PbO, K_2O and Na_2O together. Additionally, CuO (ranging from 2.0 to 4.9 wt.% and 3.6 wt.% on average) is used as the coloring agent. Copper ions can appear peacock blue or peacock green in color in alkaline oxide flux.

Table 5. The glaze formula of glazed tile production in historical records.

Information of Historical Record		Glaze Recipe of Glazed Tile				
Name	Description	PbO	SiO$_2$	FeO	CuO	PbO:SiO$_2$
Yingzao Fashi	Published in 1103 AD of the Song Dynasty The oldest technical treatise on architecture and craftmanship found in China	Lithargite 1500 g 69.9 wt.%	Luohe stone (a kind of quartzite) 500 g 23.3 wt.%	/	Copper powder 150 g 7.0 wt.%	3:1
Gongbuchangku Xuzhi	Published in 1615 AD of the Ming Dynasty It records the rules and management system of construction engineering in the Ming Dynasty	Lithargite 153 kg 73.6 wt.%	Maya stone (a kind of quartzite) 51 kg 24.5 wt.%	Ocher (a kind of hematite) 4 kg 1.9 wt.%	/	3:1

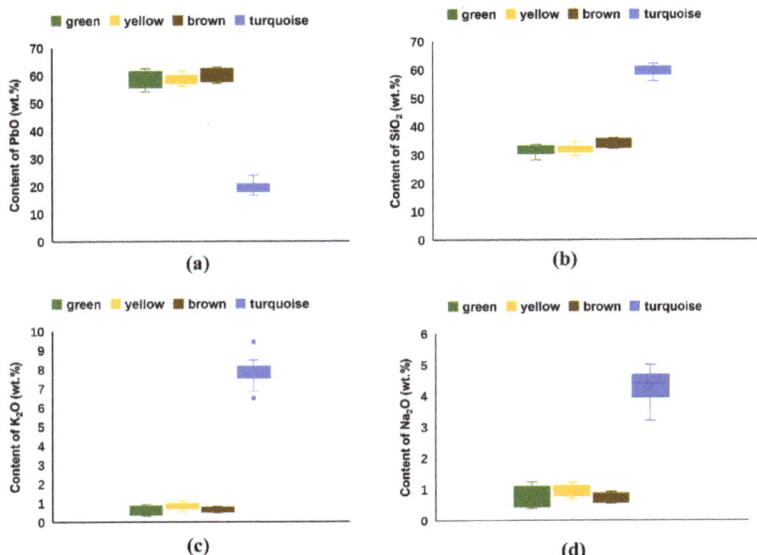

Figure 8. Plots of chemical compositions in glaze samples with different colors. (a) Content of PbO; (b) Content of SiO$_2$; (c) Content of K$_2$O and (d) Content of Na$_2$O.

In China, the turquoise glaze of the SiO$_2$-PbO-alkaline system, which was originally used for building materials of imperial palaces and temples, can be traced back to the middle period of the Tang dynasty (618–907 AD). Since the Northern Song dynasty (960–1127 AD), turquoise glaze began to be used in Chinese ceramics and its SiO$_2$-PbO-alkaline recipe became essential in the low-temperature glaze system of Chinese ceramics [21]. The famous Chinese glazes such as Fahua glazes and Cizhou polychrome alkaline glazes both have the SiO$_2$-PbO-alkaline recipe. It needs to pointed out that the maturing temperature of SiO2-PbO-alkaline recipe is normally above 1000 °C, which is higher than any typical Chinese high-lead glaze so far discussed (900–1000 °C) [12,22]. The SiO_2-PbO-alkaline is classified as a low-fired glaze in the history of Chinses glazes because it uses lead oxide as the main fluxing agent, which is totally different from the Chinese high-fired calcium glazes. The contents of alkaline (K_2O and Na_2O) in this recipe comes from nitrate and changes by different origins of the nitrate used. For example, the architectural turquoise-glazed tiles used in the Beijing Palace Museum [13] all have a similar glazing composition, of a SiO_2-PbO-K_2O system, whereas the glaze examined in this paper reveals that K_2O and Na_2O were combined to make the alkaline component. This indicates that distinct sources of nitrate were used in the turquoise glaze making of different areas.

3.2. Degradation Process

Field investigations found that due to long-term exposure to the open environment, multiple types of decay phenomena have occurred in these glazed tiles which decorate the screen walls. The glaze cracking and shedding and body salt efflorescence shown in Figure 9 are the most widespread and damaging types of decay.

Figure 9. Decay images. (**a**,**b**) Glaze cracking and shedding; (**c**) Body salt efflorescence.

3.2.1. Glaze Cracking and Shedding

Field investigations found that cracking and shedding are the most common deterioration to occur to tile glazes. This is mainly affected by the lead glaze production process itself and the exposure to the outdoor environment.

For the glazing technique, the glaze-body fit is an important factor for the stability of the glazed tile. If the thermal contraction of the glaze is greater than that of the body, the glaze will be under tension and the tensile stresses can cause cracking or 'crazing' of the glaze surface [20]. The optimum situation of the glaze-body fit is that the contraction of the glaze should be around 5–15% less than that of the body [23]. As discussed in Section 3.1.2, the glazed tiles studied in this research are decorated by lead glaze that has undergone a double firing process. The content of SiO_2 in the tile body is equivalent to that of PbO in the glaze, both of which are above 50 wt.% (for the turquoise glaze, the total content of PbO and SiO_2 is higher than 50 wt.%). Lead oxide has a relatively high thermal expansion coefficient. In the firing range of 1–100 °C, the thermal expansion coefficient of PbO is $4.2 \times 10^{-7}/°C$, while that of SiO_2 is relatively low, at $0.8 \times 10^{-7}/°C$ [24]. This means that for the lead glaze recipe, theoretically, the thermal expansion coefficient of the glaze should typically be higher than that of the body. Besides, according to Tite et al. (1998) [19], thermal expansion coefficients for typical high-lead glazes vary from 5 to $7 \times 10^{-6}/°C$.

In comparison, the expansion coefficients for non-calcareous earthenware bodies in the 0–500 °C range vary from 3 to $5 \times 10^{-6}/°C$ and from 5 to $7 \times 10^{-6}/°C$ for those made from calcareous clays (i.e., clays containing typically 15–25% of well-dispersed CaO). This means that in this study, theoretically, the thermal expansion coefficients of lead glazes might be similar or higher than their bodies (with 5.2–10.2% contents of CaO). Additionally, this theory that lead glazes have higher thermal expansion coefficients compared to their pastes has also been evidenced through the published data analysis of a range of high-lead glazed pottery [20] as well as via the glazed tiles which decorate the Beijing Palace Museum [24]. Consequently, the primary internal cause of the glaze-cracking process is the tensile stress of the lead glaze due to its higher expansion coefficient. Additionally, for those glazes with quartz inclusions and bubbles, the heterogeneity of the glaze also contributes to the cracking of the glaze.

The architectural glazed tiles are exposed to the atmospheric environment, which is more related to the problems of glazed tile conservation. The drastic changes with temperature and humidity (such as high temperature and torrential rain) can further lead to the glaze cracking because of the higher expansion coefficient of the glaze. When the tensile stresses difference at the joint of the tile body and glaze exceeds the limit, the glaze layer will even peel off. The glaze cracks not only weaken the strength of the glaze layer, but also allow the water (and salt) migration and crystallization to occur in both the tile glaze and body, which further leads to the glaze crazing and shedding.

3.2.2. Body Salt Efflorescence

Body salt efflorescent is another widespread phenomenon that occurs to the glazed tile decorations on the screen wall, and this disease normally occurs in a range of about 1–2 m from the ground. Although it is not discussed in this paper, the main body of the screen wall—the gray bricks—also suffer from the salt efflorescence disease in a similar area. As shown in Figure 9c, the tile glaze peels off, and the tile body looks loose. Additionally, some glazed tiles that seem to have intact glaze and are well preserved have also begun to undergo salt efflorescence.

The sample T-15 with intact glaze but weathered-looking paste has been scanned by a micro-CT scanner. Sequences of grayscale images were reconstructed into a 3D image, shown in Figure 10a,b. In the grayscale sliced image, the gray level is related to material density, as the brighter part is denser and the darkest part is the void. The image of the middle cross-section inside the sample is presented and rendered in Figure 10c,d. It can be seen that there are big and even connected pores with irregular shapes in the paste layer beneath the glaze layer, the biggest one showing a size of 1.6 mm. For those small holes in the paste, they either might be formed in the process of tile production or might be formed by weathering. However, for those connected holes with such big sizes above 1 mm, they demonstrate that the tile paste has suffered from salt efflorescent decay and has led to big pores in the paste, although the glaze layer of the sample is well-preserved temporarily.

To further explore the causes of the body salt efflorescent process, the moisture contents at different heights of five screen walls have been tested with non-destructive moisture measurements. The result of the Five-Dragon screen wall is shown in Figure 11. This demonstrates that the moisture content of the wall is clearly higher in the height range of approximately 1–2 m, which basically overlaps with the areas which have suffered from serious salt efflorescent decay. Additionally, the moisture content of the surface area (d = 2–3 cm) is evidently higher than that of the interior area (d = 25–30 cm).

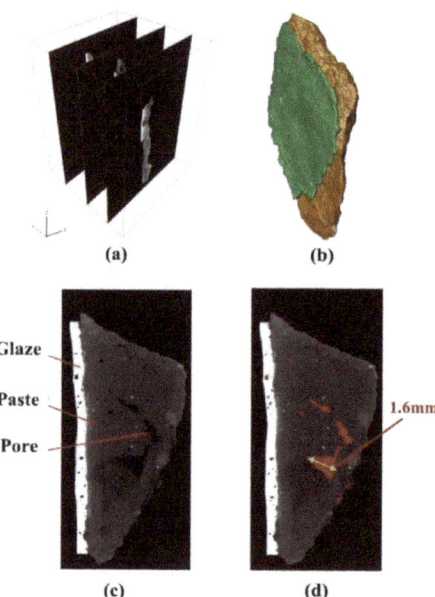

Figure 10. (**a**) The raw CT scanning grayscale slice sequence; (**b**) The 3D reconstruction model; (**c**) The middle cross-section image in grayscale; (**d**) The extracted pore area of the middle cross-section. (Note: sample T-15; the green part of 3D reconstruction model is the glaze layer of the sample, the pore area is rendered in red).

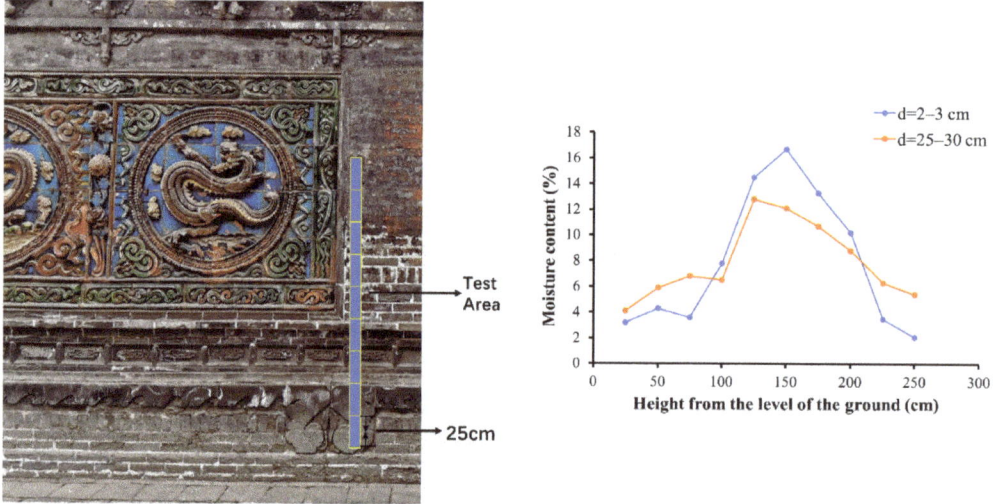

Figure 11. The moisture contents at different heights of the Five-Dragon screen wall. (d means the depth from the surface).

The samples of solute salt powder collected from the Five-Dragon screen wall and screen wall of Biping Gate were analyzed by XRD. The results show that the samples of solute salt powder mainly contain quartz (SiO_2), gypsum ($CaSO_4$) and hexahydrite ($MgSO_4 \cdot 6H_2O$). Since the sampling process inevitably brought in a small amount of tile paste, the tile pastes should be the source of SiO_2 analyzed by XRD. In addition, $CaSO_4$ and $MgSO_4 \cdot 6H_2O$ were detected and should be the main soluble salts.

This infers that the salt efflorescence of the tile body (including the gray bricks) is mainly due to the fact that the groundwater carrying the soluble salts is drawn into the screen wall through a capillary process. Then under the joint influence of thermal evaporation and capillarity, capillary water (and soluble salts) rises continuously and gradually migrates to the wall surface, finally concentrating in the height with the range of 1–2 m from the ground. When soluble salts crystallize, their volume will expand. Additionally, with the cycle of air temperature and relative humidity, the soluble salts carried by the migrated water undergo repeated crystallization. This means that when exposed to outer conditions for a long term, the soluble salts in the tile bodies are repeatedly dissolved, expanded and crystallized, which results in the looseness of the tile paste and even the formation of holes. Additionally, the infiltration of the rainwater (and soluble salt) into the tile body along the glaze cracks is another reason for the salt efflorescence. Eventually, the salt efflorescence breaks down the tile body, ultimately causing the glaze to be lost and even resulting in the shedding of the entire glazed tile.

3.3. Conservation Suggestions

3.3.1. Water Proofing

As discussed in Section 3.2.2, the salt efflorescence of the tile body (including the gray bricks) caused by groundwater (solute salt) can heavily weaken the stability of the screen wall and its glazed tile decoration. Therefore, protecting screen walls from dampness and moisture is the primary task. Specifically, the wall foundation should be exposed for water proofing treatment. The organic silicon resin materials with excellent hydrophobicity can be selected to cover and penetrate the exposed basement wall. Then the screen wall foundation should be backfilled with a mixture of lime and local soil, which is the traditional waterproof building material commonly used in China. This method can effectively prevent the foundations from absorbing groundwater. Additionally, organic silicon resin materials also must be grouted into the screen wall at a height of approximately 30 cm above the ground in order to prevent the wall root from absorbing rainwater and groundwater.

3.3.2. Desalination

Desalination treatment should be carried out on the glazed tiles and bricks of the screen walls which have deteriorated through soluble salt crystallization. In this study, desalination should be mainly applied to glazed tiles and bricks at a height of 1–2 m from the ground. Paper pulp and sepiolite specially used for cultural relics protection can be poulticed in the surface of the glazed tiles and bricks. In doing so, the soluble salt inside the glazed tiles and bricks can migrate to the externally applied material. This desalination process needs to be repeated until the surface salt content is constant.

3.3.3. Reproduction

For the areas where the glazed tiles or glaze layers have shed, we can reproduce the glazed tiles to reestablish the visual integrity of the glazed tile decoration of the screen walls on a trial basis. Besides, for the glazed tiles which have decayed seriously, it is also an effective preservation measure to move them to the museum for further protection and display and using reproduced ones to replace them in situ. The raw materials and techniques of making these Hancheng glazed tiles have been discussed in this study. Therefore, the glazed tiles can be reproduced using traditional methods. It is important that the replicas should be lightly distinguished from the original glazed tiles in both color tone and gloss.

4. Conclusions

This research has furthered the understanding of the architectural glazed tiles which decorate ancient Chinese screen walls based on scientific study. The tile pastes of the screen walls from the Hancheng Confucian Temple and the Town God's Temple in Shaanxi Province are potentially made from loess, which is not a common raw material used to produce glazed tiles. Tile glazes can be classified into two subgroups according to different colors and glaze recipes. The tile glazes with green, yellow and brown colors are typical high-lead glazes with a glaze recipe of SiO_2-Al_2O_3-PbO. On the other side the turquoise glazes are produced by a glaze recipe of SiO_2-PbO-alkaline (K_2O and Na_2O). The chemical compositions of tile glazes in this study are basically consistent with the glaze formula of architectural glazed tiles recorded in ancient literature. The microstructures show that most of the glazed tile samples have almost no paste-glaze interface, which infers that they are produced by a double-firing process. The microstructures and chemical compositions of the glazed tiles between the Ming and Qing dynasties were consistent. This indicates that the raw material selection and relevant manufacturing techniques had been optimized and kept unchanged to produce these glazed tiles in Hancheng city for a long period.

The most widespread and damaging degradation of these glazed tiles are glaze cracking and shedding and body salt efflorescence. For the deterioration of glaze cracking and shedding, the internal cause is the tensile stress of the glaze due to its higher expansion coefficient compared to tile paste. Besides, the drastic environmental changes such as high temperature and torrential rain, as well as the water (and salt) migration and crystallization within the glazed tiles, are the external factor. The salt efflorescence of the tile body normally occurs in the range of about 1–2 m from the ground. The main cause is that the groundwater carrying soluble salt is drawn into the screen wall through a capillary process, and the soluble salts undergo repeated crystallization. Therefore, preventing the permeation of water into the glazed tiles and desalination are the most urgent tasks to protect the glazed tiles which decorate screen walls from further degradation.

This research enables us to better understand the raw materials and technological choices used to produce the glazed tiles which decorate ancient Chinese screen walls and also provide insights into the causes for their key degradation processes. All these have laid a solid foundation to propose effective preservation and restoration treatments.

Author Contributions: Conceptualization, J.S., L.L. and J.-P.W.; methodology, J.S. and X.L.; formal analysis, X.L., J.-Y.L., D.Z. and J.J.; investigation, X.L. and J.S.; data curation, J.S., J.J. and D.Z.; writing—original draft preparation, J.S.; writing—review and editing, L.L. and J.-P.W.; visualization, J.S.; supervision, L.L.; project administration, J.S., L.L. and J.-P.W.; funding acquisition, J.S. and J.-P.W. All authors have read and agreed to the published version of the manuscript.

Funding: This research was funded by the Fundamental Research Funds of Shandong University (grant number: 2019HW011), the Shandong University Multidisciplinary Research and Innovation Team of Young Scholars (grant number: 2020QNQT018), and the National Natural Science Foundation of China (grant number: 51909139).

Acknowledgments: The authors would like to thank Shaojun Yan for his assistance in the SEM investigations. The scientific calculations in this paper have been completed on the HPC Cloud Platform of Shandong University.

Conflicts of Interest: The authors declare no conflict of interest.

References

1. Pang, T. Some Chinese glazed tiles of the 6th Century. *Orient. Art* **1999**, *44*, 61–68. (In Chinese)
2. Duan, H.Y.; Kang, B.Q.; Ding, Y.Z.; Dou, Y.C.; Miao, J.M. Research of technology of Qing dynasty official Glazed tile bodies in Beijing. *J. Build. Mater.* **2012**, *15*, 430–434. (In Chinese)
3. Ding, Y.Z.; Li, H.; Kang, B.Q.; Chen, T.M.; Miao, J.M. A Preliminary study on the technological process of the architectural glazed tiles during the Ming and Qing dynasties in Beijing Area. *J. Gugong Stud.* **2015**, *1*, 159–169. (In Chinese)

4. Kang, B.; Groom, S.; Duan, H.; Ding, Y.; Li, H.; Miao, J.; Lu, G. The ceramic technology of the architectural glazed tiles of Huangwa Kiln, Liaoning Province, China. In *Craft and Science: International Perspectives on Archaeological Ceramics*; Martinón-Torres, M., Ed.; Bloomsbury Qatar Foundation, Hamad bin Khalifa University Press: Doha, Qatar, 2014; pp. 213–224.
5. Kang, B.Q.; Li, H.; Miao, J.M. Research on the Qing dynasty glazed tiles with date inscriptions of the Forbidden City. *Cult. Relics South. China* **2013**, *2*, 67–71. (In Chinese)
6. Cheng, L.; Feng, S.; Li, R.; Lu, Z.; Li, G. The provenance study of Chinese ancient architectonical colored glaze by INAA. *Appl. Radiat. Isot.* **2008**, *66*, 1873–1875. [CrossRef] [PubMed]
7. Adili, D.; Zhao, J.; Yang, L.; Zhao, P.; Luo, H. Protection of glazed tiles in ancient buildings of China. *Herit. Sci.* **2020**, *8*, 37–48. [CrossRef]
8. Zhao, J.; Li, W.D.; Luo, H.J.; Miao, J.M. Research on protection of the architectural glazed ceramics in the palace museum, Beijing. *J. Cult. Herit.* **2010**, *11*, 279–287. [CrossRef]
9. Duan, H.Y.; Miao, J.M.; Li, Y.; Kang, B.Q.; Li, H. Research on diseases and damages to green-glazed tiles on Chinese traditional architecture. *J. Palace Mus.* **2013**, *166*, 114–124. (In Chinese)
10. Ding, Y.Z.; Duan, H.Y.; Kang, B.Q.; Wu, J.M.; Miao, J.M. Study on the provenance of raw materials of the body of the architectural glazed tiles from Nanjing Baoensi Pagoda. *China. Ceram.* **2011**, *47*, 70–75. (In Chinese)
11. Wu, M.Q.; Wen, Q.Z.; Pan, J.Y.; Diao, G.Y. Mass-weighted average of major chemical compositions of the Malan Loess in China. *Prog. Nat. Sci.* **1996**, *5*, 92–100. (In Chinese)
12. Wood, N. *Chinese Glazes: The Origins, Chemistry, and Recreation*, 2nd ed.; University of Pennsylvania Press: Philadelphia, PA, USA, 2011; pp. 12, 197.
13. Kang, B.Q.; Li, H.; Duan, Y.H.; Ding, Y.Z.; Zhao, L.; Lei, Y. Research on the Yuan dynasty raw materials and firing technique of turquoise-glazed tiles excavated from the Palace Museum. *J. Palace Mus.* **2018**, *1*, 191–199. (In Chinese)
14. Tian, S.; Li, Z.; Wang, Z.; Jiang, E.; Wang, W.; Sun, M. Mineral composition and particle size distribution of river sediment and loess in the middle and lower Yellow River. *Int. J. Sediment Res.* **2021**, *36*, 392–400. [CrossRef]
15. Maggetti, M. Phase analysis and its significance for technology and origin. In *Archaeological Ceramics*; Olin, J.S., Franklin, A.D., Eds.; Smithsonian Institution: Washington, DC, USA, 1982; pp. 121–133.
16. Ventolà, L.; Cordoba, A.; Vendrell-Saz, M.; Giraldez, P.; Vilardell, R.; Saline, M. Decorated ceramic tiles used in Catalan Modernist Architecture (c.1870 to c.1925): Composition, decay and conservation. *Constr. Build. Mater.* **2014**, *51*, 249–257. [CrossRef]
17. Gliozzo, E. Ceramic technology. How to reconstruct the firing process. *Archaeol. Anthropol. Sci.* **2020**, *12*, 1–35. [CrossRef]
18. Duan, H.Y.; Ding, Y.Z.; Liang, G.L.; Dou, Y.C.; Miao, J.M. The chemical composition characteristic and firing technology research on ancient building glazed tile bodies in China. *China Ceram.* **2011**, *47*, 69–72. (In Chinese)
19. Tite, T.S.; Freestone, I.; Mason, R.; Molera, J.; Vendrell-Saz, M.; Wood, N. Lead glazes in antiquity—methods of production and reasons for use. *Archaeometry* **1998**, *40*, 241–260. [CrossRef]
20. Miao, J.M.; Wang, S.W. Research on the architecture glazed tiles of Yuan, Ming and Qing dynasties. In Proceedings of the 05 International Symposium on Ancient Ceramics, Shanghai, China, 1–5 November 2005; pp. 108–115. (In Chinese)
21. Feng, M. Research on the origin of turquoise glaze technology. *Cult. Relics Cent. China* **2020**, *211*, 126–132. (In Chinese)
22. Li, J.Z. *A History of Science of Technology in China-Volume of Ceramic*; Science Press: Beijing, China, 1998; p. 474. (In Chinese)
23. Lawrence, W.G.; West, R.R. *Ceramic Science for the Potter*; Chilton Book Co.: Radnor, PA, USA, 1982; p. 182.
24. Miao, J.M.; Wang, S.W.; Duan, H.Y.; Kou, Y.C.; Ding, Y.Z.; Li, Y.; Li, H.; Kang, B.Q.; Zhao, L. Research on the elements in the vitreous glazed components used in ancient Chinese Buildings. *J. Palace Mus.* **2008**, *5*, 115–129. (In Chinese)

Article

The Technology Transfer from Europe to China in the 17th–18th Centuries: Non-Invasive On-Site XRF and Raman Analyses of Chinese Qing Dynasty Enameled Masterpieces Made Using European Ingredients/Recipes

Philippe Colomban [1,*], Michele Gironda [2], Divine Vangu [1], Burcu Kırmızı [3], Bing Zhao [4] and Vincent Cochet [5]

1. MONARIS (UMR8233), Sorbonne Université, Campus P. et M. Curie, CNRS, 4 Place Jussieu, 75005 Paris, France; divine.vangu@etu.sorbonne-universite.fr
2. XGLab S.R.L—Bruker, 23 Via Conte Rosso, 20134 Milan, Italy; michele.gironda@bruker.com
3. Department of Conservation and Restoration of Cultural Property, Faculty of Architecture, Yıldız Technical University, Yıldız Yerleşkesi B Blok, Beşiktaş, Istanbul 34349, Turkey; kirmizi@yildiz.edu.tr
4. CNRS, CRCAO, UMR8155, Collège de France, 75005 Paris, France; bing.zhao@college-de-france.fr
5. Musée National du Château de Fontainebleau, Place Charles de Gaulle, 77300 Fontainebleau, France; vincent.cochet@chateaudefontainebleau.fr
* Correspondence: philippe.colomban@sorbonne-universite.fr or philippe.colomban@upmc.fr

Citation: Colomban, P.; Gironda, M.; Vangu, D.; Kırmızı, B.; Zhao, B.; Cochet, V. The Technology Transfer from Europe to China in the 17th–18th Centuries: Non-Invasive On-Site XRF and Raman Analyses of Chinese Qing Dynasty Enameled Masterpieces Made Using European Ingredients/Recipes. *Materials* **2021**, *14*, 7434. https://doi.org/10.3390/ma14237434

Academic Editors: Žiga Šmit and Eva Menart

Received: 25 October 2021
Accepted: 1 December 2021
Published: 3 December 2021

Publisher's Note: MDPI stays neutral with regard to jurisdictional claims in published maps and institutional affiliations.

Copyright: © 2021 by the authors. Licensee MDPI, Basel, Switzerland. This article is an open access article distributed under the terms and conditions of the Creative Commons Attribution (CC BY) license (https://creativecommons.org/licenses/by/4.0/).

Abstract: Two masterpieces of the Qing Dynasty (1644–1912 CE), one in gilded brass (incense burner) decorated with *cloisonné* enamels stylistically attributed to the end of the Kangxi Emperor's reign, the other in gold (ewer offered by Napoleon III to the Empress as a birthday present), decorated with both *cloisonné* and painted enamels bearing the mark of the Qianlong Emperor, were non-invasively studied by optical microscopy, Raman microspectroscopy and X-ray microfluorescence spectroscopy (point measurements and mapping) implemented on-site with mobile instruments. The elemental compositions of the metal substrates and enamels are compared. XRF point measurements and mappings support the identification of the coloring phases and elements obtained by Raman microspectroscopy. Attention was paid to the white (opacifier), blue, yellow, green, and red areas. The demonstration of arsenic-based phases (e.g., lead arsenate apatite) in the blue areas of the ewer, free of manganese, proves the use of cobalt imported from Europe. The high level of potassium confirms the use of smalt as the cobalt source. On the other hand, the significant manganese level indicates the use of Asian cobalt ores for the enamels of the incense burner. The very limited use of the lead pyrochlore pigment (European *Naples yellow* recipes) in the yellow and soft green *cloisonné* enamels of the Kangxi incense burner, as well as the use of traditional Chinese recipes for other colors (white, turquoise, dark green, red), reinforces the pioneering character of this object in technical terms at the 17th–18th century turn. The low level of lead in the *cloisonné* enamels of the incense burner may also be related to the use of European recipes. On the contrary, the Qianlong ewer displays all the enameling techniques imported from Europe to obtain a painted decoration of exceptional quality with the use of complex lead pyrochlore pigments, with or without addition of zinc, as well as cassiterite opacifier.

Keywords: painted enamels; *cloisonné*; gold alloy; blue; yellow; green; white; China; Qing Dynasty; spectroscopy; composition

1. Aim

The attention paid to the construction of a global history [1–3] not centered on Europe or Asia has led to the symmetrical study of exchanges of all kinds between Europe and the Far East for several years, especially regarding fine arts [4–11]. The pioneering role of Chinese potters in the development of porcelain production is well known [4–14], and the enthusiasm of European elites for these ceramics from the well-established importation of blue-and-white porcelain by Portuguese then Dutch and English sea trade in the 16th and 17th centuries has been largely studied [15–17]. The role of the Jesuits living in Japan [18–22] and China [23–27] in the circulation of European science related to astronomy [22], mathematics [23], time measurement [21] and painting [26–28] is also rather well established. The role of Jesuits in the transfer of enameling technologies is less documented and only studied for a few years by research on the Imperial Palace archives [29] and the correspondence of the Society of Jesus [21–23] or diplomatic issues [30–33]. Our objective is to seek in the material of the 'Chinese' objects themselves the evidence of the use of European ingredients or recipes in the enameling procedures, similar to what was done for the first porcelains of Arita (Japan) in the late 16th century [18–20]. If shards are available, which is not the case most of the time for outstanding artefacts, it is possible to use micro-destructive methods [4,34–36] and many types of scientific analyses can be carried out in the laboratory using a variety of instruments. However, the analytical methodology for the intact invaluable objects is more restricted and difficult since it must be carried out in a perfectly non-invasive way, without any sampling and contact. The rarity and great value of the objects studied as well as their possible large size make their transfer to the laboratory facilities very expensive and impractical, inducing the necessity of on-site analysis in the exhibition halls or museum reserves with the use of mobile analytical instruments.

For several years, we have been developing procedures and models allowing the implementation and interpretation of the results obtained by mobile Raman microscopy and X-ray fluorescence devices [37–51], specifically for the identification of coloring agents and silicate matrices of the glassy materials such as the enamels. Analyses were thus carried out on the collections of enameled objects (glass, metal or porcelain) produced in France and China between the 17th and the 19th century [10,36,38,39,44–46,52,53], allowing us to identify the objects most representative of the technological change for the period mentioned.

We present here the in-depth non-invasive analysis of two masterpieces from the collections of the National Museum of the Château de Fontainebleau (Chinese Museum), particularly constituted by the Emperor Napoleon III and his wife the Empress Eugenie in the second part of the 19th century [54–56]. The two objects studied have been selected as representative of the beginning of the introduction of European recipes (a gilded brass incense burner decorated with *cloisonné* enamels and stylistically attributed to the end of the Kangxi's reign) and of the achievement of the mastering of this new technology (a gold ewer belonging to a set comprising a pair of ewers and a large gold basin with the Qianlong's reign mark decorated with both painted and *cloisonné* enamels), respectively, among the hundred Chinese outstanding artefacts studied [10,42,52,53]. Other studies reported in the literature are limited in scope and have been performed on more common objects or sherds [34,35].

2. Materials and Methods

2.1. Portable X-ray Fluorescence Spectroscopy (pXRF)

X-Ray Fluorescence analysis was performed on-site using a portable ELIO instrument (ELIO, XGLab Bruker, Milan, Italy). The set-up includes a miniature X-ray tube system with a Rh anode (max voltage of 50 kV, max current of 0.2 mA, and a 1 mm collimator), and a large area Silicon Drift Detector (SDD, 50 mm^2 active area) (ELIO, XGLab Bruker, Milan, Italy) with energy resolution of <140 eV for Mn Kα, an energy range of detection from 1 keV to 40 keV, and a maximum count rate of 5.6×10^5 cps. Measurements were carried out in the point mode with an acquisition time of 40 s, using a tube voltage of 40 kV and current of 100 µA. No filter was used between the X-Ray tube and the sample. Three measurements were made for each colored area. The working distance (distance between the sample and detector) during analysis was around 15 mm, the distance between the instrument front and the artefact being about 10 mm. Spectral signals were obtained with the optimization of the signal-to-noise ratio (SNR) by selecting the set-up parameters chosen. Information thickness during analysis of the enamel is estimated to be close to 4 μm at Si K_α, 130 μm at Cu K_α, 220 μm at Au L_α, and 2.5 mm at Sn K_α. Within the resolution of the pXRF instrument, the Fe K_β peak and the Co K_α peak are located in the same energy range. To identify the presence of Co in enamels (except when cobalt is present in traces) we can use the information obtained looking at the Fe K_α/Fe K_β ratios. In the absence of cobalt, the relative intensity between Fe K_α and Fe K_β peaks is about 6/1. Cobalt is then obvious if the superimposed peaks of Co K_α and Fe K_β exhibit a stronger intensity than that expected from the above ratio. Calculation of the local composition is also useful to detect cobalt, but the calculated 'composition' is not valid and only comparison of the counts is reliable.

XRF mapping was performed on different areas of the objects. The ELIO instrument is designed to move in front of the object measuring elemental maps up to 10 cm × 10 cm. The map is acquired in the raster scan mode with a pixel measurement time which can be configured by the user. The pixel size can be adjusted by selecting the motor step, but the spot resolution (around 1 mm) influences the spatial resolution. Good flatness of the area analyzed is required and it should be perpendicular to the axis of the instrument. Different map sizes were acquired selecting surface areas that were sufficiently flat for the tolerance of the system (plus/minus 2 mm in distance tolerance). The pixel acquisition time was set to 1 s per pixel with a pixel size of 1 mm. The ELIO software (2021, ELIO XGLab Bruker, Milan, Italy) allows the visualization of the different elements by showing the differences in intensity over the surface of a selected region of interest (ROI) in the spectrum. For element deconvolution and more advanced elaborations, the Bruker ESPRIT Reveal software (2011, Bruker AXS, Berlin, Germany) was used.

The data obtained were processed using the factory provided data reduction software, which enables automatic peak recognition supported by manual peak selection and checking. The software also enables curve fitting based on chosen elements to ensure a match between the measured spectra and theoretically predicted spectra calculated from fundamental parameters (FP). Semi-quantitative elemental compositions were also calculated for the elements of interest using FP by the instrument software when the sample thickness can be considered to fulfil the infinite thickness criteria and the material is homogeneous at the scale of the analyzed volume.

2.2. Raman Microspectroscopy

Raman analyses were carried out at the museum exhibition room with a mobile HE532 Raman set-up (HORIBA Scientific Jobin-Yvon, Longjumeau, France) as extensively described in the references [45,46,52,53]. For each colored area in the objects, at least three Raman spectra were recorded to obtain the representativeness of the collected data on a statistical basis. The reliability of the Raman spectrum starts above 80 cm^{-1} but a flat spectral background is only obtained over 500 cm^{-1}. A 200× microscope objective (~13 mm long working distance) Mitutoyo Corp., Kawasaki, Japan) was used for the analysis of the ewer while 200× and 50× (17 mm long working distance) Nikon France SAS, Champigny-sur-Marne, France) objectives were used for the study of the incense burner. These objectives provide small, focused beams (surface spot waist ~0.5 μm and 2 μm; in-depth ~2 μm and 5–10 μm, respectively, the values varying with the color), perpendicular to the sample surface, which allow the recording of spectra not/poorly contaminated by the sub-layers and/or the environment. The 200× objective is more efficient for the analysis of the individual pigment grains embedded in the glassy matrix since it provides a very small laser spot which requires a very precise focus. Thus, it allows to obtain spectra with less background than those obtained using less sophisticated objectives with lower magnification. In fact, the shape of the spectral background gives information about the color of the analyzed spot: flat for a blue area, decreasing above ~500 cm^{-1} for a red area and increasing for yellow or green areas (see further). Obviously, the power of illumination at the sample should be minimal (1 mW or less) for black or dark colored areas due to the absorption of light, although up to 10 mW is required for light colored or colorless areas of the enamels. A linear segment baseline was subtracted using Labspec® (HORIBA) software (5.25.15, 2007, HORIBA Jobin-Yvon, Longjumeau, France [36] and then the different components of the Raman spectrum were identified using Peakfitting Origin® (6.0, 1999, Microcalc Inc., Northampton, MA, USA) software. Lorentzian and Gaussian shapes were used for narrow and broad components, respectively.

3. Objects Studied

Imperial workshops were established at the Forbidden City in Beijing to satisfy the demand of Kangxi Emperor that 'new' objects similar to those given as presents by the Jesuits and Emissary of Louis XIV could be locally produced. The first imperial workshop (Zaobanchu, Office of Manufacture) opened in 1693 to manufacture *cloisonné* enamels. It is assumed that preparation of painted enamels had started with the opening of a glass workshop in 1696. This unit was headed by the German Jesuit Kilian Stumpf who is considered to have introduced the European glassmaking techniques into China as a scientist and glassblower. A dedicated enameling workshop then opened in 1716 [6,29]. At the same time, or before, another production center of painted enamels was established in Guanzhou (Custom district). Indeed, many artefacts were commissioned by the Court as tributes [11]. Technological development of the overglaze porcelain palette expanded significantly at the end of Kangxi reign to produce Famille rose, Famille verte and Famille noire porcelains [4].

3.1. The Origin of Objects

The works of art brought to Paris by the French soldiers during the sack of the Summer Palace (*Yuanming Yuan*) in Beijing in October 1860 by the Anglo-French troops in response to the execution of European persons, as well as the gifts offered by the Embassy of Siam in June 1861, were the origin of the creation of the Chinese Museum in the castle of Fontainebleau. Empress Eugenie received a part of the objects collected in China, as did Queen Victoria of the United Kingdom [54–56]. The Empress, who appreciated the sunny exposure of the ground floor of the *Gros Pavillon* (designed by A.J. Gabriel) at the Château de Fontainebleau, near the carp pond and the English Garden, had asked A. Pacard, her architect, to set up a large living room for use during summer stays. The collection was open to rare visitors from 1863. The Museum was enriched at this time with other oriental objects, in particular personal gifts from the Emperor to his wife during the feast of Saint Eugenia. Photographic details are available [57–59]. The pair of ewers (Figure 1) and the basin (see [57–59]) are believed to have been made by the imperial workshops for the Summer Palace. They were acquired on the art market and offered by Napoleon III to the Empress as a birthday present. The *cloisonné* enameled incense burner (Figure 2) is also expected to have come from a Chinese Palace.

3.2. The Visual Characteristics of the Enamels

3.2.1. Incense Burner (Assignment: Kangxi Period 1661–1722)

Figure 2 shows the incense tripod (inventory number F1448C, height: ~27 cm; diameter: ~26 cm, belonging to the collections of The National Museum of the Château de Fontainebleau (Chinese Museum)). The lower part is decorated with *cloisonné* enamels depicting multicolor lotus flowers and foliage scrolls characteristic of the influence of Buddhism on a turquoise background. The lid, rim and handles are gilded. The lid consists of golden interlacing on which a *Fô* female lion (*shi*) with a baby is placed at the top (the female lion is the symbol of education). The lid is expected to have been made by the lost wax technique, with subsequent chisel carving and gilding.

The three feet of the body are added with rivets visible inside. The handles appear to have been welded. The *cloisonné* enameled areas are turquoise, white, dark blue, red, green, light green and yellow. The center of the flowers is either yellow or green and these colors are used sparingly, which is consistent with their less availability or high cost. The use of different greens is significant. Numerous pores and cracks are visible in the enameled decoration (Figure 2). These defects are frequently encountered in such objects due to the application of *cloisonné* enameling. The *cloisonné* technique particularly involves the interposition of the glass powder between the metal sheets ('*cloison*' in French) fixed perpendicular to the object by point welding, glue or wax (before the filling of the voids in between with enamel powder) and then the firing step which requires heating/melting the enamels. The latter operation is repeated several times. This induces a significant shrinkage of the enamels, but the duration of the firing cycles is too limited to eliminate the bubbles. A surface polishing step is also performed at the end which enhances the open porosity.

Figure 1. On-site pXRF analysis of the enameled ewer (F1467C) and details of the décor (*cloisonné* and painted enamels); the bottom face exhibits an engraved Qianlong reign mark (bottom left, basin: *Qian long nian zhi*; right, ewer; *Da qing Qian long nian zhi*). Note the welding traces along the spout (arrows).

3.2.2. Ewer (3rd Quarter of 18th Century, Qianlong Mark)

Figure 1 shows the ewer analyzed (from a pair, inventory number F1467C; height: ~40 cm; base: ~27 × 20 cm^2; weight ~2750 kg, belonging to the collections of the National Museum of the Château de Fontainebleau (Chinese Museum)). One of the ewers was specifically analyzed while a few measurements were also performed on the other one. The object is gold like its pair and the basin, which especially puts these objects in an exceptional category. However, the marks are different (Figure 1). The objects appear to have been made by shaping/hammering gold foil. Traces of welds are visible in a few places (handle, spout). The interior has a greenish counter-enamel. The ewer is decorated both with *cloisonné* and painted enamels. The medallions were formed in gold body before being painted and fired. As the *cloisonné* decoration requires several fillings with enamel powder between the firing steps followed by a final polishing, we can assume that its creation was done before decorating the medallions (or the medallions were prepared before to be welded on the gold frame).

Figure 2. The incense burner (F1448C) and details of the *cloisonné* enameled décor (Kangxi reign). Note the many bubbles in the décor revealed by surface polishing.

The *cloisonné* décors consist of multicolor flowers (blue, yellow, red and green) on turquoise and blue backgrounds. The background is turquoise for the belly as usual but blue for the foot and the neck. The quality of the *cloisonné* décor is also exceptional by its surface with less bubble (Figure 1) than in the Kangxi incense burner (Figure 2). The painted décors of the small medallions depict flowers, personages and/or landscapes (four medallions on the neck, six on the central part and four on the foot) [57,58] and one large medallion on each face depicts women with a child in a garden. The fineness of the design and the variety of color tones demonstrate great technical mastery. The decor style appears to be a hybrid, combining characteristics of Chinese decor (shape of the houses) and European decor (women's faces, basket of flowers, belt and drape of the dress). The enamels are rather matt (Figure 1), except some yellow areas (Figure 3, this should involve a lower melting temperature of the enamel colored in yellow), giving the impression of a decoration on paper, which is unusual for an enamel decoration. The artist used the stamping of the metal support of the medallion for some part of the decoration (window of the pavilion, garden fence). Burrs and drips of the enameled decor on the edge of the medallions are observed (Figure 3).

Figure 3. Detailed images of the enamels of the gold ewer décor (F1467C).

4. Results and Discussion

4.1. Elemental Analysis

4.1.1. Metal Body

Figure 4 shows the XRF spectra recorded on different parts of the metal body of the two objects. The spectra recorded on the complex non-uniform areas are shown with logarithmic scale in order to make the contribution of minor components more visible.

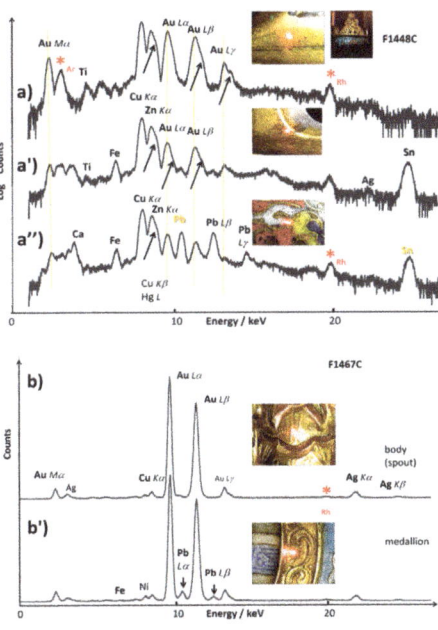

Figure 4. XRF spectra recorded on the metal parts of the incense burner (**a**,**a′**,**a″**) and the ewer (**b**,**b′**); the lightened spot on the photograph shows the analyzed area: (**a**) rim, (**a′**) rim with gilding partially lost and (**a″**) *cloison*; (**b**) spout and (**b′**) medallion rim. Main XRF peaks are labelled; arrows indicate the contribution of Hg L peaks in the spectra a, a′ and a″. Red stars (*) indicate the contribution of the instrument (Rh peaks). The logarithmic scale on the top spectra makes more visible the contribution of minor elements.

The analyses show that the rim of the incense burner (lower part) and the *cloison* foils are made of brass alloy (Figure 4a,a″). Comparison of a spot exhibiting a nice gold color (Figure 4a) with a spot that looks corroded or with gilding partially lost (Figure 4a′), shows a strong decrease of the intensity of the Au peaks, that is consistent with a gilded Cu-Zn brass (Table 1). Subtracting the 'contaminated' data due to the gilding, the composition is approximately copper (Cu) 65 wt%, zinc (Zn) 28 wt%, tin (Sn) 6 wt%, and iron (Fe) 1 wt%. The shoulders marked with arrows in Figure 4 indicate the presence of small Hg L peaks which probably result from the cold application of the gold foil using mercury as a coating agent. The measurement of the composition is likely contaminated with the contribution of the adjacent enamels rich in Pb and Ca (see further). Volatilization of lead oxide during the firing process followed by the polishing step should also have polluted the whole surface of the artefacts. However, incorporation of lead in the brass composition is also common for Chinese brass [41].

Table 1. Composition of the metal parts of the objects studied (metal wt%).

Object	Part	% Au	% Ag	% Cu	% Zn	Cu/Zn	% Ti	% Sn	% Fe	% Pb
Incense burner	Gilded body	52	-	32	15	2	1	-	-	~1
	Body (gilding lost)	17	-	52.5	22	2.5	-	7	0.5	1
	Cloison	46	-	27.5	14.5	2	-	3	0.5	~1
Ewer	Spout	84	14	1	-		-	-	-	1.5
	Medallion rim	83	11	1	-		-	-	0.5	4

The comparison of the compositions measured on different spots shows a variation in the Cu/Zn ratio. The difference concerning the measurements is assigned to the poor precision of the method and/or local heterogeneity of the alloy. Tin is also observed to some extent, as in the case of the body. This is consistent with the commonly observed fabrication of Chinese brass which involves the bringing together of different types of brass pieces that do not have identical compositions [41].

The ewer body is made of a gold (84 wt%)–silver (14 wt%) alloy with a small addition of copper (1 wt%) (Figure 4b,b′). Copper increases the hardness and mechanical strength of the alloy while silver is also considered to promote the bonding with the silicate enamels [60]. The difference in the silver content (3 wt%) is assigned to the uncertainty of the method and to the contribution of lead contamination. However, the small amount of lead present is again assigned to the pollution of the metal surface provoked by volatilization and deposition of a lead oxide film on the whole surface from the enamels during the firing step. The highest value of lead was indeed measured close to the enameled areas.

4.1.2. *Cloisonné* Enamels

The concept of 'enamel composition' is meaningful for a thick and homogeneous enamel like that of a transparent porcelain or celadon glaze. It remains relevant if the enamel is colored by ions, while it is much less when the coloring is done by pigments where the proportion varies according to the color. Therefore, the composition of the silicate matrix and the pigments should be distinguished. Painted enamels are formed of thin layers very loaded with pigment, only one or several types of pigments forming mixtures at sub-micron to micron scales (see later). The concept of 'composition' therefore no longer makes sense without specifying very precisely the color and volume of the material concerned (e.g., in the case of microdiffraction or microfluorescence on a homogeneous small volume). However, the depth probed by XRF varies according to the energy of the photons and therefore the element being measured. In addition, light elements such as sodium but also oxygen and boron are not measured in the case of pXRF. It is therefore only possible to consider the type of enamel, more precisely the nature of the fluxes detected by XRF, and in some cases their relative proportion. Measurements of alkali and earth-alkali elements correspond to the upper layers that can be corroded. Measurements are more accurate for transition metals, the in-depth penetration of XRF characteristic photons being in the order of the enamel thickness. On the other hand, the measurement of heavy elements (Pb, Sn, Sb) is 'polluted' by the contribution of the substrate [45,52,53].

Figures 5 and 6 show the representative pXRF spectra recorded on the enamels of the Kangxi incense burner and the Qianlong ewer, respectively. The silicon, potassium, calcium and lead peaks arise from the glassy silicate matrix of the enamels. The iron peak is always observed, even in the white areas, and is related to the iron impurities commonly present in silicates. Transition metals were detected as chromophores such as cobalt in the blue area, and copper in the red, green and turquoise areas. Tin is also observed in many colored areas (yellow, red, green and blue).

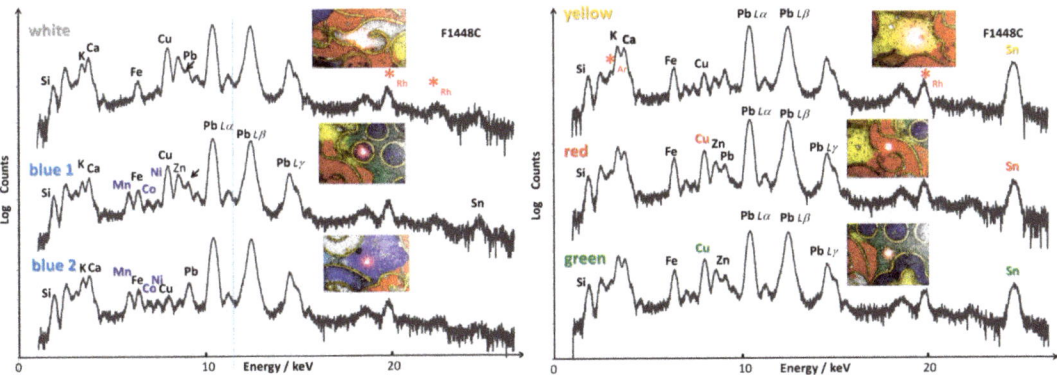

Figure 5. Representative pXRF spectra recorded on the different colored *cloisonné* enamels of the incense burner (white, blue, yellow, red and green areas). Red stars (*) indicate the contribution of the instrument (Rh peaks).

Incense burner. We now consider the pXRF spectra recorded on the enamels of the incense burner in detail (Figure 5). The Ca peak appears to be stronger than that measured for *cloisonné* and painted enamels of the ewer (Figure 6). This is confirmed by comparing local 'composition' from the net count areas (Table 2). A very small amount of tin is found in the white area, but the amount is too small to contribute to the opacification. A small Co peak was detected in the blue area, much smaller than Mn and Fe peaks which are common for the blue enamel [42,45,52,53]. This finding is related to the use of Asian cobalt ores, rich in manganese [35,36,42,61–66]. It is here worthy to mention that the resolution of the instrument does not permit to separate the residual contribution of the Fe K_β peak that superimposes the Co K_α one, leading some inaccuracy in the measurement of the Co content. Cobalt is obvious in the spectrum only if the superimposed peaks of Co K_α and Fe K_β exhibit a stronger intensity than the relative intensity ratio of these peaks. Note the very high coloration power of Co^{2+} ions in the glassy matrix, showing that ~0.5 wt% of cobalt oxide is sufficient to obtain the dark blue color [62]. Consequently, the amount of cobalt is always found to be low to very low. On the contrary, more than 5 wt% of iron oxide is required to achieve the significant coloration of a silicate type of glass [62] and a certain level of iron (~2 wt%) does not color glassy silicates significantly, especially when firing is made under a reducing atmosphere. It is also very similar in the case of manganese. Comparison of the net count areas of K_α lines of Mn and Co is rather accurate and gives a ratio between 5 and ~20 (Table 2). In the blue 1 spot, the relatively high intensity of Cu and Sn peaks are due to the contribution of the adjacent green enameled area. However, the XRF spectrum of the blue 2 spot obtained from inside the larger enameled area with cobalt points out that copper is also present in the blue enamel. The addition can be voluntary, but some cobalt ores also contain copper [60].

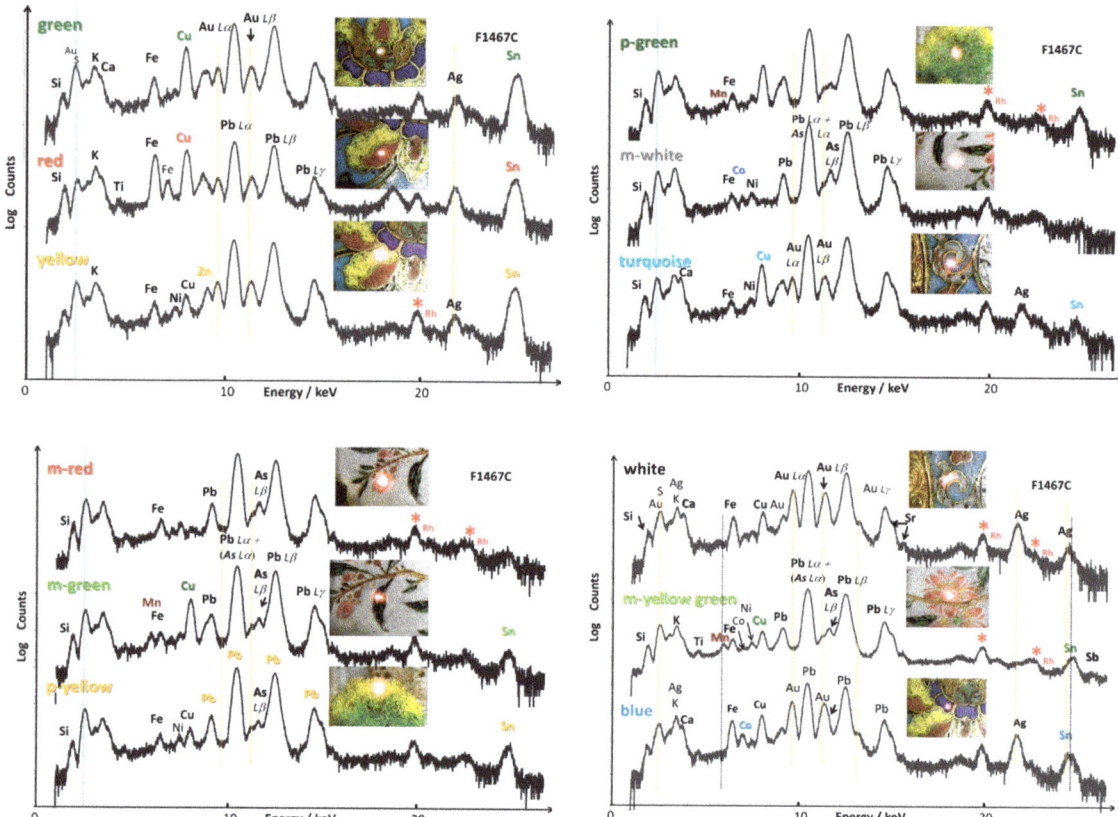

Figure 6. Representative pXRF spectra recorded on the different colored *cloisonné* (top-left: green, red and yellow; top-right: turquoise; bottom-right: white and blue) and painted (p-painted (large medallion) and m-(small) medallion) enamels of the gold ewer (white, yellow, red, green and yellow-green areas). Vertical lines serve as a guide for eyes to distinguish better the contribution of elements having peaks at very similar energy levels. Red stars (*) indicate the contribution of the instrument (Rh peaks).

In the yellow area, the presence of intense Pb and Sn peaks indicates the use of lead pyrochlore pigment (also called *Naples yellow*) [67–74]. The red area also shows the significant contribution of copper, indicating the ancient technique of red coloration in Chinese glass and glazes [75–81] where the red color is obtained by dispersion of Cu° nanoparticles. Additionally, the Fe peak seems to be greater than in the white and blue enamels, pointing out that it may have contributed to the red color in the form of hematite precipitation. The intensity of the Sn peak is rather strong, which may indicate that the yellow pigment had been mixed to adjust the hue or tin was used for the reduction of copper ions.

The Zn peak is also present but contamination of the measurement by the contribution of the *cloison* metal is certainly effective. The green enameled area shows a significant Cu peak, due to the use of Cu^{2+} ions in the silicate matrix for green coloration, the standard technique used to color an alkali-based glass turquoise.

Table 2. Elemental (local) net count areas of characteristic peaks for the blue *cloisonné* enameled areas of the incense burner (IB) and ewer (E). Ratios calculated from the net count area. Ions contributing to coloration are in bold (- not calculated). The error deduced from the study of similar heterogeneous objects is less than 15% for major elements but could be 100% for traces.

Elements (Peak)	Blue 1 IB	Blue 2 IB	Blue Flower E
Si (Kα)	889	2244	900
Pb (Lβ)	136,403	150,608	103,673
Pb/Si	153	23	115
Pb/K	67	26	15
Sn (Kα)	1895	3571	1767
K (Kα)	2010	5841	6902
Ca (Kα)	2348	3032	814
K/Ca	1	2	8
Co (Kα)	385	170	1591
Mn (Kα)	1895	2864	162
Mn/Co	5	17	0.1
Fe (Kα)	2377	2516	7182
Cu (Kα)	18,282	1429	9924
Zn (Kα)	8885	260	204
As (Kβ)	-	1770	4160
As/Co	~0	10	3
Au (Lα)		364	31,030

Table 3 compares the typical oxide compositions of Chinese *cloisonné* and Limoges painted enamels taken from the literature [38,39]. It is observed that the dispersion of the data is large [39]. However, Chinese enamels are much richer in lead oxide (15 to 40 wt% PbO that corresponds to ~10 to 30 at% Pb) than alkali-based Limoges enamels (~2 to 13 wt% PbO). The composition of the glassy matrix of the *cloisonné* enamels of the incense burner shows some similarity with the low lead glazes, such as Limoges enamels, and as previously observed for rare 17th century Chinese *cloisonné* enamels [38,39].

Table 3. Comparison of the composition range (oxide wt%) of Chinese *cloisonné* and Limoges enamels (from Ref. [38]).

Oxide	China 16th	China 17th	Limoges 17th–18th
SiO_2	40–50	45–60	60
PbO	30–40	15–22	2–13
CaO	2–7	5–15	3–5
K_2O	5–12	5–10	5
Na_2O	0.5–15	0.3–15	-

Gold ewer. Regarding the ewer (Figure 6), Au (sometimes Ag) peaks are observed for all XRF spectra recorded on the *cloisonné* enamels due to the contribution of *cloisons* (and perhaps of the gold alloy substrate).

For the blue enameled area, cobalt is again found to be responsible for the blue color, as expected. However, it is significant that the blue *cloisonné* enamel does not exhibit the Mn peak, contrary to that measured on the incense burner. This indicates that a different source such as the imported European cobalt was used in the ewers. The higher intensity of the Kα line is consistent with the use of smalt, the potassium-based glass obtained by mixing with cobalt ores [62]. Observation of a low intensity Sn peak is consistent with the addition of tin to adjust the hue. The red enamel exhibits intense Cu and Sn peaks, as also observed for the incense burner. Tin is usually added to glass in order to promote the reduction of copper ions into Cu° nanoparticles at the origin of the red color production [82]. The high intensity of Fe peaks could also suggest the use of hematite to adjust the hue. Yellow and green enamels show significant Pb and Sn peaks, indicating the use of lead–tin pyrochlore pigment plus a Cu peak for the green enamel where the Cu^{2+} ions act as the coloring agent.

Tables 2 and 4 compare the metal content (net count areas) of the blue, yellow and green enamels extracted from the fitting, respectively. These compositions are only comparative due to the intrinsic heterogeneity of the enamels, the contamination by neighboring phases and uncertainty of the method. The variable penetration depth as a function of the photon energy makes that the measured volume is very different as a function of the element [52,53,83,84]. Nevertheless, the comparison of measurements made on similar blue spots shows consistency, taking into account the contribution of neighboring phases.

Table 4. Elemental (local) near net count of characteristic peaks for the enameled yellow, green, red and white *cloisonné* and painted areas of the Kangxi incense burner (IB) and Qianlong ewer (E). Ratios calculated from the net count area. Elements contributing to coloration are in <u>bold</u>: C: *cloisonné* enamel; P: painted enamels, large medallion; m-: small medallion.

Elements	Yellow IB C	Green IB C	Yellow E C	Green E C	Red E C	Turquoise E C	White E C	m-Yellow-Green E P	m-Flower Green E P	p-Yellow E P	p-Green E P	m-White E P	m-Red E P
Si (Kα)	1793	1729	520	634	1608	1735	1245	1091	714	719	967	1141	1057
Pb(Lβ)	99,145	113,401	157,048	149,094	47,567	149,048	128,098	156,852	141,359	<u>169,959</u>	157,308	147,672	<u>161,192</u>
Pb/Si	55	66	302	235	3	86	10	144	198	236	162	129	152
Sn(Kα)	12,003	4395	14,411	9352	9017	644	400	1137	967	<u>1987</u>	1497	320	253
K(Kα)	14,919	11,607	5368	3893	5353	6144	4636	4865	4107	3790	3814	6388	5124
Ca(Kα)	272	278	5	327	239	<u>1859</u>	<u>1588</u>	1	46	10	78	1	2
K/Ca	55	42	1074	12	22	3	3	4865	90	379	48	6388	2562
Pb/K	7	10	29	38	5	24	27	38	34	45	41	23	3
Sb(Lα)	<u>6617</u>	3845	766	-	-	-	-	436	-	-	-	-	-
Mn(Kα)	102	-	153	141	142	145	-	1074 *	706	191	353	-	73
Mn/Co	0.6	0	1.1	3.7	0.1	1.9	0	7 *	5	**	**	0	**
Fe(Kα)	3809	6390	1810	3232	17506	792	4343	1010	1424	1168	-	783	<u>1627</u>
Cu(Kα)	2111	<u>13,175</u>	3253	<u>25,331</u>	<u>26,120</u>	17,790	5043	<u>17,856</u>	<u>3847</u>	2131	<u>13,823</u>	73	212
Zn (Kα)	1161	6632	524	658	909	270	614	-	215	31	639	129	20
Co(Kα)	180	428	142	38	<u>1086</u>	77	<u>146</u>	153	<u>138</u>	32	27	<u>258</u>	16
Au(Lα)	-	-	7506	3969	2762	5797	26,640	-	-	-	328	-	-
Ag(Kα)	-	-	1860	-	-	5133	-	-	-	-	-	-	-
As(Kβ)	-	-	-	-	-	-	-	7668	8473	8073	6296	<u>10,511</u>	10,392

- Not included in the fitting; * addition of brown; ** Co value too small to be significant.

The cobalt content of the blue enamel of the Kangxi incense burner is lower than that of the Qianlong ewer but the former is associated with manganese (Mn/Co net count area = 5 to ~20) while arsenic was measured for the latter and no significant amount of manganese was detected (Figures 5–7). In some spots (Blue 2 spot, Table 2) a certain level of arsenic was also detected by XRF but the absence of a characteristic As-O Raman band indicates that As remains dissolved in the silicate network.

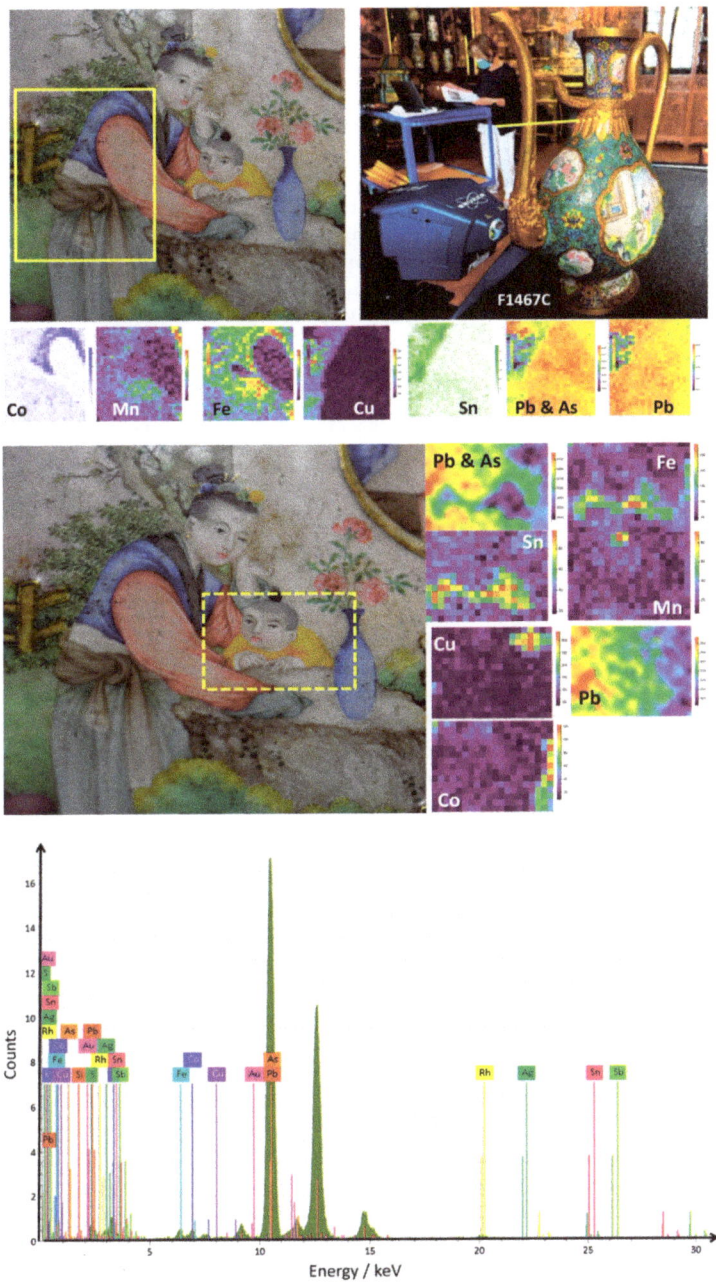

Figure 7. pXRF maps recorded on the painted décor depicting a woman with a child in the garden. The two mapped areas are delimited with a yellow line. The optical pixel image and corresponding distribution of elements (Co, Mn, Fe, Pb, Cu, Sn and Pb + As) are shown. An example of an XRF spectrum recorded on the darkest blue area of the woman's vest in the painted décor.

This indicates that some of the cobalt raw materials used contain some arsenic. Simultaneous use of cobalt ores from different origins has previously been reported [62]. These findings indicate the use of Asian Mn-rich cobalt in the blue enamel of the incense burner and European As- and Co-rich smalt for the similar enamel of the ewer [62], the potassium amount being much higher in the blue enamel of the ewer (K/Ca net count area ~1–2 for the incense burner vs. 8 for the ewer). In the ewer, as shown in Figures 6 and 7, manganese was not detected in the blue areas. The Mn/Co net count area ratios comprised between 4 and 8 are typical of measurements made on blue-and-white porcelains of the Xuande period in the Ming Dynasty [62,64–66,85,86].

The compositions of the yellow and green *cloisonné* enamels of the incense burner are rather comparable (Table 4). The detection of cobalt in the green indicates that the color was obtained by adding a yellow pigment to the blue matrix, as was characteristic of the European method. Copper ions further contribute to the coloring process. On the other hand, the *cloisonné* enamels of the ewer contain much more lead (Pb/Si net count area ratio ~50 for incense burner vs. ~250 for ewer, except for lead-poor red enamel) than those of the incense burner and show similarity with the compositions of the painted enamels (a possible argument to support the hypothesis that the last firing is made for painted enamels but also to obtain a high gloss on polished enamels). Here it should be recalled that lead-rich compositions are common for Chinese *cloisonné* enamels (Table 2; [38,39,87,88]). They also contain a significant amount of tin, indicating the use of an opacifying agent to adjust the hue and/or as an underlayer on which the other colors are put. Copper was also measured in the green enamels at a high level.

4.1.3. Painted Enamels (Ewer)

The most striking feature in the XRF spectra of the painted enamels deals with opacification based on the use of arsenic in the white enamel (Figure 6, top right and Table 4) and a small amount of antimony in the yellow-green enamel (Figure 6, bottom right) of the medallions. Cobalt was also detected as a whitening agent in the white areas opacified with arsenic. The As L_α peak cannot be separated from the Pb L_α one by the pXRF instrument and the As L_β peak is rather close to that of the Au L_β. However, the As L_β peak is sufficiently well identified in some of the pXRF spectra such as m-red (painted medallion décor, Figure 6, left-bottom), m-green (painted medallion, Figure 6, left-bottom), p-yellow (painted large medallion depicting a woman and child, Figure 6, left-bottom) and m-yellow-green (Figure 6, right-bottom). Confirmation is given by the Raman analysis (see later). However, calculation of the elemental content is difficult, except for the (homogeneous) white enamel (Table 4) which is about 5 wt% of As_2O_3. Tin was also measured in the green, yellow and yellow-green painted areas (yellow fence on the left side and green leaves around). It is also important noting that tin was used for the white hand of the child and not for the face. This could indicate that the painting of the hand and the face had been made by different artists. The distribution of lead and arsenic is wide and not directly related to the drawing, especially for the child. This is consistent with the deposit of a white arsenic and lead-rich layer as a substrate in the whole décor. The variation of the intensity reflects the difference in the thickness of this layer in relation with the variable planarity of the gold foil substrate. The poor planarity of the gold foil was also observed by X-ray radiography. Comparison with the red cloth of the woman's arm and the corresponding mapping of iron and copper elements demonstrate the absence of iron- or copper-based phases. The pink hue had thus been made with gold nanoparticles and not with the alternating techniques based on hematite or copper nanoparticles. The Mn-rich mapping spot of the child clearly corresponds to the hair bun. Manganese-free cobalt is also obvious for the blue vase.

Data measured on the similar areas are rather identical, which allows confidence in the comparison. Note that the lead content of the white area, that seems to serve as a substrate layer on which the other colors are painted, is lower than those of the colored areas. This is consistent with a deposit of the white layer first, perhaps with the preparation of *cloisonné* enamels (and their polishing). The colored décor is then added and fired at a temperature lower than the *cloisonné* enamels and painted white enamel layer, that impose a higher lead content in colored painted enamels. The highest lead content was measured for the yellow color, and we further see by Raman scattering that yellow color was obtained by lead–tin pyrochlore pigment ($Pb_2Sn_2O_6$ *Naples yellow* type) which imposes a saturation of the glass matrix by lead to preserve the dissolution of the pigment in the flux. The higher lead content involves a lowering of the melting temperature and viscosity of yellow enamel, according to the higher gloss and the smooth surface of painted yellow areas (Figure 3).

The latest generation XRF mobile instruments make it possible, if the surface to be studied is flat, to carry out maps as in Figure 7. The mapping process takes longer total measurement times than the analysis of a single spot but makes it possible to ensure the representativeness of the point measurements. An example of the spectrum recorded for a few selected pixels (blue area) is shown in Figure 7 where a well-defined Co peak is present despite the rather limited counting time imposed by the mapping procedure. In the same Figure, the comparison of the signal intensities of the woman's blue vest relating to cobalt and manganese clearly shows that the correlation between them excludes the use of cobalt ores from Asian sites used under the Ming Dynasty [62,64–66,86], and supports the use of cobalt imported from Europe (smalt), also deduced from spot analysis. Note that manganese was used in the brown belt of the woman.

The distribution of tin in the green zone is obvious, as is its non-use for the white belt obtained by an arsenate. Addition of a little copper in the green areas is obvious. The black eyes of the child were obtained with an iron-rich compound (likely a spinel). The hands are colored in white with tin (cassiterite) although the face is made with arsenic and lead (lead arsenate).

4.2. Phase Raman Identification

Figures 8–10 show representative spectra recorded with a 200× long working distance (lwd) microscope objective. Preliminary Raman identification made using a lower magnification microscope objective (50× lwd) was published in ref [52] in the course of the study of a series of Chinese enameled wares. Figures 8 and 9 show representative spectra of *cloisonné* enamels while Figure 10 shows those of painted enamels. Characteristic peak wavenumbers and phase assignments are summarized in Table 5. The Raman spectra obtained from the different colored enamels of the F1448C incense burner and F1467C ewer were baseline subtracted and spectral components were then specified with a peak-fitting process to identify the crystalline phases more clearly.

4.2.1. Silicate Matrix and Crystalline Phases

The *cloisonné* and painted enamels studied display the typical Raman signature of a glassy silicate, sometimes accompanied by some crystalline phases either coming from the raw materials used or from the addition of pigments. The Raman spectrum of a glassy silicate is mainly dominated by two broad 'bands' at about 500 and 1000 cm^{-1}, arising from the bending and stretching modes of the SiO_4 tetrahedron which polymerizes to form the crystalline or amorphous silicates [47–51]. The Si-O connectivity is interrupted with other elements (Al, K, Pb, etc.) coming from the fluxes used in the glass raw materials. In the glassy silicate signature, the symmetrical stretching mode of the SiO_4 vibrational unit dominates the Raman spectrum and the contribution of the Al-O bond (too ionic) is very poor and that of the Pb-O located at low wavenumber is suppressed by the baseline subtraction [37]. Therefore, the SiO_4 stretching mode provides a direct link between its spectral components and the SiO_4 tetrahedron with different connectivity. In this case, five components are present due to the contribution of isolated tetrahedron, and of tetrahedron connected by one, two, three or four common oxygen atoms, forming the glassy polymerized Si-O network [47–51]. Consequently, the 700–1250 cm^{-1} wavenumber range is fitted by five components. Additional bands also arise from the contribution of crystalline phases. The number of components in the bending 'band' is much higher (the symmetrical bending mode of a SiO$_4$ tetrahedron has E character and the asymmetrical mode F character that could generate 25 components) and it is not possible to assign a physical meaning to the components of the fitting.

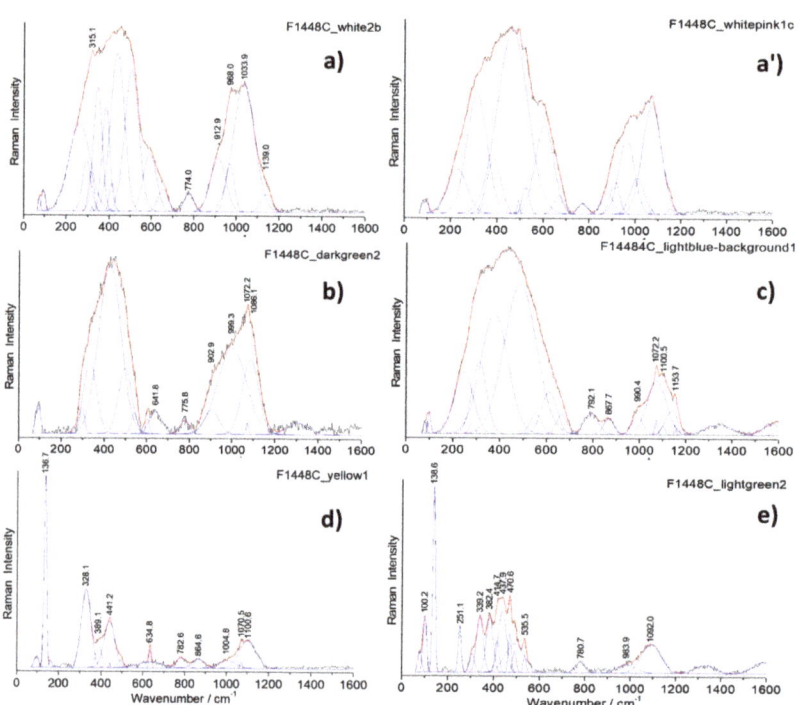

Figure 8. Representative Raman spectra recorded on the *cloisonné* enamels of the incense burner (baseline subtracted): white (**a**), white-pink (**a'**), dark green (**b**), turquoise (**c**), yellow (**d**) and light green (**e**) areas. Gaussian and Lorentzian components used to separate the contribution of the crystalline pigment from that of the glassy matrix are shown.

The quality of the spectrum collected with mobile instruments where the Rayleigh scattering is rejected by only one edge filter is less than that recorded with advanced fixed instruments equipped with a set of filters. In particular, the background of the mobile instrument is not flat, which makes its baseline subtraction partly subjective. Consequently, the fitting of the spectrum is affected by the intensity of the background and the set of SiO$_4$ stretching components should be considered mainly as a tool to determine the contribution of crystalline phases precisely.

Raman spectra recorded on white, white-pink, dark green and light blue *cloisonné* enamels of the incense burner (Figure 8) and of the turquoise and dark blue *cloisonné* enamel of the ewer (Figure 9) reveals mainly the signature of the glassy (lead earth-alkali-based) silicate: the center of gravity (and roughly the maximum) of the stretching band peaks at 1030–1050 cm^{-1}. For painted enamels (ewer) the stretching massif peaks at a slightly lower wavenumber, ~1000–1030 cm^{-1}, which indicates a more depolymerized silicate matrix, according to the higher content of lead measured (Tables 3 and 4). This should indicate a firing at a lower temperature, according to the production sequence assumed above. Small narrow peaks are, however, observed at ~975 and 1070 cm^{-1} which are assigned to alpha- and beta-wollastonite (CaSiO$_3$) precipitates, according to the large amount of calcium (Table 4). Painted enamels (Figure 10) show the similar Raman signature of the glassy silicate matrix for white and blue areas (bending and stretching broad bands) plus some additional features, such as the 461 cm^{-1} peak of alpha-quartz. This peak seems to be more frequent in the painted enamels than in the *cloisonné* ones, indicating different preparation routes.

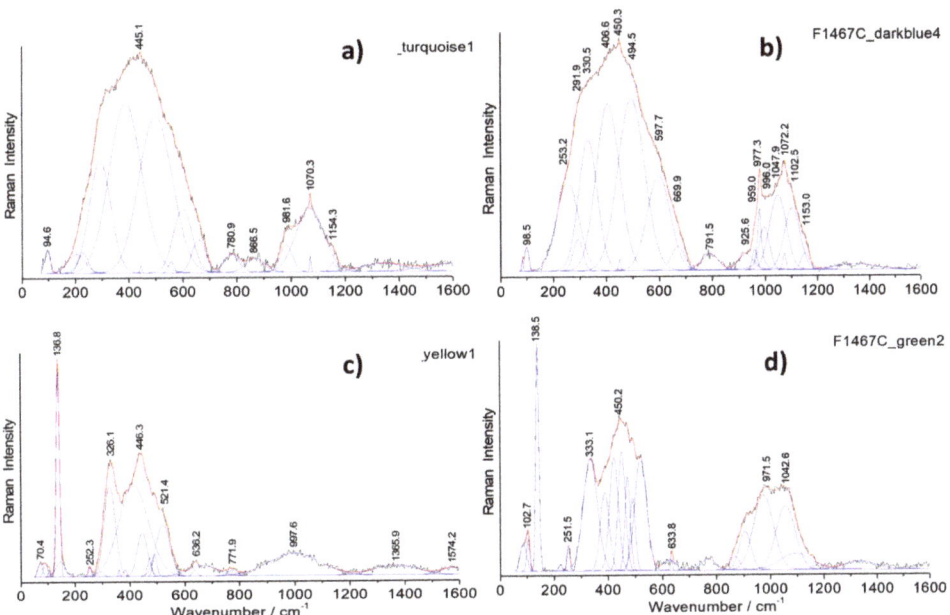

Figure 9. Representative Raman spectra recorded on the *cloisonné* enamels of the ewer (baseline subtracted): turquoise (**a**), dark blue (**b**), yellow (**c**), and green (**d**). Gaussian and Lorentzian components used to separate the contribution of the crystalline pigment from that of the glassy matrix are shown.

4.2.2. Pigments and Opacifiers

Raman analysis is particularly effective in the identification of crystal phases present in the glassy silicate matrix. In the white *cloisonné* enamel of the incense burner, the weak peak at 315 cm^{-1} is assigned to fluorite (CaF_2) as an opacifier which specifically belongs to an ancient Chinese tradition of glass opacification [38,52,81,88]. The continuous use of fluorite in Chinese glassy materials has been reported from the Tang Dynasty to the Qing Dynasty, in glass objects [81] as well as *cloisonné* enamels [38,87–89].

Regarding the blue enamels with different hues, different spectral features were observed in the artefacts studied. In the light blue-turquoise background *cloisonné* enamel of the incense burner, the typical bending and stretching bands of the glassy silicate structure are observed along with a small band at 868 cm^{-1} which can be attributed to the stretching mode of the chromate phase. This phase probably results from the presence of chromium often associated with the cobalt ore used [62,90]. The Raman spectrum of the dark blue *cloisonné* enamel of the ewer displays only the signature of a glassy silicate, indicating the use of Co^{2+} ions dispersed in the glassy matrix to obtain the dark blue color, without the precipitation of any crystalline phases. On the contrary, the painted blue enamel of the ewer strikingly shows a distinctive ~820 cm^{-1} peak with a shoulder at 788 cm^{-1} (Figure 10a,b,d) which is characteristic of the As-O symmetrical stretching mode in a lead arsenate phase. This feature is particularly assigned to lead–potassium–calcium arsenate with an apatite structure [52,53,61,63,74,91] which is formed by the reaction of lead, potassium and calcium coming from the enamel matrix with arsenic coming from the cobalt source [10,46,62,74]. In some cases, the As-O mode is of a larger intensity (Figure 10c), which indicates very small apatite grains or the formation of another kind of As-based phase (As-feldspar?) [62]. The Raman spectra recorded on red *cloisonné* enamels do not show peaks characteristic of any crystalline phases (see ref. [52]). This is consistent with coloration with copper nanoparticles.

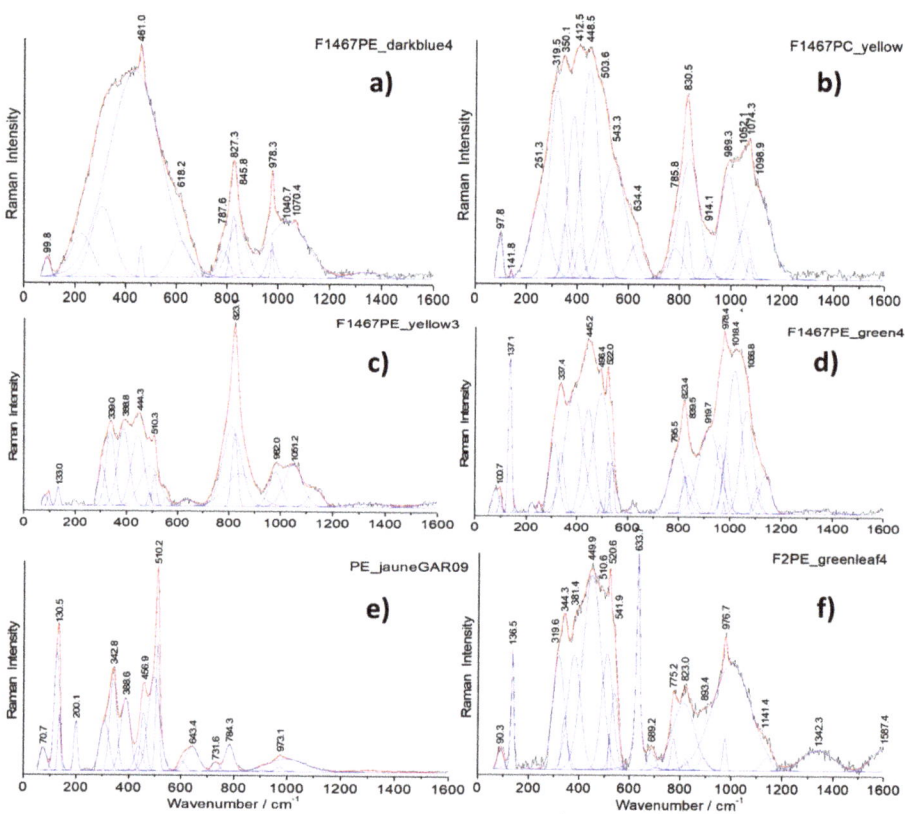

Figure 10. Representative Raman spectra recorded on the painted enamels of the ewer (baseline subtracted): dark blue (**a**), yellow (**b,c,e**), and green (**d,f**).

Raman spectra recorded on the yellow and green *cloisonné* (Figures 8 and 9) and painted enamels (Figure 10) mainly consist of a set of narrow bands, with a strong peak at ~130 cm^{-1} characteristic of lead-based pyrochlore pigment (also called *Naples yellow*) [67–74]. This mode involving Pb atoms peaks at low energy due to their heavy mass. *Naples yellow* can now be considered as a general pigment class, based on lead, antimony and/or tin (*Naples yellow* type I) with varying stoichiometry due to different routes in the production process. The pyrochlore structure may incorporate different ratios of these elements along with others such as iron, zinc and silicon (*Naples yellow* type II), forming complex solid solutions depending on the oxygen stoichiometry (i.e., the degree of oxidizing/reducing atmosphere in the firing). The availability of the raw materials and the desire to achieve different hues result in the modification of the pigment which is further affected by the glaze raw materials during the firing process. At least three types of lead pyrochlore pigment were identified in the objects according to the Raman spectra: The end member $Pb_2Sn_2O_6$ type (*Naples yellow* type I in the literature) in the yellow *cloisonné* enamels of the incense burner (Figure 8d) and the ewer (Figure 9c); a second tin-rich phase in the light green *cloisonné* enamel of the incense burner (Figure 8e) and green *cloisonné* enamel of the ewer (Figure 9d); and the tin-antimony-(zinc?) pyrochlore phase in the yellow painted enamel of the ewer (Figure 10e) as well as its green painted enamel (Figure 10f). The first type of lead pyrochlore Pb-Sn pigment is characterized by the strongest peak at 137 cm^{-1} and distinct ~330 and ~450 cm^{-1} components. The latter component is particularly assigned to the stretching mode of Sn-O (*Naples yellow* type I). The second type of pyrochlore pigment as the tin-rich phase has further components at 382 and 471 cm^{-1} along with a characteristic ~250 cm^{-1} peak while the third type as the mixed pyrochlore phase notably displays a ~510 cm^{-1} strong component which belongs to the Sb-O stretching mode. In the Raman spectra of lead pyrochlore pigment, the evident ~ca. 135 cm^{-1} peak characteristic of Pb-O mode is due to the saturation of the glassy silicate matrix with excess lead. Its position depends on the firing temperature employed [67–69]. The results are consistent with the XRF measurements. The Sn peak is clearly observed in all yellow to green areas (Figures 5 and 6, Table 4). The Sb peak was only detected for yellow-green painted medallion areas (Figure 6) as well as *cloisonné* enamels (Table 4).

Table 5. XRF and Raman results of the different colored enamels studied (sh: shoulder, w: weak, m: medium, S: strong, vS: very strong).

Enamel Color	Enamel Type	Artefact Period	Major Element (XRF)	Minor/Traces Element (XRF)	Pigment Raman Bands (cm^{-1})	Phases	European Recipe/ Ingredient
white	*cloisonné*	Kangxi		Fe,Cu	315 (w)	Fluorite	No
	cloisonné	Qianlong		Fe,Cu	-		
	painted			As,Fe,Ni,Sn	-		Yes
blue	*cloisonné* (light blue)	Kangxi		As,Mn,Fe, Co,Ni,Cu,Sn	868 (w)	Chromate	No
	cloisonné (dark blue)	Qianlong		Fe,Co,Cu,(Sn?,As?)	-	Glassy silicate matrix	No?
	painted			Co,Fe,As	788 (sh), 827 (S)	Lead arsenate apatite	Yes
yellow	*cloisonné*	Kangxi		Sn,Fe,(Cu?)	137 (vS), 328 (S), 441 (m) 635 (w), 783 (w)	Pyrochlore type I (Pb$_2$Sn$_2$O$_6$) Cassiterite	Yes
	cloisonné	Qianlong	Si,K,Ca,Pb	Sn,Fe,Ni	137 (vS), 326 (S), 446 '(S), 521 (m) 636 (w)	Pyrochlore type 1 (Pb$_2$Sn$_2$O$_6$) Cassiterite	Yes
	painted			Sn,Fe,Cu,Ni,As	130 (vS), 200 (m), 343 (S), 389 (m), 457 (m), 510 (vS)	Pyrochlore type 3	Yes
yellow-green, light green	*cloisonné*	Kangxi			139 (vS), 225, 251 (m), 339 (m), ~420 (m), 471 (m), 535 (w)	Pyrochlore type 2 (~250 cm^{-1})	Yes
	painted	Qianlong		As,Sn,Cu,Mn,Ni(Sb?)	[130–510]	Pyrochlore 3	Yes
green	*cloisonné*	Kangxi		Cu,Fe,Sn			
	cloisonné			Cu,Sn,Fe	138 (vS), 251 (w), 333 (S), 450 (S),510 (S)	Pyrochlore type 2	Yes
	painted	Qianlong		As,Cu,Sn, Fe,Cu,Mn	775 (m), 823 (m) 137 (S), 344 (vS), 450 (vS), 521 (vS) 634 (vS), 775 (m)	Lead arsenate apatite Pyrochlore type 3 Cassiterite	Yes
turquoise	*cloisonné*	Qianlong		Cu,Fe,Ni	-	Glassy silicate matrix	No
red	*cloisonné*	Kangxi		Cu,Fe,(Au?)			No?
	cloisonné	Qianlong		Fe,Cu,Sn			?
	painted			Au,Fe,Ni,As,			Yes

Other stringent features are the narrow peak doublet at ~633 and ~775 cm^{-1} characteristic of cassiterite (SnO$_2$) [39], particularly in the case of the green painted enamel of the ewer (Figure 10f). In some of the yellow and green painted enamels analyzed, the signature of lead arsenate apatite phase is also observed (Figure 10b,c,d,e). In some of these spectra, the As-O mode is larger (Figure 10b,c), which may indicate very small lead arsenate apatite grains or the formation of another As-based phase (As-feldspar?) [62].

4.3. Painting Technique

The observation of the painting technique of the Qianlong ewer at high optical magnification (Figure 11) shows dotted touches of color, which is the technique of the miniaturists in the 18th century [92,93]. Miniaturists made drawing and painting at a small scale (a few cm^2) representing landscapes or scenes including many personages, country or castle views, etc. They used lenses, fine nibs and brushes made of some polishes to achieve these miniature decorations. It is reasonable to think that similar techniques were used to paint the decoration of the enameled ewer. Only for some parts such as the hair of the human figures, the brush touch was used. The analysis of a fraction of a green paint touch is shown in Figure 10f. Here, the spot analyzed in Raman with the 200× objective is about more than ten times smaller than the paint point visible in the zoom image (Figure 11). However, in this spot at least four crystalline phases (cassiterite, lead pyrochlore type 3, wollastonite, lead arsenate apatite plus amorphous carbon) and the amorphous silicate matrix are identified in the Raman spectrum. This confirms the use of a color palette prepared by prior mixing of coloring agents to obtain a wide range of colors, as practiced in European easel paintings, in accordance with the archival texts which say that about thirty colors became available for painted enamel decorations prepared by the imperial workshops [6,94,95].

Figure 11. Images of the painted décor and zoom of the green foliage showing the use of dots of different hues and different painting techniques (spitting or brush touch).

5. Conclusions

XRF and Raman analyses enabled us to obtain a great deal of information about the coloring agents used in the different types of enamels studied as well as their glass types, despite the on-site analysis conditions with limited access time to the objects and the imperative use of non-invasive methods. The exceptional character of these objects in terms of their aesthetic quality is also attested in terms of their enameling techniques. The sparse use of new colors, such as yellow and green in the incense burner, is in perfect harmony with the imported origin of the recipes (complex lead pyrochlore type pigments, a simple lead–tin pigment being already used at least from the Ming Dynasty) used to create these colored enamels at the end of the 17th century or at the turn of the 17th–18th century. The observation of the same wavenumber at ~135 cm^{-1} for the pyrochlore pigment based on Pb-Sn in the yellow *cloisonné* enamels of the two objects and painted enamels of the ewer indicates the same temperature of preparation of the pigment, prepared before, probably around 600–700 °C which is compatible with the expected <800 °C according to the literature [20].

The complexity and mastery of the enamel decoration of the 18th century ewer as well as the extent of the color palette (Table 5) shows that the imported techniques were perfectly incorporated into the knowledge of the craftsmen of the Imperial Palace. It is evident that the lead content of the painted enamels is higher than that of the *cloisonné* ones for the ewer and much higher than that of the *cloisonné* enamels of the incense burner, except for the white background. This could also indicate that some of the recipes used in the incense burner had followed the European recipes introduced by the Jesuits. The increased XRF signal of potassium concomitant with the cobalt signal in painted enamels is a good indication of the use of smalt as a source of 'European' cobalt. This agrees with the results obtained for Japanese porcelain [18–20] as well as for paintings in China [94] and Japan [20]. The meticulousness of the dotted painting technique deposited on one or more backgrounds induces a complex stratigraphy of the enameled decoration. Although the sub-micron spatial resolution of Raman analysis allows access to grain-by-grain analyses, this is generally incompatible on-site because it takes too much measurement time to obtain such information. The lower resolution of the XRF analysis averages the related compositional data. In this case, only the availability of fragments (fragments collected during the restoration operation, sampling, shards from archaeological excavations) can lead to a more precise analysis for a better understanding of the stratigraphy of the enameled decoration in terms of composition. It is also necessary to compare the results of the analyses undertaken with the information found in the historical texts.

In conclusion, many questions still remain open. One of them is related to the functioning of the glass production workshop directed by the German Jesuit Kilian Stumpf. This workshop prepared the enamels, probably the frit, and enameled glass objects seem to have been produced at the beginning, painted enamels on copper or porcelain being produced after 1716 [95–98]. This could explain the privileged use of the opacification of lead enamels by the addition of arsenic, a classic technique of Italian glassmakers in the 17th century [99]. The use of a competing technique of opacification with cassiterite preferred by potters [99] appears to be very limited. The use of 'Italian' recipes (arsenic-based opacification, arsenic-based preparation of colloidal gold) could be linked with the venue of Italian coadjutor brothers with some expertise in the enameling techniques [100]. The highlighting of the use of European recipes in the *cloisonné* ware of the Kangxi period could indicate that the first attempts to use these European recipes were made for this type of object, as also for the 'simple' water pots as already observed [10,101]. It is necessary to analyze in detail a larger number of objects to statistically assess the use of imported recipes and their adaptation by Chinese artisans. The present study demonstrates that several phases of lead arsenate had been used for opacification, one of them being apatite. Different explanations are possible, such as that the source of arsenic-rich cobalt is different. The other one could be that the compositions of the silicate matrix are different and hence different phases were formed. Furthermore, a combination of the two phenomena is also possible. It is very likely that different pyrochlores were used simultaneously for yellows and greens (two being rich in tin, another containing antimony, plus zinc). Only µdiffraction or transmission electron microscopy analyses can provide more confident answers, but these methods require sampling.

Detailed XRF mapping of the painted area should be correlated to stylistic study. The Sn element map shows very limited use of SnO_2 for the child's hand and not for the face. This could indicate that different artists contributed to the painting of the different parts of the décor.

The present study reveals the potential of on-site non-invasive studies but also the limitations of the method.

Author Contributions: Conceptualization, P.C.; methodology, P.C., M.G. and B.K.; validation, P.C., M.G., B.K. and V.C.; investigation, P.C., D.V. and M.G.; resources, P.C., B.Z. and V.C.; writing—original draft preparation, P.C.; writing—review and editing, P.C., M.G., B.K., B.Z. and V.C. All authors have read and agreed to the published version of the manuscript.

Funding: The research in France was funded by the French Agence Nationale de la Recherche ANR EnamelFC project—19-CE27–0019-02.

Institutional Review Board Statement: Not applicable.

Informed Consent Statement: Not applicable.

Data Availability Statement: All data incorporated in the paper.

Acknowledgments: Ching-Fei Shih (NTU, Taipei) is kindly acknowledged for the critical reading of the manuscript. The authors also acknowledge Ludovic Bellot-Gurlet (SU-MONARIS, Paris), Jean-Baptiste Clais (Louvre Museum, Paris), Claire Delery (Asian Art, Guimet Museum, Paris), Yong Lei (Palace Museum, Beijing) and Pauline d'Abrigeon (Baur Fondation, Genève) for many discussions as well as Sarah Paronetto for the provision of the objects and the organization of the analytical measurements at Fontainebleau Castle.

Conflicts of Interest: The authors declare no conflict of interest.

References

1. Parthasarathi, P. Comparison in global history. In *Writing the History of the Global. Challenges for the 21st Century*; Berg, M., Ed.; Oxford University Press: Oxford, UK, 2013; pp. 69–82.
2. Schäfer, D. Technology and innovation in global history and in11 the history of the global. In *Writing the History of the Global. Challenges for the 21st Century*; Berg, M., Ed.; Oxford University Press: Oxford, UK, 2013; pp. 147–163.
3. Beaujard, P. *Les Mondes de l'océan Indien, Vol. 1, De la Formation de l'État au Premier Système-Monde Afro-Eurasien, Volume 2. L'océan Indien, au Cœur des Globalisations de l'Ancien Monde (7e-15e Siècles)*; Armand Colin: Paris, France, 2012.
4. Kingery, W.D.; Vandiver, P.B. The Eighteenth-Century Change in Technology and Style from the *Famille-Verte* Palette to the *Famille-Rose* Palette. In *Technology and Style*; Ceramics and Civilization Series; Kingery, W.D., Ed.; The American Ceramic Society: Colombus, OH, USA, 1986; Volume 2, pp. 363–381.
5. National Palace Museum (Ed.) *Special Exhibition of Ch'ing Dynasty Enamelled Porcelains of the Imperial Ateliers*; National Palace Museum: Taipei, Taiwan, 1992.
6. Shih, C.-F. Evidence of East-West exchange in the eighteenth century: The establishment of painted enamel art at the Qing Court in the reign of Emperor Kangxi. *Natl. Palace Mus. Res. Q.* **2007**, *24*, 45–94.
7. Kleutghen, K. Chinese Occidenterie: The Diversity of "Western" Objects in Eighteenth-Century China. *Eighteenth-Century Stud.* **2014**, *47*, 117–135. [CrossRef]
8. Xu, X.D. Europe-China-Europe: The Transmission of the Craft of Painted Enamel in the Seventeenth and Eighteenth Centuries. In *Goods from the East, 1600–1800 Trading Eurasia*; Berg, M., Gottmann, F., Hodacs, H., Nierstrasz, C., Eds.; Palgrave Macmillan: London, UK, 2015; pp. 92–106.
9. Palace Museum (Ed.) *Treasures from Oversea Countries, Exhibition Catalogue of Kulangsu Gallery of Foreign Artefacts from the Palace Museum Collection*; Gugong Chubanshe: Beijing, China, 2011.
10. Colomban, P.; Zhang, Y.; Zhao, B. Non-invasive Raman analyses of Chinese *huafalang* and related porcelain wares. Searching for evidence for innovative pigment technologies. *Ceram. Int.* **2017**, *43*, 12079–12088. [CrossRef]
11. Curtis, E.B. Aspects of a multi-faceted process: The circulation of enamel wares between the Vatican and Kangxi's court. *Extrême-Orient Extrême-Occident* **2020**, *43*, 29–39.
12. Wood, N. *Chinese Glazes: Their Origins, Chemistry and Recreation*; A & C Black: London, UK, 1999; pp. 194–195.

13. Kerr, R.; Wood, N. Part 12, Ceramic Technology. In *Science and Civilisation in China: Volume 5, Chemistry and Chemical Technology*; Cambridge University Press: Cambridge, UK, 2004.
14. Lili, F. *La Céramique Chinoise*; China Intercontinental Press: Beijing, China, 2011.
15. Medley, M. *The Chinese Potter: A Practical History of Chinese Ceramics*, 3rd ed.; Phaidon Press: London, UK, 1999.
16. Finlay, R. *The Pilgrim Art: Cultures of Porcelain in World History*; University of California Press: Oakland, CA, USA, 2010.
17. Castelluccio, S. *Le Goût pour les Porcelaines de Chine et du Japon à Paris aux XVIIe et XVIIIe Siècles*; Éditions Monelle Hayot: Saint-Rémy-en-l'Eau, France, 2013.
18. Montanari, R.; Alberghina, M.F.; Casanova Municchia, A.; Massa, E.; Pelagotti, A.; Pelosi, C.; Schiavone, S.; Sodo, A. A polychrome Mukozuke (1624–1644) porcelain offers a new hypothesis on the introduction of European enameling technology in Japan. *J. Cult. Herit.* **2017**, *32*, 232–237. [CrossRef]
19. Montanari, R.; Murakami, N.; Alberghina, M.F.; Pelosi, C.; Schiavone, S. The Origin of overglaze-blue enameling in Japan: New discoveries and a reassessment. *J. Cult. Herit.* **2019**, *37*, 94–102. [CrossRef]
20. Montanari, R.; Murakami, N.; Colomban, P.; Alberghina, M.F.; Pelosi, C.; Schiavone, S. European Ceramic technology in the Far East: Enamels and pigments in Japanese art from the 16th to the 20th century and their reverse influence on China. *Herit. Sci.* **2020**, *8*, 48. [CrossRef]
21. Hiraoka, R. Jesuits and Western clock in Japan's "Christian Century" (1549—c.1650). *J. Jesuits Stud.* **2020**, *7*, 204–220. [CrossRef]
22. Han, Q. The role of the French Jesuits in the seventeenth and eighteenth centuries. In *East Asian Science: Tradition and Beyond, Proceedings of the 7th International Conference on the History of Science in East Asia, Kyoto, Japan, 2–7 August 1993*; Hashimoto, K., Jami, C., Skar, L., Eds.; Kansai University Press: Osaka, Japan, 1995; pp. 489–492.
23. Landry-Deron, I. Les Mathématiciens envoyés en Chine par Louis XIV en 1685. In *Archive for History of Exact Sciences*; Springer: Berlin/Heidelberg, Germany, 2001; pp. 423–463, ⟨halshs-00676823⟩.
24. Landry-Deron, I. *La Preuve par la Chine: La «Description» de J.-B. Du Halde, Jésuite, 1735*; Editions de l'EHESS: Paris, France, 2002.
25. Jami, C. *The Emperor's New Mathematics: Western Learning and Imperial Authority in China during the Kangxi Reign (1662–1722)*; Oxford University Press: Oxford, UK, 2012.
26. Pirazzoli-t'Serstevens, M. *Giuseppe Castiglione 1688-1766: Peintre et Architecte à la Cour de Chine*; Musée des Arts de l'Asie de la Ville de Paris: Paris, France, 2007.
27. National Palace Museum (Ed.) *Giuseppe Castiglione—Lang Shining New Media Art Exhibition*; National Palace Museum: Taipei, Taiwan, 2015.
28. Montanari, R.; Alberghina, M.F.; Schiavone, S. The Jesuit painting Seminario in Japan: European Renaissance technology and its influence on Far Eastern art. *X-ray Spectrom* **2021**. [CrossRef]
29. Shih, C.-F. *Radiant Luminance: The Painted Enamelware of the Qing Imperial Court*; The National Palace Museum of Taipei: Taipei, Taiwan, 2012.
30. Zhao, B.; Simon, F. Les cadeaux diplomatiques entre la Chine et l'Europe aux XVIIe–XVIIIe siècles. Pratiques et enjeux. *Extrême-Orient Extrême-Occident* **2019**, *43*, 5–24. [CrossRef]
31. De Rochebrunne, M.L. Les porcelaines de Sèvres envoyées en guise de cadeaux diplomatiques à l'empereur de Chine par les souverains français dans la seconde moitié du XVIIIe siècle. *Extrême-Orient Extrême-Occident* **2019**, *43*, 81–92. [CrossRef]
32. Finlay, J. Henri Bertin and Louis XV's Gifts to the Qianlong Emperor. *Extrême-Orient Extrême-Occident* **2019**, *43*, 93–112. [CrossRef]
33. Guo, F. Presents and Tribute: Exploration of the Presents Given to the Qianlong Emperor by the British Macartney Embassy. *Extrême-Orient Extrême-Occident* **2019**, *43*, 143–172. [CrossRef]
34. Van Pevenage, J.; Lauwers, D.; Herremans, D.; Verhaeven, E.; Vekemans, B.; De Clercq, W.; Vincze, L.; Moens, L.; Vandenabeele, P. A Combined Spectroscopic Study on Chinese Porcelain Containing Ruan-Cai Colours. *Anal. Methods* **2014**, *6*, 387–394. [CrossRef]
35. Giannini, R.; Freestone, I.C.; Shortland, A.J. European cobalt sources identified in the production of Chinese *Famille rose* porcelain. *J. Archaeol. Sci.* **2017**, *80*, 27–36. [CrossRef]
36. Colomban, P.; Ambrosi, F.; Ngo, A.-T.; Lu, T.-A.; Feng, X.-L.; Chen, S.; Choi, C.-L. Comparative analysis of *wucai* Chinese porcelains using mobile and fixed Raman microspectrometers. *Ceram. Int.* **2017**, *43*, 14244–14256. [CrossRef]
37. Colomban, P. On-site Raman identification and dating of ancient glasses: Procedures and tools. *J. Cult. Herit.* **2008**, *9*, e55–e60. [CrossRef]
38. Kırmızı, B.; Colomban, P.; Quette, B. On-site analysis of Chinese *Cloisonné* enamels from fifteenth to nineteenth centuries. *J. Raman Spectrosc.* **2010**, *41*, 780–790. [CrossRef]
39. Kırmızı, B.; Colomban, P.; Blanc, M. On-site analysis of Limoges enamels from sixteenth to nineteenth centuries: An attempt to differentiate between genuine artefacts and copies. *J. Raman Spectrosc.* **2010**, *41*, 1240–1247. [CrossRef]
40. Colomban, P. The on-site/remote Raman analysis with mobile instruments: A review of drawbacks and success in cultural heritage studies and other associated fields. *J. Raman Spectrosc.* **2012**, *43*, 1529–1535. [CrossRef]
41. Colomban, P.; Tournié, A.; Meynard, P.; Maucuer, M. On-site Raman and XRF analysis of Japanese/Chinese Bronze/Brass Patina—The search of specific Raman signatures. *J. Raman Spectrosc.* **2012**, *43*, 799–808. [CrossRef]
42. Colomban, P.; Arberet, L.; Kırmızı, B. On-site Raman analysis of 17th and 18th century Limoges enamels: Implications on the European cobalt sources and the technological relationship between Limoges and Chinese enamels. *Ceram. Int.* **2017**, *43*, 10158–10165. [CrossRef]

43. Colomban, P. On-site Raman study of artwork: Procedure and illustrative examples. *J. Raman Spectrosc.* **2018**, *49*, 921–934. [CrossRef]
44. Colomban, P.; Lu, T.-A.; Milande, V. Non-invasive on-site Raman study of blue-decorated early soft-paste porcelain: The use of arsenic-rich (European) cobalt—Comparison with *huafalang* Chinese porcelains. *Ceram. Int.* **2018**, *44*, 9018–9026. [CrossRef]
45. Colomban, P.; Kırmızı, B.; Gougeon, C.; Gironda, M.; Cardinal, C. Pigments and glassy matrix of the 17th–18th century enamelled French watches: A non-invasive on-site Raman and pXRF study. *J. Cult. Herit.* **2020**, *44*, 1–14. [CrossRef]
46. Colomban, P.; Gironda, M.; Edwards, H.G.M.; Mesqui, V. The Enamels of the First (Soft-paste) European Blue-and-white Porcelains: Rouen, Saint-Cloud and Paris Factories: Complementarity of Raman and X-ray Fluorescence analyses with Mobile Instruments to identify the cobalt ore. *J. Raman Spectrosc.* **2021**, *52*. [CrossRef]
47. Colomban, P. Polymerisation Degree and Raman Identification of Ancient Glasses used for Jewellery, Ceramics Enamels and Mosaics. *J. Non-Cryst. Solids* **2003**, *323*, 180–187. [CrossRef]
48. Colomban, P.; Tournié, A.; Bellot-Gurlet, L. Raman Identification of glassy silicates used in ceramic, glass and jewellery: A tentative differentiation guide. *J. Raman Spectrosc.* **2006**, *37*, 841–852. [CrossRef]
49. Colomban, P.; Paulsen, O. Non-destructive Raman Determination of the Structure and Composition of Glazes by Raman Spectroscopy. *J. Am. Ceram. Soc.* **2005**, *88*, 390–395. [CrossRef]
50. Colomban, P. Non-Destructive Raman Analysis of Ancient Glasses and Glazes. In *Modern Methods for Analysing Archaeological and Historical Glass*, 1st ed.; Janssens, K., Ed.; John Wiley & Sons Ltd.: London, UK, 2012; pp. 275–300.
51. Labet, V.; Colomban, P. Vibrational properties of silicates: A cluster model able to reproduce the effect of "SiO$_4$" polymerization on Raman intensities. *J. Non-Cryst. Solids* **2013**, *370*, 10–17. [CrossRef]
52. Colomban, P.; Kırmızı, B.; Zhao, B.; Clais, J.-B.; Yang, Y. Non-invasive on-site Raman study of pigments and glassy matrix of the 17th–18th century painted enamelled Chinese metal wares: Comparison with French enamelling technology. *Coatings* **2020**, *10*, 471. [CrossRef]
53. Colomban, P.; Kırmızı, B.; Zhao, B.; Clais, J.-B.; Yang, Y.; Droguet, V. Investigation of the Pigments and Glassy Matrix of Painted Enamelled Qing Dynasty Chinese Porcelains by Noninvasive On-Site Raman Microspectrometry. *Heritage* **2020**, *3*, 915–940. [CrossRef]
54. Samoyault-Verlet, C.; Bayou, H. *Le Musée Chinois et Les Salons de l'impératrice Eugénie*; Petit Guide 132; De La Réunion Des Musées Nationaux: Paris, France, 1991.
55. Salmont, X.; Droguet, V. *Le Musée Chinois de l'impératrice Eugénie: Château de Fontainebleau*; Editions de la Réunion des Musées Nationaux: Fontainebleau, France, 2011.
56. Personne, N. *Napoléon III à Fontainebleau, Dossier n 14*; Les Amis du Chateau de Fontainebleau: Fontainebleau, France, 2017; pp. 17–19. ISSN 2103-8406.
57. Available online: https://art.rmngp.fr/en/library/artworks/paire-d-aiguieres-dans-sa-cuvette_email-cloisonne_cuivre-metal_emaille_or-metal (accessed on 17 September 2021).
58. Available online: https://www.photo.rmn.fr/C.aspx?VP3=SearchResult&IID=2C6NU0Q7XP87 (accessed on 17 September 2021).
59. Available online: https://www.proantic.com/magazine/le-musee-chinois-de-limperatrice/ (accessed on 17 September 2021).
60. Colomban, P.; Calligaro, T.; Vibert-Guigue, C.; Nguyen, Q.L.; Edwards, H.G.M. Dorures des céramiques et tesselles anciennes: Technologies et accrochage. *ArchéoSciences* **2005**, *29*, 7–20. [CrossRef]
61. Simsek, G.; Colomban, P.; Wong, S.; Zhao, B.; Rougeulle, A.; Liem, N.Q. Toward a fast non-destructive identification of pottery: The sourcing of 14th–16th century Vietnamese and Chinese ceramic shards. *J. Cult. Herit.* **2015**, *16*, 159–172. [CrossRef]
62. Colomban, P.; Kirmizi, B.; Simsek Franci, G. Cobalt and Associated Impurities in Blue (and Green) Glass, Glaze and Enamel: Relationships between Raw Materials, Processing, Composition, Phases and International Trade. *Minerals* **2021**, *11*, 633. [CrossRef]
63. Colomban, P.; Sagon, G.; Huy, L.Q.; Liem, N.Q.; Mazerolles, L. Vietnamese (15th century) blue-and-white, tam thai and "luster" porcelains/stoneware: Glaze composition and decoration techniques. *Archaeometry* **2004**, *46*, 125–136. [CrossRef]
64. Figueiredo, M.O.; Silva, T.P.; Veiga, J.P. A XANES study of cobalt speciation state in blue-and-white glazes from 16th to 17th century Chinese porcelains. *J. Electr. Spectrosc. Relat. Phenom.* **2012**, *185*, 97–102. [CrossRef]
65. Dias, M.I.; Prudêncio, M.I.; de Matos, M.A.P.; Rodrigues, A.L. Tracing the origin of blue and white Chinese Porcelain ordered for the Portugese market during the Ming dynasty using INAA. *J. Archaeol. Sci.* **2013**, *40*, 3046–3057. [CrossRef]
66. Fischer, C.; Hsieh, E. Export Chinese Blue-and-white porcelain: Compositional analysis and sourcing using non-invasive portable XRF and reflectance spectroscopy. *J. Archaeol. Sci.* **2016**, *80*, 14–26. [CrossRef]
67. Sandalinas, C.; Ruiz-Moreno, S. Lead-tin-antimony yellow-Historical manufacture, molecular characterization and identification in seventeenth-century Italian paintings. *Stud. Conserv.* **2004**, *49*, 41–52. [CrossRef]
68. Sandalinas, C.; Ruiz-Moreno, S.; Lopez-Gil, A.; Miralles, J. Experimental confirmation by Raman spectroscopy of a Pb-Sn-Sb triple oxide yellow pigment in sixteenth-century Italian pottery. *J. Raman Spectrosc.* **2006**, *37*, 1146–1153. [CrossRef]
69. Pereira, M.; de Lacerda-Aroso, T.; Gomes, M.J.M.; Mata, A.; Alves, L.C.; Colomban, P. Ancient Portuguese ceramic wall tiles (Azulejos): Characterization of the glaze and ceramic pigments. *J. Nano Res.* **2009**, *8*, 79–88. [CrossRef]
70. Rosi, F.; Manuali, V.; Miliani, C.; Brunetti, B.G.; Sgamellotti, A.; Grygar, T.; Hradil, D. Raman scattering features of lead pyroantimonate compounds. Part I: XRD and Raman characterization of $Pb_2Sb_2O_7$ doped with tin and zinc. *J. Raman Spectrosc.* **2009**, *40*, 107–111. [CrossRef]

71. Pelosi, C.; Agresti, G.; Santamaria, U.; Mattei, E. Artificial yellow pigments: Production and characterization through spectroscopic methods of analysis. *E-Preserv. Sci.* **2010**, *7*, 108–115.
72. Rosi, F.; Manueli, V.; Grygar, T.; Bezdicka, P.; Brunetti, B.G.; Sgamelotti, A.; Burgio, L.; Seccaronif, C.; Miliani, C. Raman scattering features of lead pyroantimonate compounds: Implication for the non-invasive identification of yellow pigments on ancient ceramics. Part II. In situ characterisation of Renaissance plates by portable micro-Raman and XRF studies. *J. Raman Spectrosc.* **2011**, *42*, 407–414. [CrossRef]
73. Cartechini, L.; Rosi, F.; Miliani, C.; D'Acapito, F.; Brunetti, B.G.; Sgamellotti, A. Modified Naples yellow in Renaissance majolica: Study of Pb-Sb-Zn and Pb-Sb-Fe ternary pyroantimonates by X-ray absorption spectroscopy. *J. Anal. Atom. Spectrom.* **2011**, *26*, 2500–2507. [CrossRef]
74. Colomban, P.; Maggetti, M.; d'Albis, A. Non-invasive Raman identification of crystalline and glassy phases in a 1781 Sèvres Royal Factory soft paste porcelain plate. *J. Eur. Ceram. Soc.* **2018**, *38*, 5228–5233. [CrossRef]
75. Francis, P., Jr. *Asia's Maritime Bead Trade, 300BC to the Present*; University of Hawaii Press: Honolulu, HI, USA, 2002.
76. Kingery, W.D.; Vandiver, P.B. Song Dynasty Jun (Chung) ware glazes. *Am. Ceram. Bull.* **1983**, *62*, 1269–1274.
77. Freestone, I.C.; Barber, D.J. The development of the colour of sacrificial red glaze with special reference to a Qing Dynasty saucer dish. In *Chinese Copper Red Wares*; Percival David Foundation of Chinese Art, Monograph Series No. 3; Scott, R.E., Ed.; University of London, School of Oriental and African Art: London, UK, 1992; pp. 53–62.
78. Yang, Y.M.; Feng, M.; Ling, X.; Mao, Z.Q.; Wang, C.S.; Sun, X.M.; Guo, M. Micro-structural analysis of the color-generating mechanism in Ru ware, modern copies and its differentiation with Jun ware. *J. Archaeol. Sci.* **2005**, *32*, 301–310. [CrossRef]
79. Li, Y.Q.; Yang, Y.M.; Zhu, J.; Zhang, X.G.; Jiang, S.; Zhang, Z.X.; Yao, Z.Q.; Solbrekken, G. Colour-generating mechanism of copper-red porcelain from Changsha Kiln (AD 7th–10th century), China. *Ceram. Int.* **2016**, *42*, 8495–8500. [CrossRef]
80. Sciau, P.; Noé, L.; Colomban, P. Metal nanoparticles in contemporary potters' master pieces: Lustre and red "pigeon blood" potteries as models to understand the ancient pottery. *Ceram. Int.* **2016**, *42*, 15349–15357. [CrossRef]
81. Zhou, Y.; Jin, Y.; Wang, K.; Sun, J.; Cui, Y.; Hu, D. Opaque ancient K2O-PbO-SiO2 glass of the Southern Song Dynasty with fluorite dendrite and its fabrication. *Herit. Sci.* **2019**, *7*, 56. [CrossRef]
82. Colomban, P.; Schreiber, H. Raman Signature Modification Induced by Copper Nanoparticles in Silicate Glass. *J. Raman Spectrosc.* **2005**, *36*, 884–890. [CrossRef]
83. Simsek Franci, G. Handheld X-ray fluorescence (XRF) versus wavelength dispersive XRF: Characterization of Chinese blue-and-white porcelain sherds using handheld and laboratory-type XRF instruments. *Appl. Spectrosc.* **2020**, *74*, 314–322. [CrossRef]
84. Demirsar Arli, B.; Simsek Franci, G.; Kaya, S.; Arli, H.; Colomban, P. Portable X-ray Fluorescence (p-XRF) uncertainty estimation for glazed ceramic analysis: Case of Iznik Tiles. *Heritage* **2020**, *3*, 1302–1329. [CrossRef]
85. Pinto, A. Au Coeur de la matière: Apports et limite de la science des matériaux à la question des "fausses" porcelains chinoises. *Les Cahiers de Framespa* **2019**, *31*. [CrossRef]
86. Colomban, P.; Ngo, A.-T.; Edwards, H.G.M.; Prinsloo, L.C.; Esterhuizen, L.V. Raman identification of the different glazing technologies of Blue-and-White Ming porcelains. *Ceram. Int.* **2022**, *48*, 1673–1681. [CrossRef]
87. Su, Y.; Qu, L.; Duan, H.; Tarcea, N.; Shen, A.; Popp, J.; Hu, J. Elemental analysis-aided Raman spectroscopic studies on Chinese cloisonné wares and painted enamels from the Imperial palace. *Spectrochim. Acta A Mol. Biomol. Spectrosc.* **2016**, *153*, 165–170. [CrossRef]
88. Ma, H.; Henderson, J.; Cui, J.; Chen, K. Glassmaking of the Qing Dynasty: A Review, New Data, and New Insights. *Adv. Archaeomat.* **2020**, *1*, 27–35. [CrossRef]
89. Zhou, S.Z. *Research on Painted Enamels Porcelain Ware from the Qing Court*; Wenwu Chubanshe: Beijing, China, 2008.
90. Colomban, P.; Milande, V.; Le Bihan, H. On-site Raman analysis of Iznik pottery glazes and pigments. *J. Raman Spectrosc.* **2004**, *35*, 527–535. [CrossRef]
91. Manoun, B.; Azdouz, M.; Azrour, M.; Essehli, R.; Benmokhtar, S.; El Ammari, L.; Ezzahi, A.; Ider, A.; Lazor, P. Synthesis, Rietveld refinements and Raman spectroscopic studies of tricationic lacunar apatites $Na_{1-x}K_xPb_4(AsO_4)_3$ ($0 < x < 1$). *J. Mol. Struct.* **2011**, *986*, 1–9. [CrossRef]
92. Wallert, A.; Hermens, E.; Peek, M. (Eds.) Historical Painting Techniques, Materials, and Studio Practice, The J. Paul Getty Trust. 1995. Available online: https://www.google.com/search?q=paintings+miniatures+18th+century+techniques+google+scholar&client=firefox-b-d&sxsrf=AOaemvK5FpBFDDhwHxKhx_-9UMIRaT7iZA%3A1634021495602&ei=dzBlYbKEJKiTlwT9 0oyABQ&ved=0ahUKEwjym7O2pMTzAhWoyYUKHX0pA1AQ4dUDCA4&uact=5&oq=paintings+miniatures+18th+century+techniques+google+scholar&gs_lcp=Cgdnd3Mtd2l6EAMyBAghEBU6BwgAEEcQsAM6BggAEBYQHjoICCEQFhAdEB4 6BQghEKABOgcIIRAKEKABSgQIQRgAUNNwWOezAWDAtwFoAXACeACAAdMBiAHXGJIBBzEyLjEyLjKYAQCgAQHIAQjAAQE&sclient=gws-wiz (accessed on 11 October 2021).
93. Dulac, A.-V.; Cachaud, C. La miniature à l'époque moderne/ Les objets du voyage. *Etudes Epistémé (Rev. Littér. & Civilis.)* **2019**, *36*. [CrossRef]
94. Xia, X.; Xi, N.; Huang, J.; Wang, N.; Lei, Y.; Fu, Q.; Wang, W. Smalt: An under-recognized pigment commonly used in historical period China. *J. Archaeol. Sci.* **2019**, *101*, 89–98. [CrossRef]
95. Shih, C.-F.; Hua, F. The Chinese Concept of Painted Enamels. In *The RA Collection of Chinese Ceramics: A Collector's Vision*; Jorge Welsh Books: London, UK; Lisbon, Portugal, 2021; Volume V. (in press)
96. Loehr, G. Missionary-artist at the Manchu Court. *Transactions of the Oriental Ceramic Society*, 12 August 1963.

97. Weinhold, U.; Richter, R.G. Eine Email-pretiose des Kaisers von China in der Dresdener Schatzkammer? *Dresdener Kunstblätter* **2015**, *59*, 100–109.
98. Yu, P. Fengge kushi: Kangxi Yuzhi falang caici' cezhan shougi. *Gugong Wenwu Yuekan* **2020**, *449*, 32–52. Available online: https://theme.npm.edu.tw/Academic/Book-Content.aspx?a=2599&eid=0&bid=5995&listid=2598&type=13&l=1 (accessed on 25 October 2021).
99. Colomban, P.; Kirmizi, B. Non-invasive on-site Raman study of polychrome and white enamelled glass artefacts in imitation of porcelain assigned to Bernard Perrot and his followers. *J. Raman Spectrosc.* **2020**, *51*, 133–146. [CrossRef]
100. Menegon, E. The role of the Propaganda Fide missionaries in enamel production at the Qing imperial palace in the eighteenth-century and their bItalian training. In Proceedings of the La Circulation des objets émaillés entre la France et la chine (milieu XVIIe-milieu XIXe siècle). Interactions Technologiques, Culturelles et Diplomatiques, Paris-Beijing, Online Symposium, 27–29 October 2021. CNRS-Palace Musem LIA-IRP EnamelFC.
101. Colomban, P.; D'Abrigeon, P. Non-Invasive Study of Qing Ware from Baur Fondation Collection. (work in progress).

Article

Position-Sensitive Bulk and Surface Element Analysis of Decorated Porcelain Artifacts

László Szentmiklósi [1,*], Boglárka Maróti [1], Szabolcs Csákvári [1] and Thomas Calligaro [2]

[1] Nuclear Analysis and Radiography Department, Centre for Energy Research, 1121 Budapest, Hungary; maroti.boglarka@ek-cer.hu (B.M.); csaszabolcs@gmail.com (S.C.)
[2] Centre de Recherche et de Restauration des Musées de France, Palais du Louvre, 75001 Paris, France; thomas.calligaro@culture.gouv.fr
* Correspondence: szentmiklosi.laszlo@ek-cer.hu

Abstract: Non-destructive characterization of decorated porcelain artifacts requires the joint use of surface-analytical methods for the decorative surface pattern and methods of high penetration depth for bulk-representative chemical composition. In this research, we used position-sensitive X-ray Fluorescence Spectrometry (XRF) and Prompt-gamma activation analysis (PGAA) for these purposes, assisted by 3D structured-light optical scanning and dual-energy X-ray radiography. The proper combination of the near-surface and bulk element composition data can shed light on raw material use and manufacturing technology of ceramics.

Keywords: X-ray Fluorescence Spectrometry; Prompt-gamma activation analysis; porcelain; non-destructive composition

Citation: Szentmiklósi, L.; Maróti, B.; Csákvári, S.; Calligaro, T. Position-Sensitive Bulk and Surface Element Analysis of Decorated Porcelain Artifacts. *Materials* 2022, 15, 5106. https://doi.org/10.3390/ma15155106

Academic Editor: Miguel A. G. Aranda

Received: 24 June 2022
Accepted: 20 July 2022
Published: 22 July 2022

Publisher's Note: MDPI stays neutral with regard to jurisdictional claims in published maps and institutional affiliations.

Copyright: © 2022 by the authors. Licensee MDPI, Basel, Switzerland. This article is an open access article distributed under the terms and conditions of the Creative Commons Attribution (CC BY) license (https://creativecommons.org/licenses/by/4.0/).

1. Introduction

The ancient Greek term "keramos" means clay; the derived phrase "ceramics" nowadays refers to all clay-based materials that have undergone an irreversible physical-chemical transformation during firing. Porcelain is a special class of ceramics that is of fine-grained body, usually translucent to visible light, and stands out for its whiteness. The challenge in the elemental composition analysis is common to all ceramic types due to their similar compositions.

Porcelain has a rich history of over a thousand years. The first porcelain objects were created in China [1,2] and distributed via trading routes to India, the Middle East [3–5], and later to Europe. Local production of porcelain appeared in Europe in the 16th century and became widespread in the 18th century, with the establishment of traditional manufacturing workshops [3,4]. Consequently, porcelain artifacts are abundant and in well-preserved conditions but show significant differences in terms of composition, structure, and visual appearance.

Raw materials, the mixing proportions, the formation technology, the glaze, and the firing conditions all influence the bulk material properties. Although the structure changes during the firing step of preparation and the raw material dehydrates, the characteristic bulk element composition patterns from the raw materials are preserved. Typically, it consists of 65–80 m% SiO_2, 8–22 m% Al_2O_3, and the remaining 0.5–3 m% is of various oxides (e.g., Na_2O, K_2O, MgO, CaO, Fe_2O_3, ZrO) [6]. The color palette used for decoration also varies between workshops, but the recipes remained stable over the years. Therefore, both bulk and near-surface composition offer discrimination and may be used to answer questions related to authenticity [7], classification, or manufacturing.

Most studies on porcelain artifacts are being made with surface-confined analytical techniques, such as portable XRF [8,9], Raman [10], or PIXE/PIGE. [11] In this paper, we focus on contrasting the elemental compositions at the surface and in bulk and on the methodology to collect such information. There is no single non-destructive method with penetration depth tunable in the required thickness range and adequate sensitivity; this calls for a combination of non-destructive element analysis techniques [12,13]. Here, we made use of the different information depths of X-rays and neutrons/gamma rays: the concentration difference of certain elements between the surface-confined results and the bulk indicates its presence at the surface decoration, or vice versa, in the base material.

2. Materials and Methods

Several routine instrumental element-analytical techniques [14] either require destructive sample preparation, such as powdering, homogenization, and dissolution, or they are microdestructive and leave visible traces on the object's surface after the analysis. These methods are out of scope when studying valuable heritage objects. The analytical technique based on the neutron-induced capture gamma-rays, i.e., Prompt-gamma activation analysis (PGAA) [15], and the (portable) X-ray Fluorescence Spectroscopy (XRF) [16], are both contactless and non-invasive measurement techniques for the direct, bulk- or surface-representative element analysis of solid samples, respectively. Neutrons and energetic gamma rays, unlike X-ray photons, have penetration depths as high as a few cm; this makes them appropriate for measuring the bulk composition of solid samples [17].

Our goal with this study is to benchmark the well-established X-ray Fluorescence Spectrometry (XRF) and Prompt-gamma activation analysis (PGAA) in position-sensitive applications, where both lateral and in-depth variation of the elemental concentrations are expected. Most of the PGAA data [18,19] reported in the literature are bulk concentrations, and few attempts were made for any spatial or surface-bulk discrimination. Further, the 2D XRF scanning methodology [20,21] is mostly applied to flat shreds [22–24], while in this work, we attempted to extend XRF scanning to non-flat objects by driving the positioner with a digital mesh geometry obtained by 3D optical scanning.

2.1. Benchmark Object

To benchmark analytical techniques applicable to decorated porcelain characterization, a traditional porcelain pot object (*boîte mézy*) was handcrafted at the Manufacture Nationale de Sèvres, France, using traditional techniques and raw materials (paste, glaze, flux, and pigments). The body, which has a diameter of 115 mm, was shaped from a traditional paste (70 m% kaolin) on the potter's wheel and fired. The glaze was applied by dipping, and it was again fired. It was subsequently decorated with complex patterns of high chemical contrast (Au, Co, ...), fired, and burnished. These fine details were to challenge the discrimination capabilities of the techniques. It featured a painted Colibri bird and a flower, typical motives of the 18th century, a gold inlay at the center, while a quarter of its lid was covered with traditional blue paint, *Bleu de Sèvres*. In addition, it had the reference color palette painted, allowing the identification of the pigments. The pigment palette corresponded to those used in the 19th and 20th centuries [25]: Pb-Si-B flux with colorants Cr, Co, Fe, Mn, Au, and Cu. The decorations of this object have been extensively characterized by position-sensitive PIXE at the AGLAE facility [26].

2.2. Geometry-Digitalization via 3D Structured-Light Scanning or Neutron Tomography

Preparatory to the element analysis, fragile or irregular-shaped objects were frequently 3D scanned with a RangeVision SMART [27] structured-light optical scanner (Figure 1a), or their geometries were determined using the advanced surface determination feature of the VG Studio MAX 3.2 software [28] from volumetric X-ray or neutron tomograms, to obtain 3D digital surface mesh of the objects. In addition to just displaying the visual features, we made use of these digital models during position-sensitive element analysis experiments, as discussed hereafter.

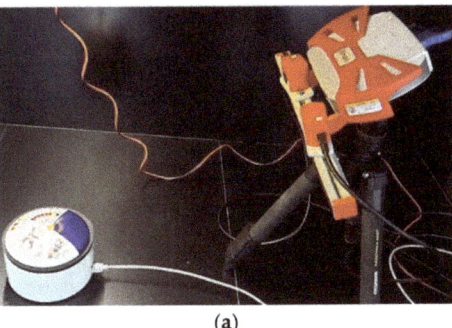  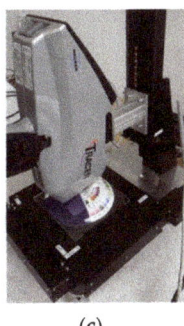

(a) (b) (c)

Figure 1. (**a**) The 3D structured-light scanning procedure, (**b**) the experimental setup of the PGAI measurement at the NIPS-NORMA facility of the BNC, and (**c**) position-sensitive xyz XRF scanning device.

2.3. Visualization of the Decoration by Dual-Energy X-ray Imaging

In medical imaging, dual-energy X-ray imaging [29] is a well-established practice. Here, we applied this approach by taking radiograms at a low-energy (35 keV) and a high-energy (200 keV) voltage setting of the X-ray generator tube to enhance sensitivity for the surface and bulk, respectively. For each setting, outlier removal, beam and dark image corrections were made, and the logarithm of the images was taken, resulting in the product of the linear attenuation coefficients (μ) and the material thickness (d) in a pixel at an (x,y) coordinate. The quantity ($\mu \times d$) differs considerably for the different energy beams, providing enhanced contrast to the otherwise very thin paint layer.

2.4. Position-Sensitive Prompt-Gamma Activation Analysis (PGAI) for Bulk Characterization

Prompt-gamma activation analysis (PGAA) is a potent in situ and contactless elemental analysis technique based on the radiative neutron capture nuclear reaction [30]. During irradiation with a well-collimated beam of slow neutrons, characteristic gamma rays up to 11 MeV energy emerge that are detected during irradiation with a perpendicularly placed gamma detector, facilitating the qualitative and quantitative elemental composition determination of the irradiated volume. The elements are identified based on their gamma-ray energies using a spectroscopic library [31]. The elemental masses within the irradiated volume are derived from the net areas of the analytical gamma-ray peaks [32,33] and recomputed to atomic or mass fractions [34]. For method validation, we used an albite standard (Centre de Recherces Pétrographiques et Géochimiques AL-1) [35] and an ancient Chinese porcelain sample characterized in the IAEA CU-2206-06 Proficiency test [36].

NIPS-NORMA [37,38] of the Budapest Neutron Centre (BNC) is the only permanent facility designed for position-sensitive elemental composition measurements based on radiative neutron capture. This extension of the PGAA technique is called Prompt-gamma Activation Imaging (PGAI) [39]. For this purpose, in addition to the setup required for the PGAA element analysis of homogeneous samples (Compton-suppressed HPGe gamma-ray detector inside a massive lead shielding), the NIPS-NORMA facility is equipped with a large, $20 \times 20 \times 20$ cm^3 sample chamber, a xyzω motorized sample stage, a computer-controlled neutron slit to adjust the neutron spot size, and an optional neutron imaging camera placed downstream of the sample chamber. These hardware components are aligned with sub mm precision to the isocenter, that is, the geometric intersection of the neutron beam and the symmetry axis of the gamma detector's collimator. This facilitates handling data from all modalities in a unified coordinate framework, directly correlating the motor positions, concentrations, and visual information without a registration process [40]. The sample positioning is based on either real-time visual feedback from the neutron imaging camera or employing a laser beam pointing along the centerline of the neutron beam (Figure 1b).

To maintain the firm placement and to avoid any damage to the upright standing sample during sample positioning, they were fixed to the sample manipulator using custom-made, disposable 3D-printed sample holders. Its upper part was developed using the exact complement of the artifact's digital geometry model, while the bottom contains an interlock to mount it on the motorized sample stage [41]. Further, the measurement geometry, including the shape of the object taken from its scanned surface mesh, can also be reproduced digitally to allow the correction for both neutron- and gamma-ray-related matrix effects by MCNP6 [42] Monte Carlo computer simulations. This methodology is discussed in detail in our earlier publication [43].

The NIPS-NORMA station can be operated with thermal and cold neutrons, depending on the status of the cold source built into the core of the Budapest Research Reactor [44]. The energy distributions of the neutrons, and consequently, the penetration depths and the elemental sensitivities of these beams differ, even for the identical sample. Therefore, some PGAA measurements were made with thermal or cold neutrons, and in some cases, two measurements were completed at the same spot using the two different beam types. This is intended to verify that the PGAA method can generate bulk-representative results independently of the exact measurement conditions.

The irradiation spots of the PGAA measurements were set by considering the surface patterns on the porcelain, followed by a corresponding paint-free area as blank. The paint-free parts showed us the compositions of the bulk porcelains, while the difference between the two spots revealed the constituents of the paint. The neutron beam spot size was adjusted to provide the required discrimination and maximize productivity at the same time.

2.5. XRF Technique as a Surface-Analytical Tool

The XRF measurements were conducted in our lab with either an OLYMPUS Delta Premium or a Bruker Tracer 5g handheld X-ray fluorescence spectrometer [45]. In this case, as the surfaces of the decorated objects are not homogeneous, the information was carried by the spatial variation of the X-ray spectrum. Therefore, a handheld XRF device must be put close to the surface, held firmly, and positioned based on a video-feedback by the operator, or can be coupled to a computer-controlled, motorized xyz sample stage to precisely position the object relative to the sampling spot or make raster scanning element mapping [24,46,47] (Figure 1c). Although the X-ray spot size of such handheld XRF devices (3–8 mm) is larger than dedicated micro-XRF scanners [48], this class of equipment is still a viable, affordable, and portable solution for many heritage-science problems, especially if combined with sophisticated image post-processing [49].

Our sample positioning device extends the macro-XRF (MA-XRF) approach to non-flat objects. An STL surface mesh from the 3D structured-light optical scan provides the elevation of the object as a function of (x,y)-coordinates in a given placement and relative to the sample stage's baseplate (Figure 2d). This allows the adjustment of the z-axis of the sample stage to maintain the close contact of the spectrometer with the sample surface. In addition, we plan to add functionality to consider the local surface normal [26] so that this approach can facilitate MA-XRF scanning of convex objects.

The OLYMPUS Delta Premium device, equipped with a 40 kV X-ray source and a silicon drift detector, has internal calibration for Soil and Mining modes. The soil option was appropriate to determine elements K, Ca, P, Ti, Cr, Mn, Fe, Ni, Cu, and Pb, while the Mining mode, which uses a dual-energy beam, was applicable to metallic elements Ti, Cr, Mn, Fe, Co, Ni, Cu, and Pb. The Bruker Tracer 5g is a research-grade pXRF that has a 50 kV Rh X-ray tube and an extremely thin, 1 mm-tick graphene entry window. The latter has higher transmission throughout the X-ray's energy spectrum and significantly improves the detection conditions for light elements. After acquiring the data, spectra were downloaded to the computer and evaluated by the bAxil software package [50].

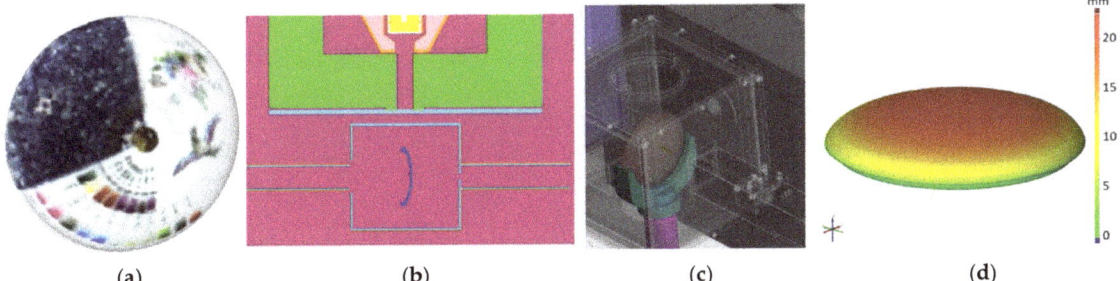

Figure 2. (**a**) the digital representation of the studied object obtained by 3D scanning, (**b**) the placement of the object in the MCNP6 simulation environment for PGAA matrix-effect correction, (**c**) the virtual model of the PGAI measurement geometry, visualizing the penetration of the pencil neutron beam, and (**d**) the elevation map for XRF scanning.

3. Results and Discussion

The 3D scanned photorealistic digital model is provided in the Electronic Supplementary Materials. These scanned geometrical data served as input data to the measurement control and simulations, as described in Ref. [43] (Figure 2b,c). The measurement positions and dual-energy radiogram are visualized in Figure 3. Relevant parts of the PGAA spectra are plotted in Figure 4. After evaluation [34], we saw many common peaks attributed to matrix elements and a few new peaks appearing, characteristic of the elements present in the surface decoration. The quantitative results are listed in Table 1. The prompt-gamma concentration data for H, Na, Al, Si, K, Ca Ti, Fe, Nd, and Gd were found to be highly reproducible at Spots 1–5 of the artifact (as labeled in Figure 3a) and agreed within the error margin. A representative element of bulk is the Si peak in Figure 4.

The different neutron beam properties did not change the analysis results either. This confirms the robustness of the PGAA in carrying out bulk analyses even if other methods are influenced by the decorations. By comparing the paint-free and decorated areas, the presence of additional elements, i.e., Cu, Co, and Au, could be confirmed.

In the case of porcelains, the flux may typically contain a high mass fraction of B_2O_3: given the high sensitivity and ppm-level of the detection limit of the PGAA to element B, this is a unique technique to quantify this element. This feature is reflected in our results: the boron concentrations show a two-fold increase at positions 3 and 4, where the flower and bird motives are situated.

Regarding the palette area, labeled as Spots A to K in Figure 3b, we overlaid these PGAA spectra and the one corresponding to the paint-free areas, and they differed only in a few spectrum regions (Figure 4). This is, on the one hand, an easy and qualitative indicator of the inorganic components of the paint, but it also proves the ability of the PGAA technique to probe the subsurface composition well. In the case of organic pigments, the indicator element is Hydrogen, as it has much higher sensitivity than the corresponding O, C, and N elements. This means that the organic nature can be confirmed this way, but the exact classification of the organic paint requires the use of another technique, e.g., handheld Raman spectrometry. We positively identified Mn in the case of measurement spot 3 (gray); blue contained Co, while the red contained elevated concentrations of Fe compared to the bulk.

Although this sample contained < 1 ppm Cd, which is the typical DL of PGAA in this matrix, it was quantified with higher accuracy in many other traditional and contemporary porcelain items we recently analyzed in our lab.

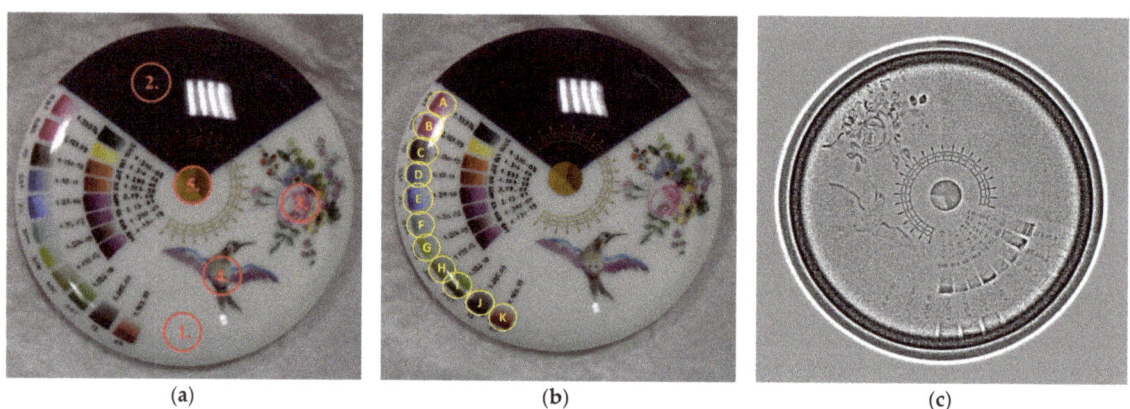

Figure 3. (**a**) PGAA measurement spots (1–5) selected for decoration, (**b**) color palette (A–K), and (**c**) dual-energy X-ray radiogram.

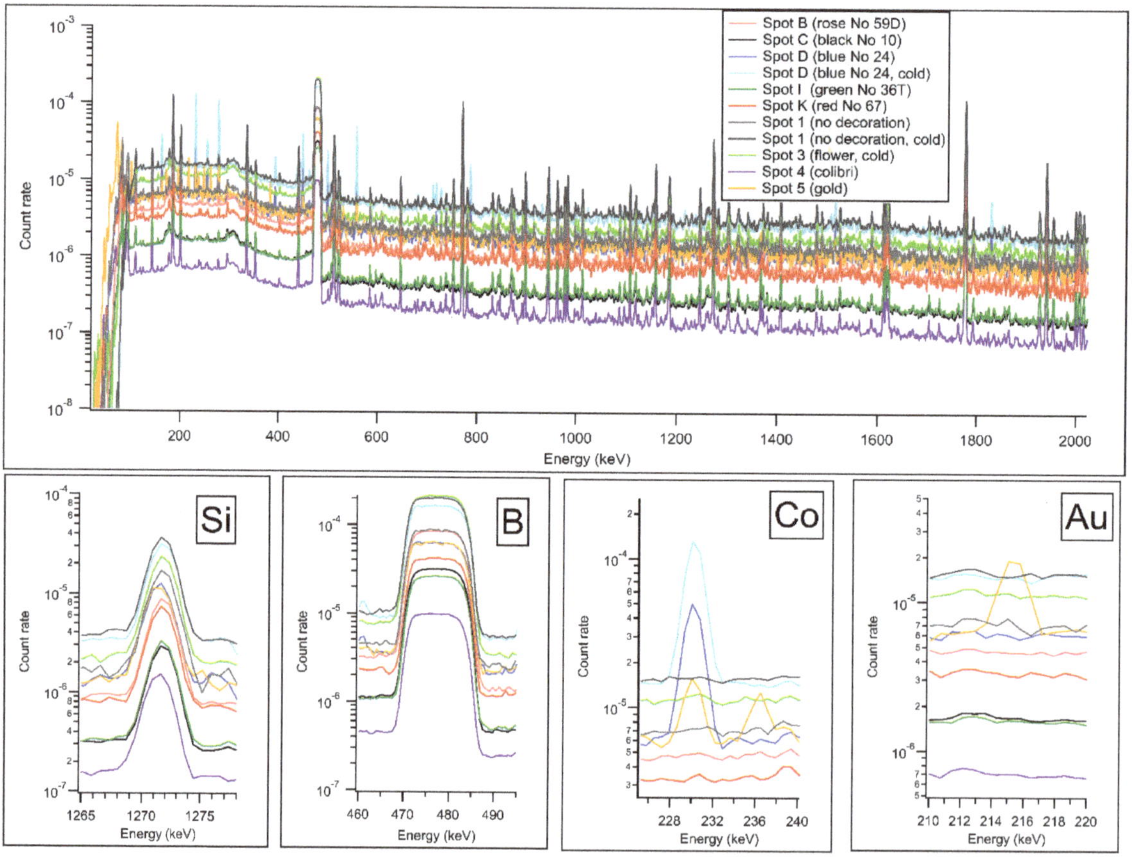

Figure 4. PGAA spectra up to 2 MeV, and the zoomed spectrum regions corresponding to bulk components Si, B, as well as elements Co and Au present in the surface decoration. PGAA data were taken with thermal neutron beam unless indicated otherwise in the legend.

Table 1. Concentrations (m%) by PGAA at measurement spots shown in Figure 3a, together with their 1-sigma relative uncertainties.

Beam	1 Thermal m%	1 Thermal unc%	1 Cold m%	1 Cold unc%	2 Thermal m%	2 Thermal unc%	2 Cold m%	2 Cold unc%	3 Thermal m%	3 Thermal unc%	3 Cold m%	3 Cold unc%	4 Thermal m%	4 Thermal unc%	5 Thermal m%	5 Thermal unc%
H	130 ppm	9	100 ppm	6	130 ppm	6	100 ppm	6	220 ppm	10.	210 ppm	6	100 ppm	5	260 ppm	7
B	60 ppm	0.9	63 ppm	0.8	59 ppm	0.8	58 ppm	1.0	121 ppm	0.9	106 ppm	0.9	110 ppm	0.9	62 ppm	1.0
Na	0.48	3.0	0.46	2.0	0.57	2.9	0.54	3.1	0.63	3.0	0.52	2.3	0.58	2.9	0.57	2.9
Mg							1.0	9.7					1.3	8.1		
Al	16.1	1.6	17.2	1.5	15.7	1.6	15.5	1.6	16	1.6	16	2.1	16	2.0	15.8	1.8
Si	29	1.6	28	1.6	29	1.6	29	1.7	29	1.7	29	1.9	28	1.7	29	1.7
K	2.4	1.9	2.4	1.9	2.4	2.0	2.3	1.9	2.4	1.9	2.5	2.0	2.4	2.3	2.5	3.1
Ca	2.5	3.0	2.7	2.5	2.6	3.1	2.5	2.6	2.6	2.9	2.6	2.8	2.4	2.6	2.6	3.6
Ti	110 ppm	16.9	120 ppm	9.0			120 ppm	10.6	80 ppm	22.0	140 ppm	11.7	120 ppm	10.4		
Mn											100 ppm	9	100 ppm	4.8		
Fe	0.24	9.3	0.24	4.1	0.20	12.1	0.25	6.5	0.23	8.0	0.21	7.3	0.22	4.9		
Cu									0.11	9.0						
Co					0.24	3.2	0.24	2.8			50 ppm	17	30 ppm	14.2	0.059	3.8
Nd			20 ppm	16.8							30 ppm	19				
Gd	3.9 ppm	6.2	4 ppm	9.4	3.8 ppm	6.9	3.4 ppm	7.3	4 ppm	13.0	3.7 ppm	6.2	1.8 ppm	7.0	4.1 ppm	5.2
Au															0.21	3.5

With XRF, whose spectra are plotted in Figure 5 and data listed in Table 2, there is a clear difference already between measurement points 0 (bare bulk without glaze) and 1 (white area covered with glaze) and an apparent negative bias of the major components Al. Si at decorated regions was observed, proving the presumption that the painted areas allowed the X-rays to penetrate less into the bulk. The overall scatter of the data was far larger than those of PGAA. However, due to the limited sampling depth, the XRF was more sensitive to indicate the surface variations and determined Sn that PGAA did not report. The detection limits for some elements were found to be better for XRF than for PGAA, but this also differs on the thickness of the bulk relative to the decorated layer. All green color shades contained Cr, highest in spot 8, while in the light green, one finds both Cr and Co. The gray paint contained Mn, Co, and Ni that PGAA was not able to detect. The dual-energy X-ray radiogram successfully revealed the thickness variation of the brushstrokes (Figure 3c).

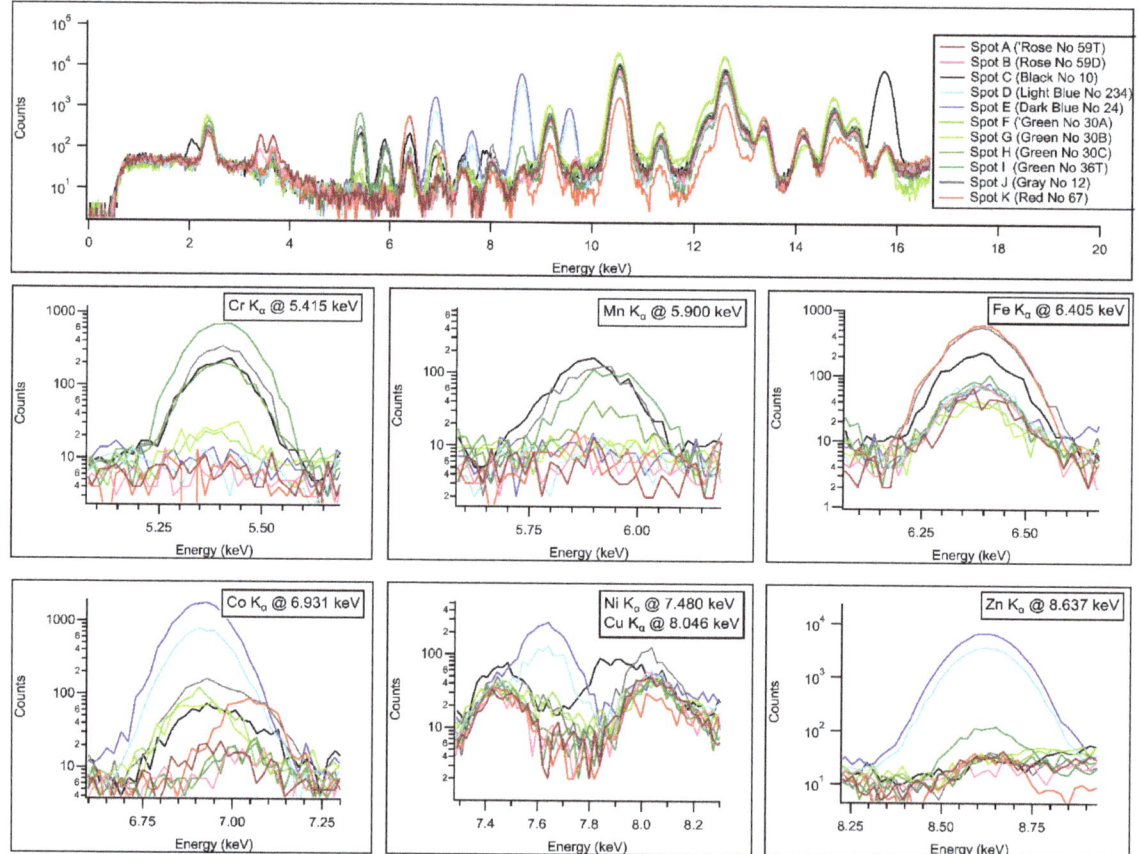

Figure 5. The K_α analytical lines of Cr, Mn, Fe, Co, Ni, Cu, and Zn in the 5–9 keV region of the XRF spectra for the color palette.

Table 2. Mass percentage (m%) concentrations by pXRF at measurement spots shown in Figure 3a, together with their 1-sigma relative uncertainties.

	0 m%	0 unc %	1 m%	1 unc %	2 m%	2 unc %	3 m%	3 unc %	4 m%	4 unc %	5 m%	5 unc %
Al	15.4	2.6	8.28	3.3	6.59	3.9	2.38	7.6	3.71	5.7	4.57	7.4
Si	29.5	1.0	42.2	0.7	35.9	0.8	22.6	0.9	20.1	1.0	4.6	2.4
K	1.74	1.7	3.61	1.1	2.36	1.3	1.97	1.0	0.616	3.2		
Ca	2.89	1.0	3.04	1.0	2.78	1.1	2.49	1.2	1.42	1.4		
Ti	0.089	14.7	0.0501	22.2	0.0711	14.9	0.0768	15.0	0.0534	18.4		
Mn							0.0116	30.2	0.0822	8.4		
Fe	0.312	2.9	0.229	3.1	0.156	3.8	0.12	4.2	0.409	2.2	0.365	7.1
Co					3.64	0.5			0.0652	5.1	6.81	1.2
Cu	0.0107	10.3	0.0104	10.6	0.0166	7.8	0.0171	7.0	0.0191	6.3		
Sn	0.0435	3.9	0.0319	4.7	0.0366	4.4	0.133	1.5	0.0382	4.2		
Au											40.5	0.5

The PGAA spectrum taken at the central gilding indicated not only the presence of gold but also showed a strong correlation with the cobalt spectrum (see the correlated peaks at 230 and 236 keV of the golden-colored line in the Co plots of Figure 4), proving that underneath the gold layer the Sevres blue paint is present. This is also justified by the radiogram, as well as the anti-correlation of the matrix peaks Al, Si, K, and Ca with the thickness of the gilding/blue paint in the XRF spectrum (Figure 6).

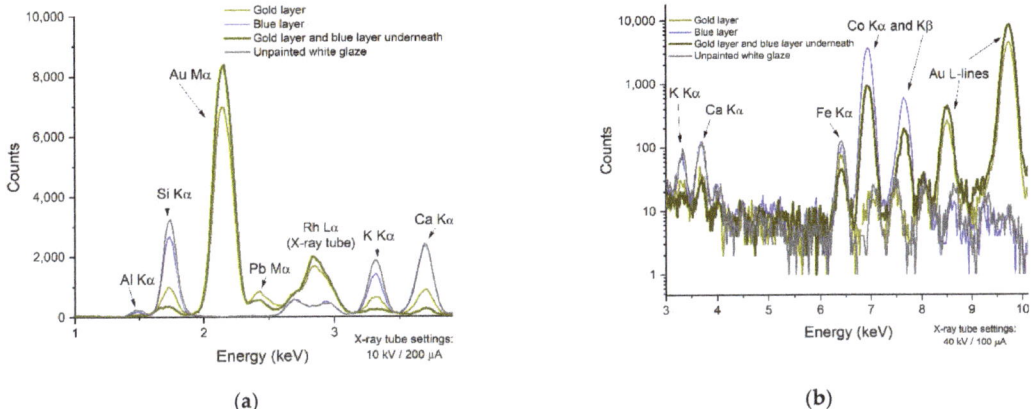

Figure 6. Two XRF spectra recorded at the 10 kV (**a**) and 40 kV (**b**) voltage settings at the central gilded area. Note the suppression of matrix peaks (Si, K, Ca) by the increasing thickness of golden/Sevres blue layers relative to the unpainted areas.

4. Conclusions

In addition to the main components of porcelain, such as Si, Al, K, Ca, Na, Fe, and Mg, PGAA could quantify multiple trace elements, mainly gadolinium, titanium, and in some pigments, cobalt, manganese, cadmium, and neodymium. In most of the cases, there were clear differences between the paint-free spots and decorated areas. Based on the PGAA results, we could differentiate between the decorated and non-decorated parts and proved that the blue paint contained cobalt, the pink and brown were manganese-based paints, the red paint contained Fe, the green paint used Cr, and the thin golden layers were also well detectable with Co-blue positively identified underneath.

The results of the pXRF measurements both supported and complemented the PGAA results. By choosing appropriate, glaze-free measurement points, we could achieve almost identical porcelain bulk compositions. This mostly remained true even if we measured the different surface patterns on the porcelains by PGAA. Most of the differences between the two techniques' results could be explained by the differing probing volumes and the detectability conditions of the elements. Our data are compatible with the results of the detailed PIXE element mapping published by our French collaborators [26].

Overall, we can conclude that the PGAA and pXRF methods are complementary and help us to gain both surface and bulk-related information on the samples. In addition, the techniques can confirm the other one's results, if not quantitatively, at least qualitatively, contributing to the comprehensive interpretation of the measurement results and fully non-destructive but detailed characterization of valuable artifacts.

Supplementary Materials: The following supporting information can be downloaded at: https://www.mdpi.com/article/10.3390/ma15155106/s1, Figure S1: The digital geometry model of the investigated Sevres porcelain, obtained via 3D optical scanning.

Author Contributions: Conceptualization, methodology, writing—original draft preparation, visualization, supervision, project administration, funding acquisition L.S.; formal analysis, investigation, data curation, writing—review and editing: B.M. and S.C.; resources, conceptualization, writing—review and editing: T.C. All authors have read and agreed to the published version of the manuscript.

Funding: This research was part of project No. 124068 that was implemented with financial support from the National Research, Development and Innovation Fund of Hungary, financed under the K_17 funding scheme. L.S. acknowledges the financial support of the János Bolyai Research Fellowship of the Hungarian Academy of Sciences. We also thank the Infrastructure Upgrade Initiative of the Hungarian Academy of Sciences (Grant No. IF-8/2020) for funding the purchase of the Bruker Tracer 5g pXRF spectrometer and some components of the sample stage. The fruitful collaboration within the Joint Research Activity of the IPERION CH European program (GA 654028) is also highly appreciated.

Acknowledgments: We highly appreciate the contributions of V. Jonca, head of creation and production, and O. Dargaud, head of the research service at the Cité de la céramiques, Sèvres, for providing the test sample, V. Szilágyi for the fruitful discussions and proofreading, and Z. Kis for measuring the radiograms.

Conflicts of Interest: The authors declare no conflict of interest.

References

1. Leung, P.L.; Luo, H. A Study of Provenance and Dating of Ancient Chinese Porcelain by X-ray Fluorescence Spectrometry. *X-ray Spectrom.* **2000**, *29*, 34–38. [CrossRef]
2. He, L. *Chinese Ceramics: A New Comprehensive Survey from the Asian Art Museum of San Francisco*; Rizzoli: New York, NY, USA, 2006; ISBN 978-0847819737.
3. Pierson, S. The Pilgrim Art: Cultures of Porcelain in World History by Robert Finlay. *China Rev. Int.* **2012**, *19*, 582–586. [CrossRef]
4. Chaffers, W. *Marks and Monograms on European and Oriental Pottery and Porcelain, with Historical Notices of Each Manufactory; over 3500 Potters' Marks and Illustrations*, 7th ed.; Reeves and Turner: Nashville, TN, USA, 1908.
5. Valenstein, S.G. *A Handbook of Chinese Ceramics*; Metropolitan Museum of Art: New York, NY, USA, 1989. [CrossRef]
6. Dondi, M.; Ercolani, G.; Melandri, C.; Mingazzini, C.; Marsigli, M. Chemical Composition of Porcelain Stoneware Tiles and Its Influence on Microstructural and Mechanical Properties. *InterCeram Int. Ceram. Rev.* **1999**, *48*, 75–83.
7. Neelmeijer, C.; Roscher, R. PIXE-RBS Survey of a Meissen Porcelain Snuff Box: First Version or Not? *X-ray Spectrom.* **2012**, *41*, 93–97. [CrossRef]
8. Simsek Franci, G.; Colomban, P. On-Site Identification of Pottery with PXRF: An Example of European and Chinese Red Stonewares. *Heritage* **2021**, *5*, 88–102. [CrossRef]
9. Fornacelli, C.; Volpi, V.; Ponta, E.; Russo, L.; Briano, A.; Donati, A.; Giamello, M.; Bianchi, G. Grouping Ceramic Variability with PXRF for Pottery Trade and Trends in Early Medieval Southern Tuscany. Preliminary Results from the Vetricella Case Study (Grosseto, Italy). *Appl. Sci.* **2021**, *11*, 11859. [CrossRef]
10. Colomban, P.; Ngo, A.T.; Fournery, N. Non-Invasive Raman Analysis of 18th Century Chinese Export/Armorial Overglazed Porcelain: Identification of the Different Enameling Techniques. *Heritage* **2022**, *5*, 233–259. [CrossRef]
11. Neelmeijer, C.; Pietsch, U.; Ulbricht, H. Eighteenth-Century Meissen Porcelain Reference Data Obtained By Proton-Beam Analysis (PIXE-PIGE). *Archaeometry* **2014**, *56*, 527–540. [CrossRef]

12. LeMoine, J.B.; Halperin, C.T. Comparing INAA and PXRF Analytical Methods for Ceramics: A Case Study with Classic Maya Wares. *J. Archaeol. Sci. Rep.* **2021**, *36*, 102819. [CrossRef]
13. Torrisi, L.; Venuti, V.; Crupi, V.; Silipigni, L.; Cutroneo, M.; Paladini, G.; Torrisi, A.; Havránek, V.; Macková, A.; La Russa, M.F.; et al. RBS, PIXE, Ion-Microbeam and SR-FTIR Analyses of Pottery Fragments from Azerbaijan. *Heritage* **2019**, *2*, 1852–1873. [CrossRef]
14. Carter, S.; Clough, R.; Fisher, A.; Gibson, B.; Russell, B. *Atomic Spectrometry Update: Review of Advances in the Analysis of Metals, Chemicals and Materials*; Royal Society of Chemistry: London, UK, 2021; Volume 36, ISBN 0324532008.
15. Molnár, G.L. *Handbook of Prompt Gamma Activation Analysis*; Springer: Berlin/Heidelberg, Germany, 2004; pp. 1–423.
16. Hegewisch, M.; Daszkiewicz, M.; Schneider, G. (Eds.) *Using PXRF for the Analysis of Ancient Pottery*; Universität Berlin und der Humboldt-Universität zu Berlin: Berlin, Germany, 2021; ISBN 9783981968590.
17. Bode, P. Kilogram Sample Analysis by Nuclear Analytical Techniques: Complementary Opportunities for the Mineral and Geosciences. *Minerals* **2021**, *11*, 443. [CrossRef]
18. Kasztovszky, Z. Application of Prompt Gamma Activation Analysis to Investigate Archaeological Ceramics. *Archeometria Műhely* **2007**, *2*, 49–54.
19. Dias, M.I.; Prudêncio, M.I.; Kasztovszky, Z.; Maróti, B.; Harsányi, I.; Flor, P. Nuclear Techniques Applied to Provenance and Technological Studies of Renaissance Majolica Roundels from Portuguese Museums Attributed to Della Robbia Italian Workshop. *J. Radioanal. Nucl. Chem.* **2017**, *312*, 205–219. [CrossRef]
20. Domoney, K. X-ray Fluorescence (XRF) Analysis of Porcelain: Background Paper. *Anal. Methods* **2017**, *9*, 2371–2374. [CrossRef]
21. Capobianco, G.; Sferragatta, A.; Lanteri, L.; Agresti, G.; Bonifazi, G.; Serranti, S.; Pelosi, C. MXRF Mapping as a Powerful Technique for Investigating Metal Objects from the Archaeological Site of Ferento (Central Italy). *J. Imaging* **2020**, *6*, 59. [CrossRef] [PubMed]
22. De Pauw, E.; Tack, P.; Verhaeven, E.; Bauters, S.; Acke, L.; Vekemans, B.; Vincze, L. Microbeam X-ray Fluorescence and X-ray Absorption Spectroscopic Analysis of Chinese Blue-and-White Kraak Porcelain Dating from the Ming Dynasty. *Spectrochim. Acta Part B At. Spectrosc.* **2018**, *149*, 190–196. [CrossRef]
23. Lin, C.; Meitian, L.; Youshi, K.; Changsheng, F.; Shanghai, W.; Qiuli, P.; Zhiguo, L.; Rongwu, L. The Study of Chemical Composition and Elemental Mappings of Colored Over-Glaze Porcelain Fired in Qing Dynasty by Micro-X-ray Fluorescence. *Nucl. Instrum. Methods Phys. Res. Sect. B Beam Interact. Mater. At.* **2011**, *269*, 239–243. [CrossRef]
24. Alberti, R.; Crupi, V.; Frontoni, R.; Galli, G.; La Russa, M.F.; Licchelli, M.; Majolino, D.; Malagodi, M.; Rossi, B.; Ruffolo, S.A.; et al. Handheld XRF and Raman Equipment for the in Situ Investigation of Roman Finds in the Villa Dei Quintili (Rome, Italy). *J. Anal. At. Spectrom.* **2017**, *32*, 117–129. [CrossRef]
25. Brongniart, A. *Traité des Arts Céramiques ou des Poteries Considérées dans Leur Histoire, Leur Pratique et Leur Théorie*; Wentworth Press: Paris, France, 1844.
26. Calligaro, T.; Arean, L.; Pacheco, C.; Lemasson, Q.; Pichon, L.; Moignard, B.; Boust, C.; Bertrand, L.; Schoeder, S.; Thoury, M.; et al. A New 3D Positioner for the Analytical Mapping of Non-Flat Objects under Accelerator Beams. *Nucl. Instrum. Methods Phys. Res. Sect. B Beam Interact. Mater. At.* **2020**, *467*, 65–72. [CrossRef]
27. RangeVision SMART Scanner. Available online: https://rangevision.com/en/products/smart/ (accessed on 23 June 2022).
28. Volume Graphics VG Studio Max. Available online: http://www.volumegraphics.com (accessed on 23 June 2022).
29. Fredenberg, E. Spectral and Dual-Energy X-ray Imaging for Medical Applications. *Nucl. Instrum. Methods Phys. Res. Sect. A Accel. Spectrometers Detect. Assoc. Equip.* **2018**, *878*, 74–87. [CrossRef]
30. Bohr, N. Neutron Capture and Nuclear Constitution. *Nature* **1936**, *137*, 344–348. [CrossRef]
31. Révay, Z.; Firestone, R.B.; Belgya, T.; Molnár, G.L. Prompt Gamma-Ray Spectrum Catalog. In *Handbook of Prompt Gamma Activation Analysis*; Molnár, G.L., Ed.; Kluwer Academic Publishers: Dordrecht, The Netherlands; Boston, MA, USA; London, UK, 2004; pp. 173–364.
32. Fazekas, B.; Belgya, T.; Dabolczi, L.; Molnár, G.; Simonits, A. HYPERMET-PC: Program for Automatic Analysis of Complex Gamma- Ray Spectra. *J. Trace Microprobe Tech.* **1996**, *14*, 167–172.
33. Simonits, A.; Östör, J.; Kálvin, S.; Fazekas, B. HyperLab: A New Concept in Gamma-Ray Spectrum Analysis. *J. Radioanal. Nucl. Chem.* **2003**, *257*, 589–595. [CrossRef]
34. Révay, Z. Determining Elemental Composition Using Prompt γ Activation Analysis. *Anal. Chem.* **2009**, *81*, 6851–6859. [CrossRef]
35. Govindaraju, K. 1995 working values with confidence limits for twenty-six crpg, anrt and iwg-git geostandards. *Geostand. Newsl.* **1995**, *19*, 22. [CrossRef]
36. Shakhashiro, A.; Trinkl, A.; Törvenyi, A.; Zeiller, E.; Benesch, T.; Sansone, U. *Report on the IAEA-CU-2006-06 Proficiency Test on the Determination of Major, Minor and Trace Elements in Ancient Chinese Ceramic*; International Atomic Energy Agency (IAEA): Seibersdorf, Austria, 2006.
37. Szentmiklósi, L.; Kis, Z.; Belgya, T.; Berlizov, A.N. On the Design and Installation of a Compton–Suppressed HPGe Spectrometer at the Budapest Neutron-Induced Prompt Gamma Spectroscopy (NIPS) Facility. *J. Radioanal. Nucl. Chem.* **2013**, *298*, 1605–1611. [CrossRef]
38. Kis, Z.; Szentmiklósi, L.; Belgya, T. NIPS–NORMA Station—A Combined Facility for Neutron-Based Nondestructive Element Analysis and Imaging at the Budapest Neutron Centre. *Nucl. Instrum. Methods Phys. Res. Sect. A Accel. Spectrometers Detect. Assoc. Equip.* **2015**, *779*, 116–123. [CrossRef]

39. Belgya, T.; Kis, Z.; Szentmiklósi, L.; Kasztovszky, Z.; Kudejova, P.; Schulze, R.; Materna, T.; Festa, G.; Caroppi, P.A. First Elemental Imaging Experiments on a Combined PGAI and NT Setup at the Budapest Research Reactor. *J. Radioanal. Nucl. Chem.* **2008**, *278*, 751–754. [CrossRef]
40. Szentmiklósi, L.; Kis, Z.; Maróti, B. Integration of Neutron-Based Elemental Analysis and Imaging to Characterize Complex Cultural Heritage Objects. In *Handbook of Cultural Heritage Analysis*; D'Amico, S., Venuti, V., Eds.; Springer: Berlin/Heidelberg, Germany, 2022; Volume 1, pp. 239–272.
41. Szentmiklósi, L.; Maróti, B.; Kis, Z.; Janik, J.; Horváth, L.Z. Use of 3D Mesh Geometries and Additive Manufacturing in Neutron Beam Experiments. *J. Radioanal. Nucl. Chem.* **2019**, *320*, 451–457. [CrossRef]
42. Goorley, T.; James, M.; Booth, T.; Brown, F.; Bull, J.; Cox, L.J.; Durkee, J.; Elson, J.; Fensin, M.; Forster, R.A.; et al. Features of MCNP6. *Ann. Nucl. Energy* **2016**, *87*, 772–783. [CrossRef]
43. Szentmiklósi, L.; Kis, Z.; Maróti, B.; Horváth, L.Z. Correction for Neutron Self-Shielding and Gamma-Ray Self-Absorption in Prompt-Gamma Activation Analysis for Large and Irregularly Shaped Samples. *J. Anal. At. Spectrom.* **2021**, *36*, 103–110. [CrossRef]
44. Belgya, T. Prompt Gamma Activation Analysis at the Budapest Research Reactor. *Phys. Procedia* **2012**, *31*, 99–109. [CrossRef]
45. Kasztovszky, Z.; Maróti, B.; Harsányi, I.; Párkányi, D.; Szilágyi, V. A Comparative Study of PGAA and Portable XRF Used for Non-Destructive Provenancing Archaeological Obsidian. *Quat. Int.* **2018**, *468*, 179–189. [CrossRef]
46. Lins, S.A.B.; Manso, M.; Lins, P.A.B.; Brunetti, A.; Sodo, A.; Gigante, G.E.; Fabbri, A.; Branchini, P.; Tortora, L.; Ridolfi, S. Modular Ma-Xrf Scanner Development in the Multi-Analytical Characterisation of a 17th Century Azulejo from Portugal. *Sensors* **2021**, *21*, 1913. [CrossRef]
47. Shugar, A.N. Handheld Macro-SRF Scanning: Development of Collimators for Sub-Mm Resolution. In *Proceedings of the Fourth International Symposium on Analytical Methods in Philately*; Lera, T., Barwis, J.H., Eds.; Smithsonian National Postal Museum: Washington, DC, USA, 2020; pp. 13–20.
48. Tsuji, K.; Matsuno, T.; Takimoto, Y.; Yamanashi, M.; Kometani, N.; Sasaki, Y.C.; Hasegawa, T.; Kato, S.; Yamada, T.; Shoji, T.; et al. New Developments of X-ray Fluorescence Imaging Techniques in Laboratory. *Spectrochim. Acta Part B At. Spectrosc.* **2015**, *113*, 43–53. [CrossRef]
49. Yang, J.; Zhang, Z.; Cheng, Q. Resolution Enhancement in Micro-XRF Using Image Restoration Techniques. *J. Anal. At. Spectrom.* **2022**, *37*, 750–758. [CrossRef]
50. Brightspec BAxil XRF Evaluation Software Package. Available online: https://www.brightspec.be/brightspec/?q=node/31 (accessed on 23 June 2022).

Article

Corrosion Layers on Archaeological Cast Iron from Nanhai I

Minghao Jia, Pei Hu and Gang Hu *

School of Archaeology and Museology, Peking University, Beijing 100871, China; 2001110770@stu.pku.edu.cn (M.J.); 1801110803@pku.edu.cn (P.H.)
* Correspondence: hugang@pku.edu.cn

Abstract: Archaeological iron objects were excavated from the Nanhai I ship from the Southern Song Dynasty that sunk in the South China Sea. Most of these artifacts were severely corroded and fragmented. In order to understand their current corrosion state and guide their restoration and protection, optical microscopy, scanning electron microscopy, micro-laser Raman spectroscopy, infrared spectroscopy and X-ray diffraction were all selected for analysis. It was clear that the archaeological iron material was hypereutectic white iron with a carbon content of about 4.3–6.69%, and had experienced low-melt undercooling. There were many internal cracks formed by general corrosion that extended to the iron core, which tended to make the material unstable. At the interface between the iron and rust, there was a black dense layer enriched with chlorine, and a loose yellow outer layer. The dense layer was mainly composed of magnetite, akaganeite and maghemite, while the rust of the loose layer was composed of lepidocrocite, goethite, feroxyhite, maghemite and hematite. The major phases of all corrosion products were akaganeite and lepidocrocite. Numerous holes and cracks in the rust layer exhibited no barrier ability to the outside electrolyte, hence the iron core formed many redox electrochemical sites for general corrosion with the rust. Meanwhile, the dense rust located close to the iron core was broken locally by an enriched chlorine layer that was extremely detrimental to the stability of the archaeological iron. Using electrochemical impedance spectroscopy, it could be determined that the rust layers had no protective effect on the internal iron core under conditions of simulated seawater, and these rust layers even accelerated the corrosion. A mechanism for the rust growth as a result of laboratory testing was proposed to explain the entire corrosion process. In view of the desalination preservation treatment that had been applied for ten years, it was not recommended to maintain a single desalination operation. The archaeological rusted iron of the Nanhai I ship that was excavated from the marine environment should be properly stabilized and protected using corrosion inhibition and rust transformation for iron oxyhydroxides, since the rust structure and the internal iron core retain well together.

Keywords: Nanhai I; archaeological iron; corrosion layers; conservation

Citation: Jia, M.; Hu, P.; Hu, G. Corrosion Layers on Archaeological Cast Iron from Nanhai I. *Materials* **2022**, *15*, 4980. https://doi.org/10.3390/ma15144980

Academic Editors: Žiga Šmit and Eva Menart

Received: 20 June 2022
Accepted: 16 July 2022
Published: 18 July 2022

Publisher's Note: MDPI stays neutral with regard to jurisdictional claims in published maps and institutional affiliations.

Copyright: © 2022 by the authors. Licensee MDPI, Basel, Switzerland. This article is an open access article distributed under the terms and conditions of the Creative Commons Attribution (CC BY) license (https://creativecommons.org/licenses/by/4.0/).

1. Introduction

The Nanhai I is a wooden merchant ship from the Southern Song Dynasty (1127–1279 AD), which was sunk in the South China Sea. It was unearthed in Guangdong province and is now a collection in the Maritime Silk Road Museum [1]. It conforms to the hull structure of a "Fu Ship (Fujian-style freighter)" from the typical types of Chinese ships. It is the oldest, largest and best-preserved shipwreck ever found for ancient international trade. The ship dates to the mid-13th century, as a result of the copper coins and china commodities that were unearthed from it [2]. The Song Dynasty was a period of rapid development in maritime trade. Developments in shipbuilding technology and navigation knowledge further expanded the scope of marine trade during this period, and communication with overseas became more frequent. Cargo from the Nanhai I reflects the trade situation during the Song Dynasty, filling gaps in the physical data of Sino-foreign maritime trade in academic research. The cargo salvaged from the shipwreck is of a wide variety and is on a large scale, retaining most of the products from that time. It provides us with new ideas for studying the maritime metals trade and navigation history from the Song Dynasty. Therefore, there are good reasons to conduct pre-conservation guided research on excavated archaeological artifacts, especially metal objects from the Nanhai I.

In ancient trade, metal products dominated the international market, and were important commodities of market circulation. Steel or iron can be used to manufacture weapons and labor tools. It was an indispensable military material that was closely related to the country's politics, economy and military at that time [3]. Therefore, the study of metal artifacts is beneficial, since it reveals the influence of ancient China in the historical development of Asia and the world. A large number of daily necessities and commercial products which were made of iron or steel, such as pots, nails and long slabs, were excavated from the Nanhai I. Finding a method to stabilize the material under favorable conditions against further corrosion has been a serious problem. Further corrosion of the iron objects often occurs after extraction from their marine environment, and their exposure to a new environment tends to transform their metal cores into corrosion products, even when they are stored in stable climatic conditions [4]. The presence of chlorides raises serious problems about the protection of archaeological artifacts recovered from marine environments. Currently, objects are treated to remove chloride ions using alkaline solutions. The purpose of this operation is to ensure dechlorination on the one hand, and on the other hand to slow down the corrosion rate of metal materials. These treatments have proven their efficiency in most cases, but it still is difficult to get a clear indicator that an artifact shows any corrosion recovery after treatment [5]. The cycles of these treatments are generally several months or several years, based on the volume and corrosion state of the metal artifacts. Therefore, protection methods should be regularly tested and optimized according to the correlation between the existing corrosion products and their metal cores. Analysis is conducive to the efficient development of conservation methods.

The complex properties of archaeological materials make every step of the protection operation a meticulous one. Meanwhile, the rationale for a preservation operation should only be designed and assessed after a complete characterization of the corroded archaeological metal. Metal artifacts unearthed from marine environments have been immersed in a high-chlorine electrolytic environment for hundreds of years. The corrosion state is significantly different in the structure of the rust layers compared to that of metal monuments from an outdoor dry-wet cycle [6]. Few data are dedicated to identifying marine corrosion rust at the microscopic scale, especially for archaeological artifacts [7–9]. Researchers believed that the distribution of different components in the rust layer had remarkable effects on future corrosion. In marine environments, the long-term immersion of iron is prone to thick rust forming, with many different phases: oxides, oxyhydroxides, sulphides, sulphates, chlorides, etc. Furthermore, the prior phases of rusts are usually a complex mixture of goethite (α-FeOOH), akaganeite (β-FeOOH), lepidocrocite (γ-FeOOH) and magnetite (Fe_3O_4) [10]. Recent articles have shown that the formation of akaganeite requires the simultaneous presence of high concentrations of dissolved Fe(II) and chloride [11]. This condition is extremely common in the sea. Therefore, the existence of akaganeite can directly lead to a variety of localized corrosion phenomena. The rust composition and distribution in archaeological irons are always more complicated than simulations of them in the laboratory, especially under natural marine corrosion conditions of more than 800 years. Determining the state of rust is beneficial to infer its corrosion mechanism and guide the protection of archaeological iron for conservationists.

This article reported on the metallic phase of iron materials from the Southern Song Dynasty of China, and analyzed their rusted state from marine corrosion over hundreds of years, in order to provide useful information for preventing further degradation of iron artifacts after salvation, and ensuring their preservation. These results also contribute to a better understanding of archaeologically complex corrosion systems, which will lead to improved diagnosis and conservation of archaeological and cultural heritage pieces. Therefore, this paper applies multi-scaled, comprehensive methods to study the corrosion of archaeological iron extracted from marine environments. Iron fragments covered with rust from the Nanhai I ship were used for analysis in this research. The rust was mainly analyzed using optical stereo microscopy, micro-Raman spectroscopy and electrochemical impedance spectroscopy, combined with SEM, XRD and FTIR, to evaluate current corrosion characteristics and to find the corrosion causes during the marine period to predict the future corrosion situation for conservationists.

2. Materials and Methods

2.1. Archaeological Iron Objects

The iron samples were collected from the Maritime Silk Road Museum of Guangdong. The objects were all completely covered with rust and mud after being salvaged from a marine environment; the silt was gradually removed by spraying. The samples had been dechlorinated in pH = 9–10 sodium hydroxide solution for about 10 years in the archaeological laboratory. The water quality is regularly updated every month by adding about 4 g/t sodium hydroxide to a new soaking tank, and then transferring the iron objects into it. Before this experiment, the concentrations of Cl^-, NO_3^- and SO_4^{2-} were determined using ion chromatography to be about 26.93, 4.17 and 4.45 mg/L, respectively. During pretreatment cleaning with deionized water, the rust layer was gradually exposed. The surface appearance and approximate dimensions are shown in Figure 1a.

Figure 1. Archaeological iron artifacts from the Nanhai I (**a**); working electrode (**b**).

2.2. Sample Preparation

Rust powder: a razor and brush were used to scrape the rust powder from the iron, then an agate mortar and pestle was used to grind the rust to a uniform and fine size. The rust was previously powdered and screened to a particle size of less than 125 μm.

Cross-section of iron core and rust layers: the cross-section was cut from an edge of the archaeological iron and then dried at room temperature for 3 days. The object was embedded in epoxy resin for mechanical grinding with SiC paper (grade 180–2000) under absolute ethanol. The sample was polished with diamond paste, in order to obtain a smooth cross-section.

Working electrode: the electrode was cut from along the above cutting edge, in order to ensure the continuity and similarity of the rust layers and the iron core. The sample was cut into a small piece of 10 mm × 10 mm, and embedded in epoxy resin to protect the cut edges for the electrochemical impedance test. The details are shown in Figure 1b.

2.3. Morphology Observation

An optical stereo microscope (Eclipse LV100ND, Nikon, Tokyo, Japan) equipped with a digital camera (DS-Ri2, Nikon, Tokyo, Japan) was used to observe the metallographic structure of the iron core and the appearance of the rust layers.

A scanning electron microscope (Quanta 200F, FEI Company, Hillsboro, OR, USA) was used to show the morphology of the sample surface in detail under an accelerating voltage of 10 kV.

2.4. Determination of Rust Composition

X-ray diffraction was recorded by a diffractometer (X'pert-3 Power, PANanalytical, Almelo, The Netherlands) with a Cu anode, in order to determine the crystalline structure of the rust. The generator voltage used was 40 kV, while the tube current was 40 mA. Angular scanning was performed from 5° to 80°, with a scan step size of 0.013°.

FTIR transmittance analysis (Tensor 27, Bruker, Karlsruhe, Germany) was prepared by KBr-matrix pellets to define chemical structures of rust in the ranges from 400 to 4000 cm^{-1} at 4 cm^{-1} resolution. A powder sample of 0.2 mg was mixed with 200 mg of dry KBr (>99% FTIR grade, Sigma-Aldrich, Burlington, MA, USA), in order to be milled and pressed into pellets.

Micro-Raman spectroscopy was performed on the cross-sectional sample using a micro-Raman spectroscope (Thermo Scientific DXRxi, Thermo Fisher Scientific, Waltham, MA, USA) that was equipped with a microscope (OLYMPUS BX51, Olympus Corporation, Tokyo, Japan) to define different rust components. Raman excitation was provided by a frequency-doubled Nd:YAG laser operating at 532 nm, with a power of about 0.2 mW and with a probe diameter of about 1 mm. All spectra were calibrated using the 519.5 cm^{-1} line of a silicon wafer.

2.5. Electrochemical Impedance

Electrochemical impedance spectroscopy of rusted and naked archaeological iron objects was conducted using an electrochemical workstation (CS-350, CorrTestTM, Wuhan, China). Measurements were carried out in a three-electrode cell with 1.5 g/L NaCl and 1.5 g/L Na_2SO_4 mixed together to form an electrolyte to simulate sea water. The three-electrode cell included a saturated calomel reference electrode filled with saturated KCl solution which served as a reference electrode; a platinum auxiliary electrode with an exposure surface of 10 mm × 10 mm served as a counter electrode; and the iron objects with an exposure surface of 1 cm^2 functioned as working electrodes, as described above. Before testing, the electrodes were kept in the solution for 30 min to stabilize the free corrosion potential. An open-circuit potential was applied with frequencies ranging from 100 kHz to 10 mHz, and a sinusoidal perturbation signal with a 10-millivolt amplitude was used. The obtained data were interpreted based on an equivalent circuit using Zview (Zview2, Scribner Associates Inc., Southern Pines, NC, USA), in order to obtain the fitting parameters.

3. Results and Discussion

3.1. Cracks in the Archaeological Iron

After grinding and polishing, the cross-sectional sample in Figure 2 shows a significant corrosion state without any corrosive agent. The cross-section had a length of 0.78 cm and a width of 0.36 cm, which was approximately a rectangular plane. General corrosion and long cracks extending from the rust layer to the iron core were clearly observed, and the corroded state of the upper surface was most obvious. There was a worryingly large crack in the upper left corner, which may be caused by the combined effect of general corrosion and crevice corrosion. Due to the internal stresses of corrosion products [12], the core gradually expanded and resulted in the peeling off of small parts from the metal object. The cracks in the upper right corner were densely distributed and intersecting, extending in different directions to the core; these conditions are not conducive to stable storage of the archaeological iron. This embedded sample would continue to be used for the analysis of the cast state of the iron core and for the composition of the rust layer on its upper surface.

Figure 2. Digital and microscopic graphs of the cross-section of archaeological iron.

3.2. Metallographic Studies of the Archaeological Iron

In Figure 3, the silvery white strips in the cross section are primary cementite, and the matrix is eutectic ledeburite. The archaeological sample was a typical hypereutectic white cast iron with high carbon content in the range of 4.3–6.69 wt.%. The elemental analysis was further tested by EDS, and the composition was 5.31% C, 0.11% Si, 0.15% Mn, 0.08% P, 0.05% S and 94.3% Fe, which agreed with the metallographic data.

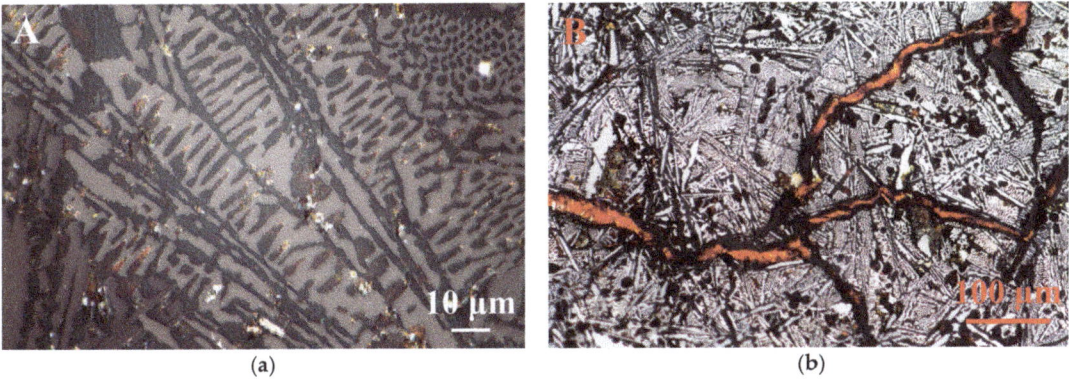

Figure 3. Metallographic diagrams of an archaeological iron artifact from the Nanhai I. Primary cementite and eutectic ledeburite (**a**); cracks of the iron (**b**).

When the hypereutectic molten iron began to cool, the long and coarse primary cementite was produced. It could grow freely in the liquid, thus it became shaped as strips or flakes. Gradually, the temperature of the metal liquid continued to drop, and a eutectic ledeburite phase formed. Due to local cooling, primary cementite and eutectic ledeburite also arranged in different cooling directions. Generally, two major phases of eutectics were encountered in white cast iron with high carbon content under different undercoolings. They could be described as ledeburite eutectic which had dendrite branches, while another phase was plate-like eutectic. Besides, plate-like eutectic in white cast iron became the main constituent of the microstructure with an increased degree of melt undercooling (ΔT~35–40 °C), while ledeburite eutectic often appeared at lower undercooling (ΔT~5–25 °C). This transformation does not really depend on the starting alloy's composition (hypo- or hyper-eutectic) [13]. The ledeburite in the metallographic images of the archaeological iron section was mostly dendritic. Based on this, it could be inferred that the iron object from the Southern Song Dynasty experienced slow cooling after being cast into a mold.

3.3. Composition of the Rust Powder

The basic components of the rust layer were identified by FTIR. Figure 4 shows that the main components of the powdered rust were akaganeite (β-FeOOH) and lepidocrocite (γ-FeOOH), with a small amount of maghemite (γ-Fe_2O_3). The absorption peaks were at 833 cm^{-1} and 1627 cm^{-1}, corresponding to the bending of O-H bonds and vibrations of Fe-O bonds in β-FeOOH [14]. A peak at 1026 cm^{-1} was attributed to O-H bending in γ-FeOOH [15]. γ-Fe_2O_3 and β-FeOOH overlapped at the characteristic peak of 648 cm^{-1} [16]. As for the peak at 3377 cm^{-1}, it represented a common peak from stretching vibrations of the −OH bond [14].

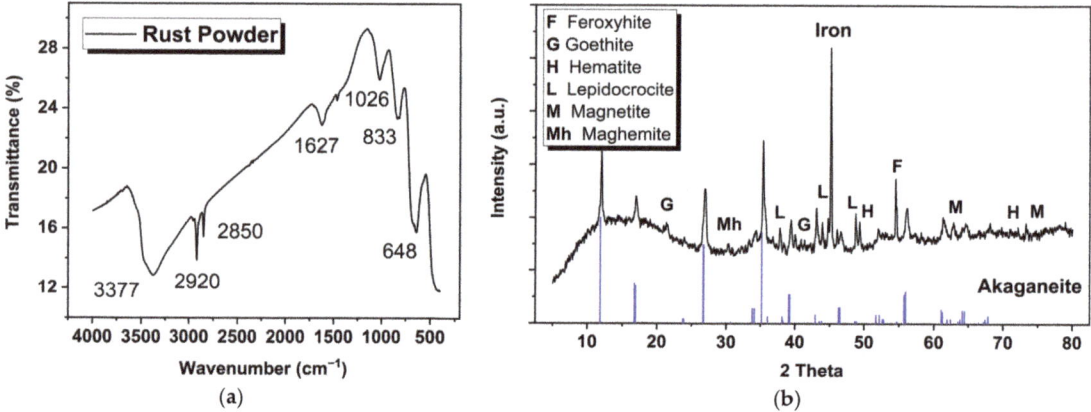

Figure 4. FTIR (a) and XRD (b) diagrams of rust powder.

An XRD was used to analyze the composition of rust powder, combined with data from infrared spectroscopy. The XRD spectrum in Figure 4b was compared individually using Jade software (Jade 9.0, MDI, Livermore, CA, USA) and PDF-card references (α-FeOOH: #00-029-0713; β-FeOOH: #00-042-1315; γ-FeOOH: #00-008-0098; δ-FeOOH: #00-013-0087; α-Fe_2O_3: #00-001-1053; γ-Fe_2O_3: #00-039-1346; Fe_3O_4: #00-019-0629). The peaks of the XRD data and the provided standard spectrum of akaganeite clearly corresponded to each other, which may be regarded as the main phase of rust powder. Meanwhile, a significant signal of pure iron appeared, proving that there were some tiny iron particles that peeled off with rust due to corrosion. The peak intensities of other rust components were weak, but a small amount of goethite (α-FeOOH), lepidocrocite, feroxyhite (δ-FeOOH), hematite (α-Fe_2O_3), maghemite and magnetite could be confirmed according to the PDF information.

Ferric oxyhydroxides have four common types; goethite is an electrochemically stable phase, whereas others are all active phases that may be transformed into relatively stable magnetite [9]. The results of FTIR and XRD analyses indicated that a large amount of β-FeOOH was present in the rust, which was associated with the high concentration of chloride in the South China Sea. However, in this work, α-FeOOH was only detected in the XRD with a weak intensity, and with a scarce signal in the IR spectra. Most iron-based compounds involved in corrosion reactions are prone to transforming into β-FeOOH and γ-FeOOH in marine environments, resulting in small amounts of α-FeOOH in rust. Generally, formation of β-FeOOH requires chloride ions in order to stabilize the structure of its crystal [17]. Moreover, β-FeOOH is the most detrimental for continuous corrosion of archaeological iron among all the ferric oxyhydroxides [18]. Therefore, the composition of the rust powder preliminarily indicated that the rust phase of this archaeological iron had not yet reached stability, and that the subsequent stabilization treatment was indeed required.

3.4. Structure of the Rust Layers

Figure 5 shows details of rust layer A, which is marked in Figure 2. The surface rust of archaeological iron had an obvious dividing line, separating two layers: a yellow outer layer that was mostly loose and porous, and an inner layer near the iron core that was almost dominated by black dense rust. Large cracks ran through the entire rust to the iron core, such that the rust layer could not completely protect the internal metal, and tended to peel off. The iron element was almost evenly distributed throughout the rust. The chlorine element was evenly distributed in the outer rust, and formed a highlight band that coincided with the position of the dense rust layer.

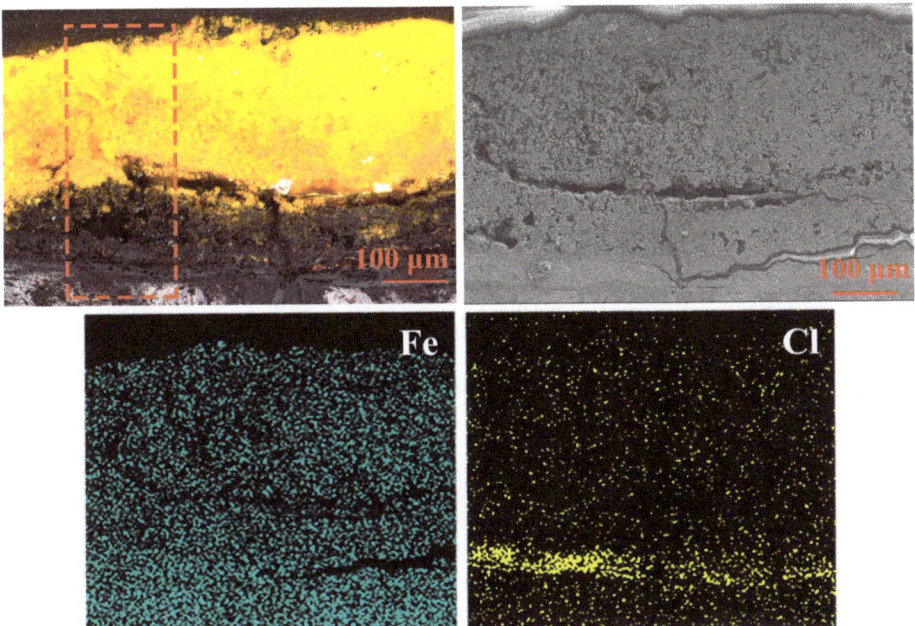

Figure 5. Graphs of the microstructure and element distribution of rust layer A (red square for Raman analysis).

Micro-Raman analysis revealed eight points, as shown in Figure 6. The oxides and hydroxides of iron compounds were identified, layer by layer, according to the literature's values [10,19,20]. The cross-section of rust layer A indicated four distinct layers based on color: a yellow outermost layer dotted with light-yellow spots (I); a yellow layer (II); a black layer dotted with yellow spots (III); and a black innermost layer which was closely connected with the iron core (IV). Raman spectra showed that the outermost rust was mainly formed by hematite (α-Fe_2O_3) and maghemite (γ-Fe_2O_3), with characteristic peaks at 214, 280 and 386 cm^{-1} to α-Fe_2O_3, and a relatively weak peak at 1608 cm^{-1} to γ-Fe_2O_3. The light-yellow dots in the outer rust showed a strong peak at 393 cm^{-1}, with less intense peaks at 298, 683 and 1302 cm^{-1} that were attributed to the formation of goethite (α-FeOOH). The strongest peak was close to 249 cm^{-1}, and a second peak in the Raman shift of 380 cm^{-1} was related to the lepidocrocite phase (γ-FeOOH) that coexisted with α-FeOOH phase. The presence of feroxyhite (δ-FeOOH) was confirmed by its characteristic bands at a single peak of 698 cm^{-1} at the junction between the black and yellow rust. Akaganeite (β-FeOOH), whether in marine environments or soils, is usually hygroscopic and is easily converted into other oxides, especially magnetite (Fe_3O_4) [21]. Furthermore, chloride ions are often present in the lattice of β-FeOOH, which can lead to a complete loss of artifacts [10]. In this particular object, akaganeite had bands at 137, 299 and 386 cm^{-1}, while magnetite was assigned the Raman bands at 362 and 694 cm^{-1}. The two kinds of rust had obvious signals near the black dense layer, where chlorine was enriched.

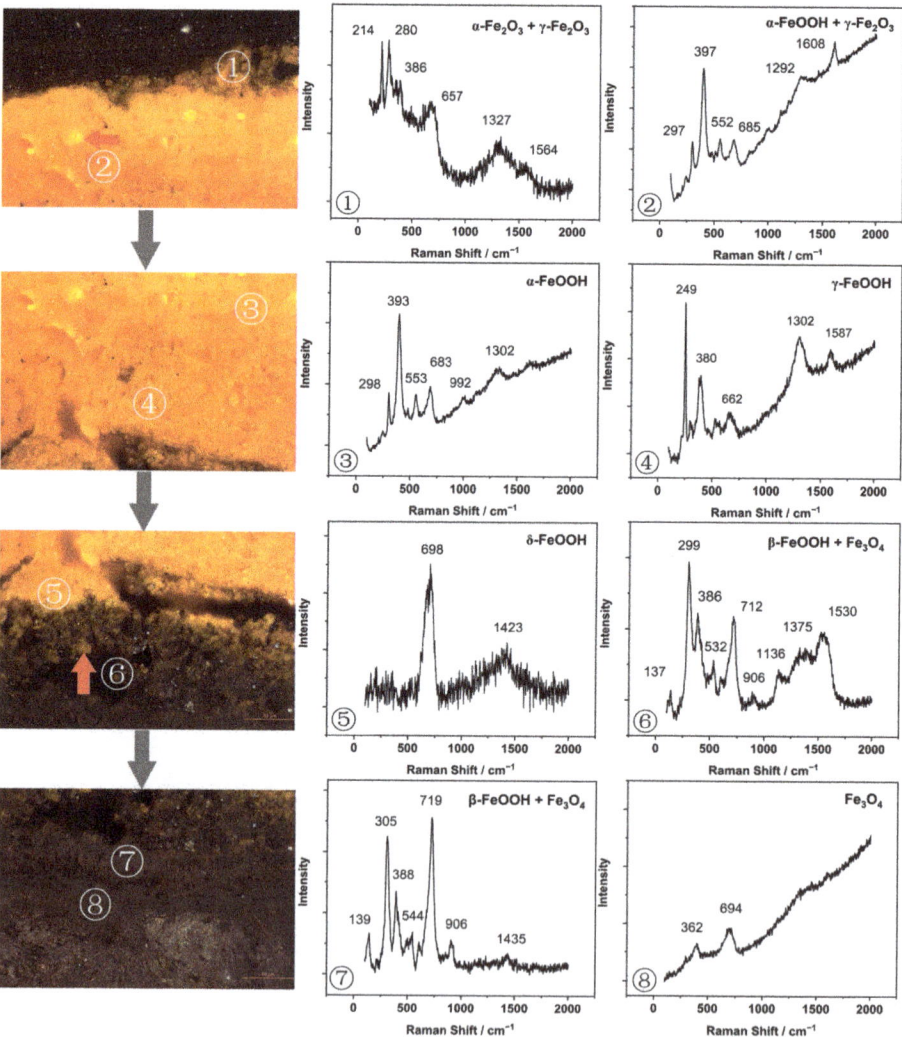

Figure 6. Raman spectra of rust layer A (red square in Figure 5). ①–⑧ were the main points for Raman analysis.

Area B in Figures 7 and 8 was the rust layer on the upper right of the iron sample. There were several wide and hollow lines at the interface between rust and iron, which did not tightly wrap around the internal metal to protect it against the external environment. The overall structure was similar to that of layer A, except that the black rust gradually occupied the yellow outer layer. Maghemite, goethite and lepidocrocite were the dominant phases in the yellow loose rust layer, and the black dense layer where chlorine gathered was still mainly composed of akaganeite and magnetite.

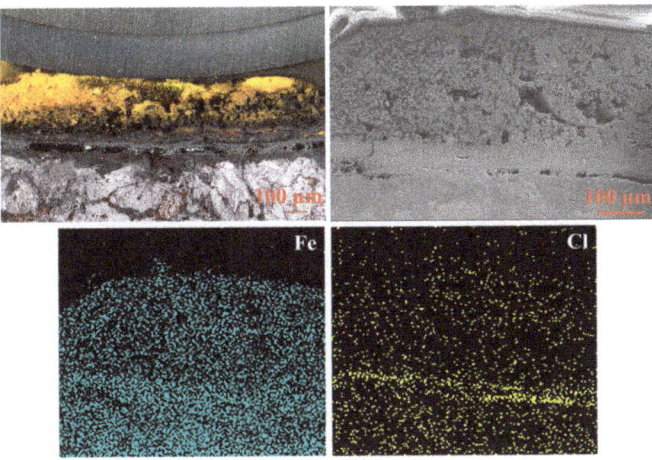

Figure 7. Graphs of the microstructure and element distribution of rust layer B.

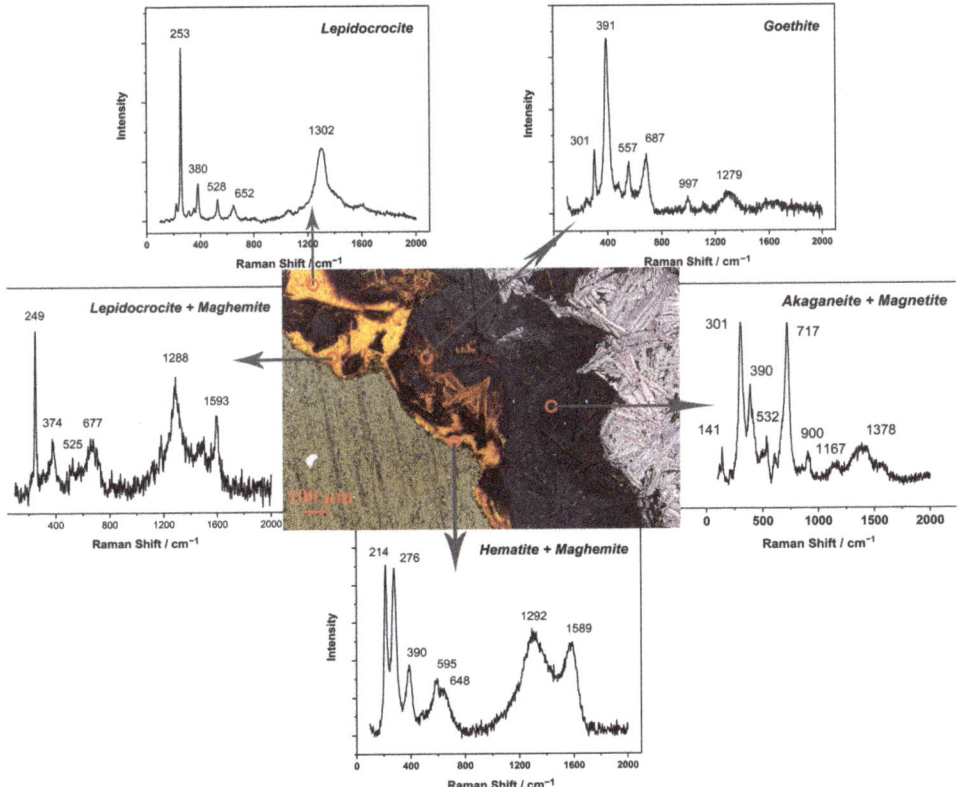

Figure 8. Raman spectra of rust layer B.

Therefore, there were two obvious rust layers on the upper surface of the iron. The loose yellow rust had a tendency to exfoliate, and the dense black rust had a lot of cracks, hence the inner iron core still had channels that were in contact with the external environment, making it difficult for the material to remain stable.

3.5. Corrosion Evaluation in Simulated Seawater

Electrochemical impedance spectroscopy (EIS) is an effective analysis tool that can be used to obtain valuable information about the nature of the rust layers formed on the marine-corroded archaeological iron.

Figure 9 shows the Nyquist and Bode plots of naked (removing the rust) and rusted archaeological iron, respectively. For the rusted surface, the Nyquist plot is composed of a depressed capacitive semi-arc in the high frequency range, and a long tail in the low frequency region that represents typical Warburg impedance. This indicates that the electrochemical corrosion process on the rust/iron interface in the $NaCl/Na_2SO_4$ solution was mainly controlled by the diffusion process. The charge transfer resistance of the iron rust dominates at the medium-low frequency region, while the rust resistance dominates at the high frequency region [22]. The construction of rust layers could be described using the equivalent circuit in Figure 10b. After the rust surface was completely removed, the archaeological sample reacted with the simulated seawater directly. Only one double-layer capacitive semicircle was observed in the Nyquist plot, and its equivalent circuit is shown in Figure 10a. R_s represents the electrolyte resistance, C_r the rust capacitance, R_r the rust resistance, C_{dl} the double-layer capacitance, R_{ct} the charge transfer resistance, and W the Warburg resistance or barrier diffusion impedance. Diffusion impedance was regarded as the diffusion of corrosive electrolyte through pores in the rust layer that acted as a diffusion barrier. The constant phase angle element (CPE) was introduced to describe the C_{dl} and C_r in the fitting circuits due to the rough and uneven surfaces of the sample [23].

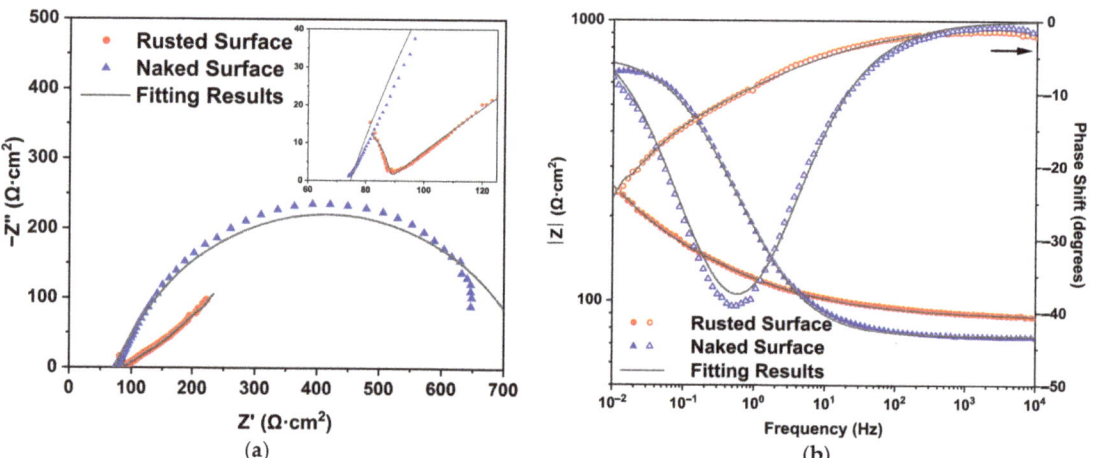

Figure 9. Nyquist (**a**) and Bode (**b**) plots of naked and rusted archaeological iron in simulated seawater.

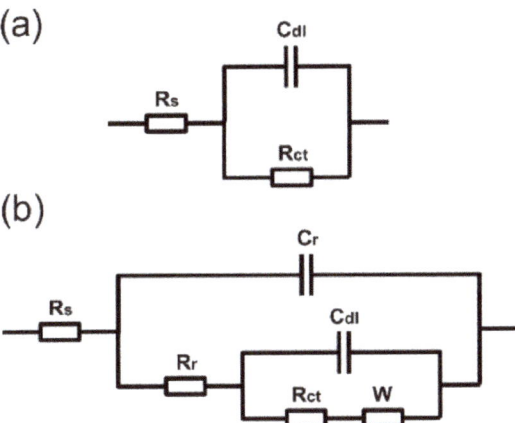

Figure 10. Equivalent circuits of naked (**a**) and rusted (**b**) archaeological iron in simulated seawater.

The total impedance for the circuit in Figure 10a could be expressed by the following equation:

$$Z = R_s + \frac{1}{Y_{dl}(j\omega)^{n_{dl}} + \frac{1}{R_{ct}}} \tag{1}$$

The electrochemical impedance for the circuit in Figure 10b could be expressed by the following equation:

$$Z = R_s + \frac{1}{Y_r(j\omega)^{n_r} + \frac{1}{R_r + 1/[Y_{dl}(j\omega)^{n_{dl}} + 1/(R_{ct} + Z_W)]}} \tag{2}$$

where Y_r and n_r, and Y_{dl} and n_{dl} are constants representing the elements C_r and C_{dl}, respectively. Z_W is the Warburg resistance, represented as follows [18]:

$$Z_W = A_W \cdot (j\omega)^{-0.5} \tag{3}$$

where A_W is the modulus of Z_W.

The equivalent circuits of Figure 10a,b were used to fit the Nyquist and Bode plots of rusted and naked iron, respectively. Fitting data are listed in Table 1. R_{ct} of the rusted iron was 17.01 Ω·cm², and was much smaller than that of the naked iron; this was attributed to the reduction of the rust that accelerated the cathodic reaction. The lower R_{ct} value indicated that the corrosion products formed on the archaeological object were less compact, less continuous and porous [22]. R_r is usually used to evaluate the corrosion resistance of rust [17]. Therefore, it could be that the rust layer has poor barrier behavior, and can not effectively prevent the corrosion of the iron core, which is consistent with the porous structure of the rust layer mentioned previously.

Table 1. Fitting data of impedance parameters in equivalent circuits.

Samples	R_s (Ω·cm²)	Y_{dl} (mF·cm^{-2}·s^{-n})	n_{dl}	R_{ct} (Ω·cm²)	R_r (Ω·cm²)	Y_r (μF·cm^{-2}·s^{-n})	n_r	A_W (Ω·cm²·s$^{-0.5}$)	χ^2
Naked	75.28	1.80	0.74	675.70					0.056
Rusted	42.10	8.32	0.31	17.01	42.65	0.012	0.92	0.41	0.011

3.6. Mechanism of the Rust Growth

The rust growth and iron corrosion mechanism of this archaeological iron could be proposed from the iron surface to the outer rust layer. Since the rust growth of archaeological iron was not an independent process of a single rust phase, it was better to describe the rust growth according to environmental changes over multiple periods.

When the Nanhai I became shipwrecked, the cargo on the ship sank into the sea. The iron products became immersed in a salt-rich water environment and began corroding. The corrosion reactions during this initial stage could have occurred as follows [24]:

The iron objects functioned as an anode region:

$$Fe \rightarrow Fe^{2+} + 2e^- \tag{4}$$

As the ship sank, a large amount of air would have been brought in instantly, and attached to the surface of the cargo in the form of bubbles. Hence, the interface between iron and sea water would be a cathodic reduction due to the oxygen dissolved in the thin water film. Meanwhile, countless electrochemical cells involving cathodic and anodic areas began reacting, and scattered on the iron surface to form numerous corrosion sites:

$$O_2 + 2H_2O + 4e^- \rightarrow 4OH^- \tag{5}$$

Then, the interface would quickly turn into normal seawater with a high chloride concentration, with the consumption and bursting of tiny bubbles. The chloride ions were more readily adsorbed than oxygen in competition for surface sites, and consequently became deposited on the metal with high retention, promoting the formation of ferrous chloride and accelerating the corrosion of iron. Moreover, the accompanying hydrolysis reaction began to create a weakly acidic environment around it [25]:

$$Fe^{2+} + 2Cl^- \rightarrow FeCl_2 \tag{6}$$

$$FeCl_2 + 2H_2O \rightarrow Fe(OH)_2 + 2Cl^- + 2H^+ \tag{7}$$

Mud and sand would gradually envelop and cover the archaeological objects with surrounding water flow, thereby blocking the update and circulation of dissolved oxygen and reserving the chloride ions around the iron. The high chloride concentration and local acidic conditions at the interface increased the chance formation of FeOOH, especially akaganeite and lepidocrocite [15]:

$$4Fe(OH)_2 + O_2 \rightarrow 4\beta, \gamma - FeOOH + 2H_2O \tag{8}$$

Meanwhile, some ferrous ions were not chosen to become compounds with chloride ions, instead forming hydrated ions in solution or becoming oxidized to $Fe(OH)_3$, which has very low solubility. The intermediate corrosion products $FeOH^+$ and $Fe(OH)_3$ could be converted into γ-FeOOH and δ-FeOOH, with small amounts of oxygen remaining in the mud and sand [24]:

$$Fe^{2+} + H_2O \rightarrow FeOH^+ + H^+ \tag{9}$$

$$2FeOH^+ + O_2 + 2e^- \rightarrow 2\gamma - FeOOH \tag{10}$$

$$2Fe(OH)_2 + O_2 + H_2O \rightarrow 2Fe(OH)_3 \tag{11}$$

$$Fe(OH)_3 \rightarrow \gamma, \delta - FeOOH + H_2O \tag{12}$$

Eventually, the archaeological iron became wrapped and buried tightly under the silt in deep sea, forming an anaerobic and weakly acidic environment. Fe_3O_4 could be reduced and converted rapidly from γ-FeOOH and β-FeOOH [26]. Ferrous ions and electrons could pass through an Fe_3O_4 layer due to its good conductivity. Moreover, Fe_3O_4 would have been prone to accumulate at the interface and form a dense layer on the surface of iron under the cathodic reaction. The reduction of the γ-FeOOH and β-FeOOH layer would continue its transformation into the cathode area of the Fe_3O_4 layer. Meanwhile, the chloride ions would gradually cause local breaks in the oxide/oxyhydroxide film on the iron objects, especially at rough and uneven areas, thereby making the corrosion process of the archaeological iron slow but unstoppable in its marine environment for more than 800 years:

$$3\beta,\gamma-\text{FeOOH}+H^++e^- \rightarrow Fe_3O_4+2H_2O \qquad (13)$$

$$8\beta,\gamma-\text{FeOOH}+Fe^{2+}+2e^- \rightarrow 3Fe_3O_4+4H_2O \qquad (14)$$

After being excavated out of seawater, the archaeological iron was immersed in a weakly alkaline solution and then transported to the laboratory for research. For preservation and display, this iron object was placed at room temperature in order to gradually remove internal moisture. Since the oxygen concentration and humidity changed significantly on the rust surface during the drying stage, it became prone to forming γ-FeOOH again as well as α-FeOOH. At this stage, γ-FeOOH and α-FeOOH could be generated by oxidizing the precipitation of $Fe(OH)_2$ derived from the outer rust, and only a fraction of FeOOH was generally adsorbed on the surface of the outermost layer. The structure of newly generated FeOOH was loose and did not cover the object tightly, resulting in the formation of pores in yellow rust. Hence, when the archaeological iron was removed from its dechlorination solution or rinsed with distilled water, it became easily detached. After immersion in alkaline solution, FeOOH in the dehydration stage was more likely to form maghemite and hematite [21]:

$$4Fe(OH)_2+O_2 \rightarrow 4\alpha,\gamma-\text{FeOOH}+2H_2O \qquad (15)$$

$$Fe(OH)_3 \rightarrow \alpha,\gamma-\text{FeOOH}+H_2O \qquad (16)$$

$$2\alpha,\gamma-\text{FeOOH} \rightarrow Fe_2O_3 \cdot H_2O \rightarrow Fe_2O_3+H_2O \qquad (17)$$

4. Conclusions

Archaeological iron artifacts excavated from the Nanhai I ancient ship were corroded and fragmented on a large scale. The metallographic structures of these objects belonged to hypereutectic white cast iron with a carbon content of 4.3–6.69 wt.%, and experienced low-melt undercooling. There were many cracks in the iron core that were caused by general corrosion, which was the direct cause of the fragmentation.

Furthermore, the upper rust of the archaeological iron that was most corroded could be distinguished as two layers. The outer layer was loose yellow rust mainly composed of α-Fe_2O_3, γ-Fe_2O_3, α-FeOOH, γ-FeOOH and δ-FeOOH. The dense, black rust layer close to the iron core was gathered β-FeOOH and Fe_3O_4, full of chlorine that threatened to break this dense layer. Meanwhile, there were also many cracks in the rust layers that extended to the iron surface, resulting in a very low resistance against the seawater. This indicated that the rust was a poor barrier for the internal metal. These two features were the important reasons behind the observed general corrosion of the archaeological iron.

There was a reasonable mechanism proposed to explain the growth and transformation of each rust layer, as well as to explain the reason why the rust layer cracked and lost its protective effect; the mechanism combines corrosion conditions and reactions from the initial stage of being submerged in seawater with the condition of the artifacts before laboratory protection.

In view of the desalination treatment that had been applied for ten years, it was not recommended to maintain a single desalination operation. For better conservation of these archaeological iron pieces, they should be properly stabilized and protected using corrosion inhibition and rust transformation for iron oxyhydroxides. Moreover, the preservation status should be monitored through a daily routine. These results also contribute to a better understanding of archaeologically complex corrosion systems, which may lead to improved diagnoses and conservation of archaeological and cultural heritage pieces by conservationists and archaeologists.

Author Contributions: Conceptualization, M.J. and G.H.; methodology, P.H.; validation, M.J. and P.H.; investigation, M.J.; resources, G.H.; writing—original draft preparation, M.J.; supervision, G.H.; project administration, G.H.; funding acquisition, G.H. All authors have read and agreed to the published version of the manuscript.

Funding: This research was funded by the National Key Research and Development Program of China (State assignment No. 2020YFC1522100).

Institutional Review Board Statement: Not applicable.

Informed Consent Statement: Not applicable.

Data Availability Statement: The data presented in this study are available in the article.

Acknowledgments: Thanks to Li Chen in the School of Physics, Peking University, for her support in the SEM and EDS work in this study. Thanks to Huabin Xie at the Archives of Xiamen University for his help with the graphic design in this paper. Thanks to Jian Sun at the National Centre for Archaeology and to Yong Cui at the Guangdong Provincial Institute of Cultural Relics and Archaeology for their support in providing archaeological iron from the Nanhai I that was used in this study. Thanks to Dongbo Hu at the School of Archaeology and Museology, Peking University, for guiding the experimental design and article writing.

Conflicts of Interest: The authors declare no conflict of interest.

References

1. Hao, X.L.; Zhu, T.Q.; Xu, J.J.; Wang, Y.R.; Zhang, X.W. Microscopic study on the concretion of ceramics in the "Nanhai I" shipwreck of China, Southern Song Dynasty (1,127-1,279 AD). *Microsc. Res. Techniq.* **2018**, *81*, 486–493. [CrossRef] [PubMed]
2. Wang, Y.; Xiao, D. The excavation of the "Nanhai No. 1" shipwreck of the Song Dynasty in 2014. *Archaeology* **2016**, *12*, 56–83.
3. Wan, X.; Mao, Z.P.; Zhang, Z.G.; Li, X.H. Analysis of iron pans and iron nails unearthed from Nanhai I shipwreck. *China Cult. Herit. Sci. Res.* **2016**, *2*, 46–51.
4. Selwyn, L.S.; Sirois, P.J.; Argyropoulos, V. The corrosion of excavated archaeological iron with details on weeping and akaganeite. *Stud. Conserv.* **1999**, *44*, 217–232.
5. Remazeilles, C.; Neff, D.; Kergourlay, F.; Foy, E.; Conforto, E.; Guilminot, E.; Reguer, S.; Refait, P.; Dillmann, P. Mechanisms of long-term anaerobic corrosion of iron archaeological artefacts in seawater. *Corros. Sci.* **2009**, *51*, 2932–2941. [CrossRef]
6. Chen, H.; Cui, H.; He, Z.; Lu, L.; Huang, Y. Influence of chloride deposition rate on rust layer protectiveness and corrosion severity of mild steel in tropical coastal atmosphere. *Mater. Chem. Phys.* **2021**, *259*, 123971. [CrossRef]
7. Remazeilles, C.; Neff, D.; Bourdoiseau, J.A.; Sabot, R.; Jeannin, M.; Refait, P. Role of previously formed corrosion product layers on sulfide-assisted corrosion of iron archaeological artefacts in soil. *Corros. Sci.* **2017**, *129*, 169–178. [CrossRef]
8. Gao, X.L.; Han, Y.; Fu, G.Q.; Zhu, M.Y.; Zhang, X.Z. Evolution of the Rust Layers Formed on Carbon and Weathering Steels in Environment Containing Chloride Ions. *Acta Metall. Sin-Engl.* **2016**, *29*, 1025–1036. [CrossRef]
9. Kamimura, T.; Hara, S.; Miyuki, H.; Yamashita, M.; Uchida, H. Composition and protective ability of rust layer formed on weathering steel exposed to various environments. *Corros. Sci.* **2006**, *48*, 2799–2812. [CrossRef]
10. Saleh, S.A. Corrosion Mechanism of Iron Objects in Marine Environment an Analytical Investigation Study by Raman Spectrometry. *Eur. J. Sci. Theol.* **2017**, *13*, 185–206.
11. Remazeilles, C.; Refait, P. On the formation of beta-FeOOH (akaganeite) in chloride-containing environments. *Corros. Sci.* **2007**, *49*, 844–857. [CrossRef]
12. Bazan, A.M.; Galvez, J.C.; Reyes, E.; Gale-Lamuela, D. Study of the rust penetration and circumferential stresses in reinforced concrete at early stages of an accelerated corrosion test by means of combined SEM, EDS and strain gauges. *Const. Build. Mater.* **2018**, *184*, 655–667. [CrossRef]
13. Mazur, A.; Gasik, M.M.; Mazur, V.I. Thermal analysis of eutectic reactions of white cast irons. *Scand. J. Metall.* **2005**, *34*, 245–249. [CrossRef]
14. Veneranda, M.; Aramendia, J.; Bellot-Gurlet, L.; Colomban, P.; Castro, K.; Madariaga, J.M. FTIR spectroscopic semi-quantification of iron phases: A new method to evaluate the protection ability index (PAI) of archaeological artefacts corrosion systems. *Corros. Sci.* **2018**, *133*, 68–77. [CrossRef]
15. Liu, Y.W.; Wang, Z.Y.; Wei, Y.H. Influence of Seawater on the Carbon Steel Initial Corrosion Behavior. *Int. J. Electrochem. Sci.* **2019**, *14*, 1147–1162. [CrossRef]
16. Labbe, J.P.; Ledion, J.; Hui, F. Infrared spectrometry for solid phase analysis: Corrosion rusts. *Corros. Sci.* **2008**, *50*, 1228–1234. [CrossRef]
17. Nishimura, T.; Katayama, H.; Noda, K.; Kodama, T. Electrochemical behavior of rust formed on carbon steel in a wet/dry environment containing chloride ions. *Corrosion* **2000**, *56*, 935–941. [CrossRef]
18. Li, Q.X.; Wang, Z.Y.; Han, W.; Han, E.H. Characterization of the rust formed on weathering steel exposed to Qinghai salt lake atmosphere. *Corros. Sci.* **2008**, *50*, 365–371. [CrossRef]

19. Rocca, E.; Faiz, H.; Dillmann, P.; Neff, D.; Mirambet, F. Electrochemical behavior of thick rust layers on steel artefact: Mechanism of corrosion inhibition. *Electrochim. Acta* **2019**, *316*, 219–227. [CrossRef]
20. Bernard, M.C.; Joiret, S. Understanding corrosion of ancient metals for the conservation of cultural heritage. *Electrochim. Acta* **2009**, *54*, 5199–5205. [CrossRef]
21. Calero, J.; Alcantara, J.; Chico, B.; Diaz, I.; Simancas, J.; de la Fuente, D.; Morcillo, M. Wet/dry accelerated laboratory test to simulate the formation of multilayered rust on carbon steel in marine atmospheres. *Corros. Eng. Sci. Technol.* **2017**, *52*, 178–187. [CrossRef]
22. Dhaiveegan, P.; Elangovan, N.; Nishimura, T.; Rajendran, N. Electrochemical Characterization of Carbon and Weathering Steels Corrosion Products to Determine the Protective Ability Using Carbon Paste Electrode (CPE). *Electroanal* **2014**, *26*, 2419–2428. [CrossRef]
23. Han, D.; Jiang, R.J.; Cheng, Y.F. Mechanism of electrochemical corrosion of carbon steel under deoxygenated water drop and sand deposit. *Electrochim. Acta* **2013**, *114*, 403–408. [CrossRef]
24. Hu, J.Y.; Cao, S.A.; Xie, J.L. EIS study on the corrosion behavior of rusted carbon steel in 3% NaCl solution. *Anti-Corros. Method. Mater.* **2013**, *60*, 100–105. [CrossRef]
25. Zheng, L.G.; Yang, H.Y. Influence of Organic Inhibitors on the Corrosion Behavior of Steel Rebar inside Mortar Specimens Immersed in Saturated NaCl Solution. *Acta Phys.-Chim. Sin.* **2010**, *26*, 2354–2360.
26. Tanaka, H.; Mishima, R.; Hatanaka, N.; Ishikawa, T.; Nakayama, T. Formation of magnetite rust particles by reacting iron powder with artificial alpha-, beta- and gamma-FeOOH in aqueous media. *Corros. Sci.* **2014**, *78*, 384–387. [CrossRef]

Article

Archaeological and Chemical Investigation on Mortars and Bricks from a Necropolis in Braga, Northwest of Portugal

Ana Fragata [1,*], Carla Candeias [1], Jorge Ribeiro [2], Cristina Braga [2], Luís Fontes [2], Ana Velosa [3] and Fernando Rocha [1]

1. GeoBioTec—GeoBioSciences, GeoTechnologies and GeoEngineering Research Unit, Geosciences Department, University of Aveiro, 3810-193 Aveiro, Portugal; candeias@ua.pt (C.C.); tavares.rocha@ua.pt (F.R.)
2. Lab2PT—Landscapes, Heritage and Territory Laboratory Research Unit, Institute of Social Sciences, University of Minho, 4710-057 Braga, Portugal; jribeiro@uaum.uminho.pt (J.R.); cristinabraga@era-arqueologia.pt (C.B.); luisfontes1959@gmail.com (L.F.)
3. RISCO—Research Centre for Risks and Sustainability in Construction, Department of Civil Engineering, University of Aveiro, 3810-193 Aveiro, Portugal; avelosa@ua.pt
* Correspondence: afragata@ua.pt

Citation: Fragata, A.; Candeias, C.; Ribeiro, J.; Braga, C.; Fontes, L.; Velosa, A.; Rocha, F. Archaeological and Chemical Investigation on Mortars and Bricks from a Necropolis in Braga, Northwest of Portugal. *Materials* 2021, 14, 6290. https://doi.org/10.3390/ma14216290

Academic Editors: Žiga Šmit and Eva Menart

Received: 11 September 2021
Accepted: 19 October 2021
Published: 22 October 2021

Publisher's Note: MDPI stays neutral with regard to jurisdictional claims in published maps and institutional affiliations.

Copyright: © 2021 by the authors. Licensee MDPI, Basel, Switzerland. This article is an open access article distributed under the terms and conditions of the Creative Commons Attribution (CC BY) license (https://creativecommons.org/licenses/by/4.0/).

Abstract: This investigation intends to study and characterize the mortars and bricks from walls and floors used in the funerary nucleus of the archaeological site of Dr. Gonçalo Sampaio Street (Braga, Portugal), associated with the Via XVII necropolis of the *Bracara Augusta* Roman city. The diversity of the funeral structures and their exceptional state of conservation make this sector of the necropolis an unprecedented case and a reference site in the archaeology of Braga, a determinant for its conservation and musealization. Nineteen mortars samples were analysed by X-ray Fluorescence. The results showed clear chemical composition differences among coating and floor mortars (CFM), masonry mortars (MM) and bricks (B) groups of samples. The chemical affinity between CFM from the V to IV centuries, CFM from the IV to V centuries, MM from brick walls (IV–V centuries), MM from stone walls (V–VII centuries) and B from the IV to V centuries samples were confirmed by statistical analyses. Their composition was distinctly related to the use of different raw materials, according to their chronological context; in mortars, according to their function in the structures; and in some samples, from contamination.

Keywords: mortars; roman bricks; *Bracara Augusta*; necropolis; funerary nucleus; archaeological sites; chemical characterization

1. Introduction

Mortars are an anthropic material made of binder and aggregates, with a fundamental role in the construction of ancient buildings. The study of Roman mortars and bricks from archaeological sites provides important information on the composition and execution techniques of those highly durable materials. In Portugal, Roman mortars from Beja-Pisões, Braga, Conimbriga, Marvão-Ammaia and Tróia [1–5] and bricks and clayed ceramic materials (CCM) [6–8] were investigated, regarding the preservation, conservation, and archaeological perspectives. Specifically, there are some investigations on necropolis archaeological sites from Jordan [9], France [10,11] and Portugal [12] from an archaeological perspective, and in Italy [13,14] and Spain [15] from a materials characterization perspective.

The archaeological funerary site of Dr. Gonçalo Sampaio Street, excavated by the Archaeology Unit of the University of Minho (Braga, Portugal), is an excellent case study of preserved mortars in masonry constructions (stone and brick), but also in different levels of *opus signinum*, applied in the diverse documented graves. Additionally, it allowed the analysis of the lateritic materials used in the structures.

2. Archaeological Setting

Bracara Augusta, the city. The story of *Bracara Augusta* began in Roman times. The city was founded by Emperor *Augustus* around 16–15 BC in the NW of *Hispania* [16] (Figure 1). Convent capital of the *Tarraconensis* province, the city became the capital of the new province of the *Gallaecia* under Diocletian. During the V and VI centuries, waves of Germanic tribes swept into the former Roman territories, amongst them the Suevi and the Visigoths. The Suevic Kingdom adopted Braga as its capital in 411 AD, having been absorbed by the Visigothic Kingdom in 585 AD [17]. The political changes and the disturbances generated did not prevent the city from maintaining a remarkable economic and constructive dynamism.

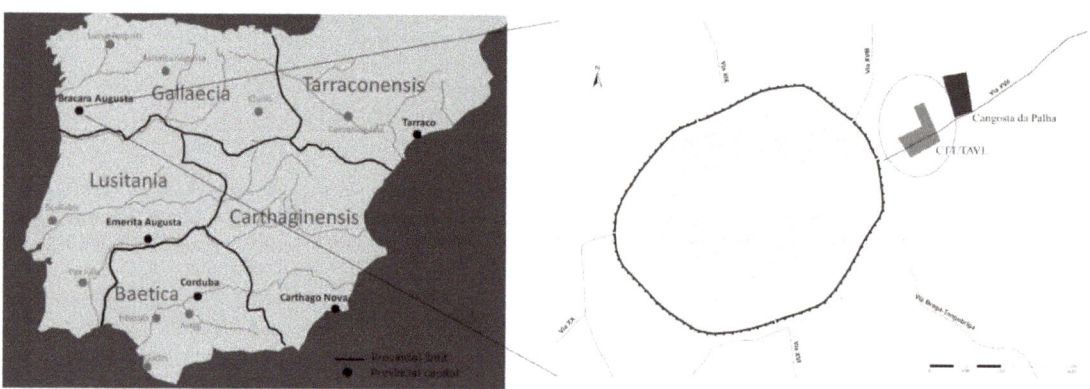

Figure 1. Location of the studied area of Via XVII necropolis in the Roman urban layout, on the Iberian Peninsula. Adapted from [6,17].

Bracara Augusta, the necropolis. The city had several necropolis spaces that, as usual in Roman cities, were located outside the walls, associated with main and secondary roads. Thus, in Braga, the Maximinus necropolis (Via XX and XVI), the Rodovia necropolis (*Via Bracara Augusta—Emerita Augusta*), the Campo da Vinha nuclei (Via XIX/ XVIII?) and the necropolis of Via XVII (Via *Bracara Augusta—Asturica Augusta*), revealed a recent building intervention, forming the archaeological nucleus of Dr. Gonçalo Sampaio Street and the nucleus of Cangosta da Palha [18] and references therein (Figure 1).

The study of the Roman necropolises of Braga resulted from the archaeological record accumulated over nearly 40 years of research in the city, in the context of the *Bracara Augusta* project [16], under the responsibility of the Archaeology Unit of the University of Minho. Several studies resulted from this research, among which one dedicated to the theme of the necropolis nucleus of Via XVII [19], and another one related to the funerary topography of the Via XVII necropolis in Late Antiquity [18]. The Via XVII necropolis, located in one of the East exits of *Bracara Augusta*, known since the 1940s, is undoubtedly the most studied ancient funerary context in the city, of which 12 distinct nuclei are known, dating from the Roman period to the Suevo–Visigothic era.

A nucleus of the Via XVII *Necropolis.* Between the end of 2007 and May 2009, a building from the beginning of the 20th century, and its surrounding area, with a total area of 5600 m^2, was subjected to a major urban rehabilitation project, in the context of which a major archaeological intervention took place. The need to build an underground car park implied the excavation of the area and the archaeological work for over two years. The excavation work of this nucleus was completed in November 2016.

The work carried out led to the identification of a large necropolis area, a section of the Via XVII and a glass workshop, which was active between the Lower Roman Empire and the Suevo–Visigothic period. Regarding the identified necropolis area (integrated into the Via XVII necropolis), it is a large burial space with 129 incineration graves, 94 *ustrinae*, 65 burials, 4 mausoleums and 6 funerary enclosures, with a wide chronology of use, between the first century and the Late Antiquity (V–VII centuries) [17,20].

Nowadays, the only nucleus of a necropolis accessible in Braga occupies an area of reduced dimensions, in which five burial graves were discovered, identified (from west to east, and from north to south) as LXIII, XLIX, CCX, LXXXV and LVII, presenting differentiated constructive orientations and techniques, dated from the IV to VII centuries. There is also a rectangular enclosure built with masonry walls, which partially overlaps the graves XLIX and LXIII, and a stone masonry wall oriented SE–NO, which seemed to delimit the area to the west [20] (Figure 2). The funerary enclosure was assumed to be constructed between the V and VII centuries, later than the graves XLIX and LXIII, from the IV to V centuries. The chronology was attributed through the relative dating methodology, based on the analysis of the materials collected in the intervention, namely ceramics, glasses and metals, considering their evolution according to the stratigraphic levels [21].

Grave CCX   Grave LXXXV

Grave LVII   Funerary enclosure

Grave XLIX   Grave LXIII

Figure 2. *Cont.*

Stone wall

Figure 2. Funerary enclosure, graves and stone wall of the funerary nucleus.

Description of the Structures Identified

A schematic representation of the graves LXIII, XLIX, CCX, LXXXV and LVII, the funerary enclosure and the stone wall can be found in Figure 3.

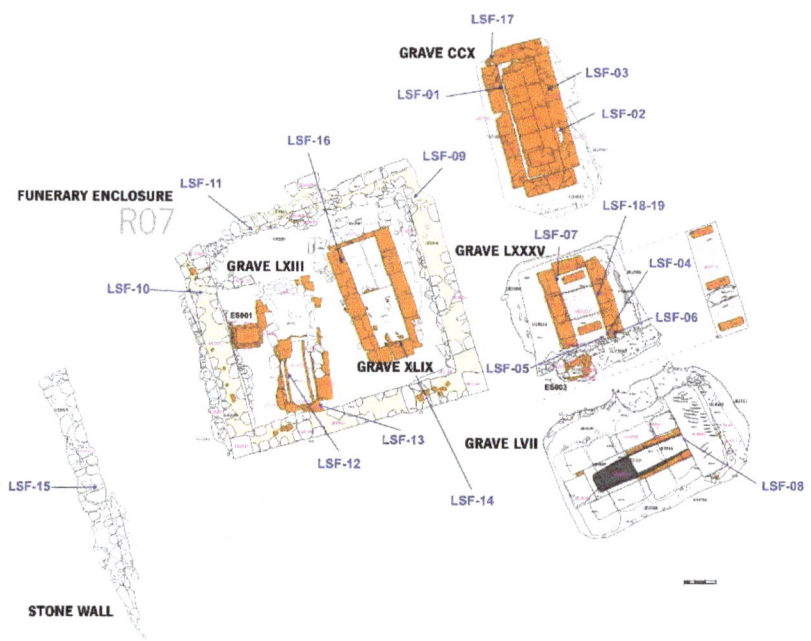

Figure 3. Schematic representation of the funerary nucleus and sampling areas. Adapted from [21].

Grave LXIII is formed by a rectangular box, oriented N–S, with an implantation area of 3.94 m^2 and 2.08 × 0.54 × 0.54 m. It is entirely made of bricks, mostly of the *lydion* type (0.45 × 0.30 × 0.05 m), and it sits directly on the granitic alterite, adapted for this purpose. The size, configuration and the absence of spoils of the grave point to it being a burial grave dating from the IV to V centuries.

Grave XLIX is located to the east of LXIII grave, and has a roughly rectangular box, oriented N/NW–S/SE, with an implantation area of 5.4 m^2 and internal measures of 2.08 × 0.60 − 0.67 × 0.9 m. Its construction is also made of brick masonry, of the *lydion* type, being delimited on the surface by a back row of granitic elements. The base consists of large well-cut granite slabs, covered with a reddish mortar, such as *opus signinum*, penetrating the interstices of the stones. The proposed chronology is of the same period as the LXIII grave.

Grave CCX is located NE of the above graves, with an NNO–SSE orientation. It was discovered in 2009 but only fully excavated in 2016. It is a box with walls and a base made of brick, mostly of the *lydion* typology, covered with a mortar-like *opus signinum*. On the bottom, at each end of the box, *lydion* bricks establish a platform for laying the stretcher in such a way that allows the collection of the moorings. The cover is made of six granitic stones with different dimensions, with the flat face facing inwards. Its proposed chronology is the same as from LXIII and XLIX graves.

Grave LXXXV was located at the south of CCX grave with an NNO–SSE orientation. It was discovered in 2009 but only fully excavated in 2016. It had an implantation area of 6.9 m^2 and an interior span of 2.10 × 0.59 × 0.54 m. This grave is in a brick box, with the interior coated with an *opus signinum*-type mortar, closed with granitic elements. On the structure's leveling embankment, a brick box was documented. It was closed with large granitic blocks, with the joints sealed with a mortar-like *opus signinum*, occasionally linked to the burial signal. It was dated from the V–VII centuries.

Grave LVII is located to the south of the LXXXV grave. It is a box construction with a rectangular plan, oriented OSO–ENE, made from granitic masonry, well squared and with smooth flanks. It has an area of deployment of 8.8 m^2 and an interior span of 2.5 × 0.60 × 0.60 m. The cover is made of monolithic granite slabs, sealed with a mortar-like *opus signinum*. Inside the grave, a rare rectangular container made of lead (Pb) was found, which was considered a unique archaeological feature in the Iberian Peninsula, despite other examples known [10,11]. In the excavation, the occurrence of water infiltration and possibly the contamination of the grave materials was observed. The presence of fine materials was also observed. As for its chronology, it was suggested between the V and VII centuries.

The west wall of the enclosure separated the necropolis (II century), located to the west of the funerary nucleus, showing some deterioration and other nuclei with different characteristics.

Lastly, the funerary enclosure is a quadrangular building, quite flat, made of granitic stone, situated about 28 m north of the Roman road Via XVII with an NNO–SSE orientation and occupying an area of 15.12 m^2, with 3.98 × 3.80 m. The perimeter walls are preserved, which rest on footings made of small stones arranged in an irregular shape. It is made of asymmetrical masonry, showing two-row blocks with irregular size and shape. At the SE and SW limits, the walls sit on parallelepiped granitic ashlars. The laying mortar, as well as that of the joints, has clay characteristics and a yellowish color. The enclosure contains two previous graves (LXIII and XLIX), possibly functioning as an appropriation of important funerary space. It was dated between the V and VII centuries.

The value of this set, with its diversity of burials and rarity, particularly regarding the lead coffin identified in the grave LVII, led to its *in situ* conservation and future musealization.

The graves' material samples were taken from the different types of constructions identified according to a multidisciplinary approach, archaeological and geochemical, by the evidence of typo chronologies to determine whether certain characteristics are related to their use, the existing binders, to characterize geochemically the materials and to compare them with other cases studied.

The aim of this study was to characterize the mortars and bricks to find differences and/or similarities considering the type of mortars/function in the structures and the period of construction. Two types of mortars were considered: the coating and floor mortars, with a reddish color that can be associated with ceramic inclusions, and the masonry mortars with a light brown tone, in general, with some samples being brown but all without ceramic inclusions.

3. Materials and Methods

3.1. Samples Collection

For this study, 19 mortar and brick samples were collected from different structures of the funerary nucleus, graves, funerary enclosure and the dividing wall. Sampling was undertaken considering the structure's function, the macroscopic characteristics, and the available material without destroying the funerary site (Table 1 and Figure 3). The coating and floor mortars (CFM) from the graves, between the IV–V and V–VII centuries, were removed from the internal renders of brick walls (samples SF01 and SF07), the stone wall (SF08), the top of the covers (SF02 and SF06), the uppermost layer (SF04) and the lowermost layer (SF05). These last two were from the grave floor. The masonry mortars (MM) were taken from bedding mortars of graves' brick walls (SF03, SF12, SF16 and SF19) and stone walls (SF09, SF10, SF11 and SF15), attributed to different chronologies, II, IV–V and V–VII centuries. Brick samples from the walls showed different chronological contexts between the IV–V (SF13, SF14 and SF17) and the V–VIII (SF18) centuries.

3.2. Chemical Analysis by X-ray Fluorescence (XRF)

The present study focuses on the chemical composition of the collected mortars and bricks. The chemical analyses were performed on crushed samples by X-ray fluorescence (XRF) carried out by a Panalytical Axios spectrometer PW4400/40 X-ray (Marvel Panalytical, Almelo, The Netherlands) operating on an Rh tube under argon/methane at the GeoBioTec/University of Aveiro laboratories. For major and minor elemental analysis, the Omnian 37 and Pro-Trace2021 software were used, respectively. Loss on Ignition (LOI) was determined by heating the samples at 1000 °C with an electric furnace for 3 h. The major elements analysed with a detection limit of 1%, were: aluminum oxide (Al_2O_3), calcium oxide (CaO), iron oxide (Fe_2O_3), potassium oxide (K_2O), magnesium oxide (MgO), manganese oxide (MnO), sodium oxide (Na_2O), phosporus oxide (P_2O_5), silicon oxide (SiO_2), sulfur oxide (SO_3) and titanium oxide (TiO_2). For the trace elements analysed, the detection limits were: arsenic (As) = 4.06 mg/kg, barium (Ba) = 6.90 mg/kg, bromine (Br) = 0.78 mg/kg, chloride (Cl) = 20 mg/kg, chromium (Cr) = 1.96 mg/kg, copper (Cu) = 2.84 mg/kg, niobium (Nb) = 0.84 mg/kg, neodymium (Nd) = 2.00 mg/kg, nickel (Ni) = 2.00 mg/kg, lead (Pb) = 1.72 mg/kg, tin (Sn) = 3.02 mg/kg, strontium (Sr) = 0.72 mg/kg, vanadium (V) = 2.78 mg/kg, yttrium (Y) = 0.86 mg/kg, zinc (Zn) = 1.28 mg/kg and zirconium (Zr) = 0.80 mg/kg. The precision and the accuracy of analyses and procedures were monitored using internal standards, adopting quality control blanks and certified reference material. The confidence limits of the results were 95% and the relative standard deviation was between 5% and 10%.

3.3. Chemical Statistical Analysis

Variables were processed using IBM SPSS® statistics v25. The normality of the data was verified by the Shapiro–Wilk test ($p > 0.05$). In order to confirm the groups and determine the statistically significant differences ($p < 0.05$), ANOVA, cluster, discriminant analyses, Tukey's test, Student's t-test and k-means were used.

Table 1. Description of the mortars and bricks samples from the funerary nucleus.

	Sample	Location Description	Brick or Mortar Function in the Structure/Construction Technique	Construction Phase (Century)	Color/Consolidation/Conservation State
Masonry mortars (MM)	SF03	Grave CCX; brick wall	Bedding mortar	IV-V	Greyish Brown/no/cohesion loss
	SF09	Funerary enclosure; East stone wall	Bedding mortar	V-VII	Light brown/no/cohesion loss
	SF10	Funerary enclosure; West stone wall	Bedding mortar	V-VII	Light brown/no/cohesion loss
	SF11	Funerary enclosure; North stone wall	Bedding mortar from foundations	V-VII	Light brown/no/cohesion loss
	SF12	Grave LXIII (inside funerary enclosure); brick wall	Bedding mortar	V-VII	Light brown/no/cohesion loss
	SF15	Stone wall	Bedding mortar	II	Brown/no/cohesion loss
	SF16	Grave XLIX (inside funerary enclosure); brick wall	Bedding mortar	IV-V	Light brown/no/cohesion loss
	SF19	Grave LXXXV; brick wall	Bedding mortar	V-VII	Greyish brown/no/cohesion loss
Coating and floor mortars (CFM)	SF01	Grave CCX; brick wall	*Opus signinum*; coating mortar—from the wall	IV-V	Reddish/cohesion loss
	SF02	Grave CCX; brick wall	*Opus signinum*; coating mortar with brick fragments from the top of the cover (may act as impermeabilization render)	IV-V	Pink or reddish brown/paraloid 10%/cohesion loss
	SF04	Grave LXXXV	*Opus signinum*; floor mortar, uppermost layer	V-VII	Reddish/cohesion loss
	SF05	Grave LXXXV	Preparatory layer; floor mortar, lowermost layer	V-VII	Reddish/no/cohesion loss
	SF06	Grave LXXXV; brick wall	*Opus signinum*; coating mortar from the top of the cover (may act as impermeabilization render)	V-VII	Orange or reddish/no/cohesion loss
	SF07	Grave LXXXV; brick wall	*Opus signinum*; coating mortar from the wall	V-VII	Reddish/cohesion loss
	SF08	Grave LVII; stone wall	*Opus signinum*; coating mortar from the wall	V-VII	Reddish brown/no/cohesion loss

Table 1. Cont.

	Sample	Location Description	Brick or Mortar Function in the Structure/Construction Technique	Construction Phase (Century)	Color/Consolidation/Conservation State
Bricks (B)	SF13	Grave LXIII (inside funerary enclosure); brick wall	Brick from cover	IV–V	Orange/no
	SF14	Grave XLIX (inside funerary enclosure); brick wall	Brick from the wall	IV–V	Orange/no
	SF17	Grave CCX; brick wall	Brick (*lydion* type) from the wall	IV–V	Orange/no
	SF18	Grave LXXXV; brick wall	Brick from the wall	V–VII	Orange/no

4. Results and Discussion

The mortar and brick samples' chemical composition is detailed in Tables 2 and 3. Coating and floor mortars group (CFM) include all *opus signinum* mortars and SF05 sample from the preparatory layer of LXXXV grave's floor, showing higher mean values of Al_2O_3, Fe_2O_3, P_2O_5, SO_3, As, Br, Cl, Cr, Nb, Ni, Sn, Y and Zn than masonry mortars (MM). Bricks (B) were characterized with higher content on Al_2O_3 and Fe_2O_3 and K_2O, As, Nb, Ni, Sn, Y and Zr and lower content on CaO, MgO, MnO, Na_2O, P_2O_5, SO_3, TiO_2, Ba, Cl, Cu, Nd, Pb, Sr and V when compared to CFM and MM samples. The mortars and bricks' low CaO content [1,22] support the idea of the use of locally available raw materials, as the Braga region is dominated by granites [23]. Additionally, in mortars, the high SiO_2 concentration is directly related to the low CaO and MgO contents, confirming the absence of calcareous/dolomitic aggregates (Table 2). The Fe_2O_3 mortars content did not reveal a significant variation (5.37–7.61%, except for SF02 with the lowest content among mortars—3.51%). Among mortars, the higher Fe_2O_3 content (between 5.54% and 7.61%) in *opus signinum* ones (cocciopesto mortars) from CFM (except SF01 and SF02, with lower content) can be related to the presence of red ceramic fragments or powder, which may result, although in lower quantity, from the aggregate. The higher aluminate content in those mortars confirms this idea, as well as the most abundant presence of ceramic fragments in SF04, SF06, SF07 and SF08). However, MM displayed relatively high contents of Fe_2O_3 suggesting that they came from the aggregate in this case.

Brick samples, ranging from 6.33 to 8.52% of Fe_2O_3, also did not reveal significant variation ($p > 0.05$); although, a relatively higher content could have been the result of Fe_2O_3 application to facilitate bricks' firing by increasing the heating storage capacity. The lowest LOI content was observed in bricks SF14 and SF17, with 1.62 and 1.36%, respectively, which can be related to their high firing temperature and higher kaolinitic content. The chemical composition of the raw clay used in bricks is relatively uniform, as the variability means of Al_2O_3 (30.17%) and SiO_2 (52.07%) exhibit low variance, with a standard deviation of 2.01–2.65%, respectively (Table 2). Low content of CaO, Na_2O and TiO_2 was found in all brick samples, although the first two can be associated with contamination by lime mortars and soluble salts [24].

The high content of Cu and Pb in mortars can possibly be the result of the considerable degree of exposure to the modern construction materials (concrete and Portland cement) (e.g., [25]) that were used above this archaeological site. Considerably higher contents of Cu and Pb were observed in mortar samples SF01 (Pb = 180 mg/kg), SF03 (Cu = 210 mg/kg), SF09 (Cu = 150 mg/kg, Pb = 190 mg/kg) and SF15 (Cu = 87.5 mg/kg, Pb = 220 mg/kg), which are closer to those modern materials (Table 3), as result of contamination. A remarkable Pb content of 17,890 mg/kg was found in sample SF08, a coating mortar from grave LVII, that might be related to contamination due to a leaden coffin found in the grave with signs of corrosion related to water infiltration.

The Cl and Ba contents, although low, found in bricks (Table 3) may result from contamination. As with clayed bricks, those elements are not expected due to the firing process. The presence of Cl in those samples (except in SF14, in which Cl content was below the detection limit) can be ascribed to infiltration and/or capillary rise that may transport these soluble salts through those porous materials [26].

Table 2. Samples major elements chemical composition, in %.

Group		Al_2O_3	CaO	Fe_2O_3	MgO	K_2O	MnO	Na_2O	P_2O_5	SiO_2	SO_3	TiO_2	LOI
Masonry mortars (MM)	Min	20.77	0.51	5.37	1.43	5.80	0.06	0.33	0.38	50.07	0.03	1.11	4.78
	Max	23.99	0.96	7.23	1.66	6.73	0.11	0.89	1.82	56.57	0.11	1.19	6.54
	Mean	22.88	0.75	6.16	1.53	6.30	0.08	0.62	0.94	53.80	0.07	1.15	5.31
	SD	1.29	0.19	0.58	0.07	0.31	0.02	0.22	0.54	1.88	0.03	0.03	0.56
Coating and floor mortars (CFM)	Min	18.28	0.35	3.51	0.90	3.82	0.07	0.28	0.36	42.26	0.05	0.65	3.61
	Max	29.78	0.99	7.61	1.45	6.61	0.11	0.75	4.53	63.83	0.32	1.25	7.89
	Mean	25.96	0.54	6.35	1.20	4.94	0.08	0.45	1.86	51.41	0.17	1.09	5.35
	SD	4.52	0.21	1.48	0.21	1.07	0.01	0.22	1.33	7.00	0.11	0.21	1.38
Bricks (B)	Min	28.27	0.12	6.33	0.54	3.70	0.02	0.14	0.31	50.38	0.02	1.10	1.36
	Max	32.83	0.33	8.52	0.82	3.85	0.04	0.20	1.03	56.01	0.12	1.23	6.04
	Mean	30.17	0.23	7.02	0.69	3.80	0.03	0.17	0.68	52.07	0.05	1.18	3.63
	SD	2.01	0.09	1.02	0.11	0.07	0.01	0.03	0.35	2.65	0.05	0.06	2.48

Min: minimum; Max: maximum; Sd: Standard deviation; LOI: Loss of ignition.

Table 3. Samples trace elements chemical composition, in mg/kg.

Group		As	Ba	Br	Cl	Cr	Cu	Nb	Nd	Ni	Pb [1]	Sn	Sr	V	Y	Zn	Zr
Masonry mortars (MM)	Min	5.9	720.0	6.8	120.0	21.9	32.3	15.7	61.1	8.1	46.9	10.4	140.0	56.4	13.0	57.5	370.0
	Max	21.4	900.0	16.9	370.0	89.8	210.0	18.4	80.8	9.8	220.0	21.4	180.0	70.1	18.7	78.9	410.0
	Mean	10.9	820.0	10.5	222.5	38.2	85.7	17.1	69.1	8.8	94.5	12.8	163.8	64.7	15.1	66.6	386.3
	SD	4.3	61.4	3.2	98.8	20.5	58.4	0.9	5.5	0.6	64.8	3.6	14.9	4.0	1.7	5.6	13.2
Coating and floor mortars (CFM)	Min	5.4	400.0	3.9	160.0	28.4	27.9	11.7	24.3	6.3	43.5	8.1	67.6	27.3	10.8	46.0	180.0
	Max	28.1	820.0	62.0	610.0	65.4	160.0	28.8	91.5	16.1	180.0	34.1	150.0	90.0	71.5	110.0	460.0
	Mean	14.9	565.7	22.5	357.1	50.8	67.8	21.3	62.1	11.5	81.6	19.7	85.6	62.1	35.0	72.2	348.6
	SD	7.5	136.3	17.9	166.2	11.0	41.5	6.8	20.0	3.4	46.7	7.7	26.9	18.0	19.7	21.1	90.5
Bricks (B)	Min	11.9	270.0	6.3	40.0	46.5	20.3	27.5	38.6	9.9	42.6	19.8	35.6	51.7	30.1	56.7	370.0
	Max	34.7	340.0	16.7	280.0	54.3	31.4	34.4	56.9	13.6	63.6	35.1	89.8	67.6	45.2	74.7	590.0
	Mean	22.4	307.5	10.8	146.7	50.4	28.6	32.0	48.5	12.5	51.6	28.1	52.6	56.7	39.3	65.3	465.0
	SD	8.2	32.7	4.4	99.8	2.8	4.8	2.6	6.6	1.5	7.9	5.7	21.8	6.4	6.1	7.2	93.4

SD: Standard deviation; Min: minimum; Max: maximum; LOI: Loss of ignition. [1] The Pb content of SF08 (17,890 mg/kg) sample, from coating and floor mortars group was not included in the statistical analysis.

The sample's chemical content was explored using multivariate cluster analyses. The samples major elements dendrogram revealed two major clusters (Figure 4a): Group 1 samples, which included all masonry mortars (MM) and three coating and floor mortars (CFM) samples (SF01, SF05 and SF02), a set enriched in SiO_2 (50.07–63.83%), MgO (1.43–1.58%) and K_2O (5.80–6.75%). The higher MgO content in group 1 (1.43–1.58%, except sample SF02), when compared to group 2, can be related to the need to increase plasticity through the addition of more plastic clays (poorer in kaolinite, richer on illite/smectite); however, it can also be ascribed to lower quality (less kaolinitic) of raw materials used, or to mortars degradation. The outlier SF02 sample, removed from a structure previously consolidated with paraloid, showed the highest SiO_2 (63.83%) and the lowest K_2O (5.45%) and Al_2O_3 (18.28%) content among group 1. Group 2 includes all brick (B) samples and the remaining CFM samples (SF04, SF06, SF07 and SF08—all *opus signinum* from the V to VII centuries), a set of samples enriched in Al_2O_3. The higher Al_2O_3 (28.27–32.83%) content can be ascribed to Al-rich kaolinitic clays, which highlighted the careful selection of locally (Braga region is very rich in kaolin deposits) raw materials on those materials. SF08 was revealed to be an outlier in group 2, with the lowest SiO_2 and K_2O contents among the group samples, which may be the result of previous contamination with fine materials. By considering the variable cluster analysis of all the samples, Al_2O_3 and SiO_2 concentrations were distinguished from the other major elements analysed, forming a distinctive cluster (Figure 4b). A subgroup of Group A included Fe_2O_3 and K_2O, with bricks showing lower K_2O and higher Fe_2O_3 than mortars.

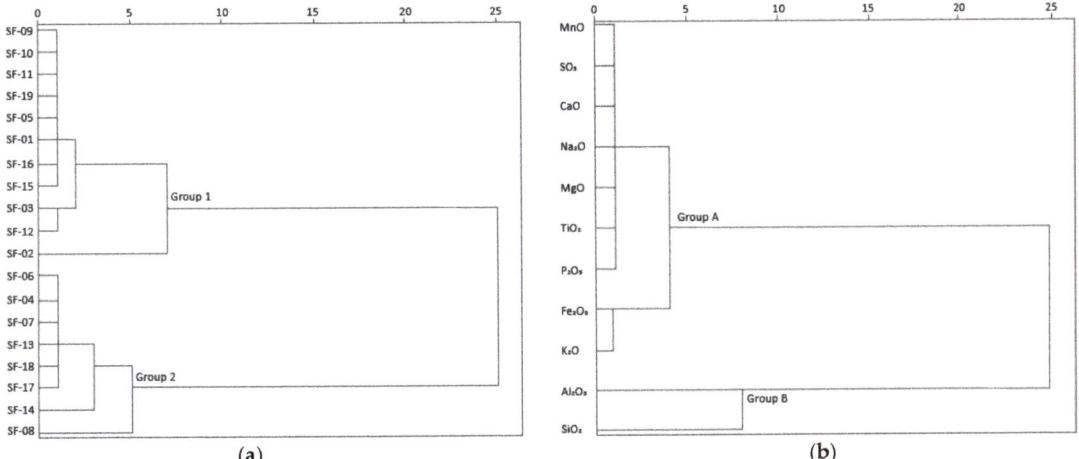

Figure 4. Cluster analysis of mortars and bricks samples major element concentrations: (a) samples, and (b) variables.

Cluster analysis of CFM, MM and B groups' major elements content was also performed individually. On CFM samples, two clusters were identified, one cluster composed by SF08 with an affinity to SF04, SF06, and SF08 (V–VII centuries), with higher Fe_2O_3 and Al_2O_3 (that can be ascribed to the higher inclusion of brick fragments) than the other cluster with SF05 (V–VII centuries), and SF01 and SF02 (IV–V centuries). The cluster analyses of MM group, samples SF03 and SF12, revealed a distinct composition, with lower SiO_2 and Al_2O_3 content, from SF15. In the B group, sample SF14 presented a slight distinct composition from the remaining brick samples, with higher SiO_2, MgO and TiO_2 and lower P_2O_5 content.

The diagram CaO + MgO vs. SiO_2 + Fe_2O_3 + Al_2O_3 can be roughly related to the presence of binder and aggregate [27], respectively, since these mortars do not present calcareous aggregates. Mortars formed two groups (Figure 5): a group composed by all

MM and one CFM (SF01) with a good compositional homogeneity, in the upper part of the diagram, with less binder (2.0–2.6%) and higher aggregate (80.9–84.4%) content; another group was composed by the remaining CFM samples, despite SF05 and SF07, with binder content similar to MM samples (1.9 and 1.8%, respectively). Sample SF08 revealed the lowest binder and aggregate content among mortars, possibly due to contamination with fine materials. Mortars displayed higher CaO (0.35–0.99%) and MgO (0.90–1.66%) content than bricks CaO (0.12–0.33%) and MgO (0.11–0.54%). The higher MgO content in MM samples may be the result of the application of Mg-enriched clays (non-pure kaolin) to increase mortars' plasticity (pure kaolins are low plastic clays). Additionally, in CFM samples (*opus signinum* mortars with ceramic powder or fragments), an increased content of SiO_2 + Al_2O_3 + Fe_2O_3 was observed when compared to masonry mortars, suggesting the use of pozzolanic materials and, therefore, the hydraulic characteristics of these mortars. Although a good degree of hydraulicity of the mortar could be attained using finely grounded bricks, if produced with poor clay raw materials, as was this case, the pozzolanic reaction could not occur due to the low amounts of amorphous materials [22,28]. Bricks showed lower binder content (0.8–1.1%) than mortars, and their aggregate content clearly falls into two distinct groups: one more similar to CFM mortars (SF18 and SF13) and another with the highest aggregate content (SF14 and SF17). The binder and aggregate proportion allowed to distinguish CFM (*opus signinum* mortars) from the V to VII centuries, MM from stone walls (inside funerary enclosure) from the V to VII centuries, MM and CFM (both from brick walls) from the IV to V centuries, and bricks (except *lydion* brick).

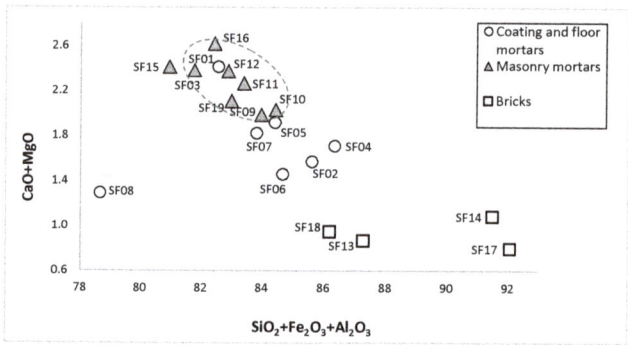

Figure 5. Mortars and bricks CaO + MgO vs. SiO_2 + Fe_2O_3 + Al_2O_3 binary diagram.

The binary diagram TiO_2/Fe_2O_3 vs. SiO_2/CaO showed a homogeneous group composed of MM samples, with low SiO_2/CaO ratio, differing from CFM samples with lower TiO_2/Fe_2O_3 ratio and bricks with higher SiO_2/CaO ratio (Figure 6). The similar TiO_2/Fe_2O_3 ratio (0.16–0.17%) of the CFM (SF04, SF07, SF08 and SF06) samples from the V to VII century construction phase suggested the use of the same raw materials. Similarly, MM samples from V to VII (SF09, SF10, SF11 and SF19) and from the IV to V centuries (SF03, SF12 and SF16) seemed to use different raw materials, according to the construction phase of the structures to which they belong. Sample SF15, a unique MM sample from the II century stone wall, was clearly separated from MM samples with affinity to CFM samples due to its higher content in Fe_2O_3 (7.13%), which may arise from the aggregate. The MM showed a more homogeneous chemical composition than CFM, and their lower content in Fe_2O_3 and higher content in TiO_2 and MgO can be attributed to the raw materials used (binder and aggregates): a less pure lime may have been used, with higher content in magnesium combined with the fact that local materials with a particular composition may have been used. The high content of Fe_2O_3 in CFM (Table 2) can be attributed to the presence of ceramic particles, which may also come, although in smaller quantities, from the aggregate. Among CFM samples, the results pointed out to lower percentage of ceramic particles in samples SF01 and SF05 (Figure 5). Observing the percentage of aluminates in CFM samples, this idea was reinforced, as well as the more abundant presence of ceramic particles in samples SF04, SF06, SF07 and SF08, forming a distinct group (Figure 5). Bricks were more dispersed than mortars and showed the highest SiO_2/CaO ratio (160–430%). The *lydion* brick sample (SF-17) differed in composition from other brick samples, showing the lowest TiO_2/Fe_2O_3 ratio and CaO content (0.12%) and sample SF13 with the lowest SiO_2/CaO content. Considering the relation in the binary diagram TiO_2/Fe_2O_3 vs. SiO_2/CaO, associated with the raw materials used, compositional differences were observed between CFM (*opus signinum* mortars from brick walls) from the V to VII centuries, MM (from stone walls inside the funerary enclosure) from the V to VII centuries, and MM (from brick walls) from the IV to V centuries.

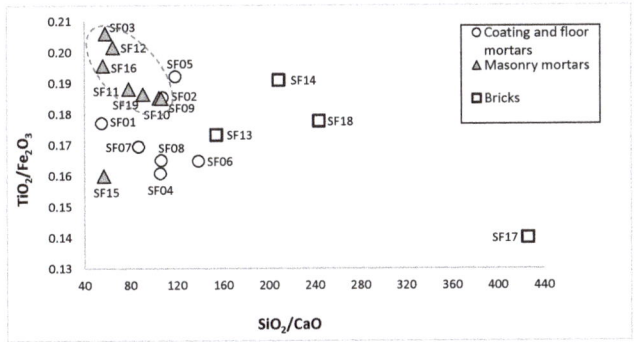

Figure 6. Mortars and bricks TiO_2/Fe_2O_3 vs. SiO_2/CaO binary diagram.

The CaO/MgO ratio showed some variability (Figure 7), which in some cases was not directly linked to the variation in calcium oxide, suggesting the use of binders from different sources. As a result, the binder materials used in MM samples SF09, SF10, SF11 and SF19 (V–VII centuries) and for CFM samples SF04, SF05, SF06, SF07 and SF08 (V–VII centuries) could have been the source. The CFM samples (SF01 and SF02) and MM samples (SF03, SF12, SF16) from the IV to V centuries showed a similar CaO/MgO ratio (between 0.58 and 0.70), suggesting the same material source. The MM sample SF15, from the II century, could have used the same binder materials source and different from samples from the IV–V centuries. The Al_2O_3/SiO_2 ratio clearly differentiates two groups: one mainly composed of MM samples with lower Al_2O_3 content, and a group mainly composed of CFM samples (SF04, SF06, SF07 and SF08), suggesting a more careful selection of raw materials than other samples, as well the presence of brick particles on mortars composition. Bricks showed less homogeneous composition than mortars, although some similarities in the raw materials used could be found between SF14 and SF18 and between SF13 and SF17 samples. From the binary diagram CaO/MgO vs. Al_2O_3/SiO_2 analyses, chemical differences were observed between CFM (*opus signinum* from brick walls) from the V–VII centuries, MM (from stone walls inside the funerary enclosure) and MM (from brick wall) both from the V–VII centuries, CFM (coating mortars from brick walls) from IV–V centuries and MM from the brick wall.

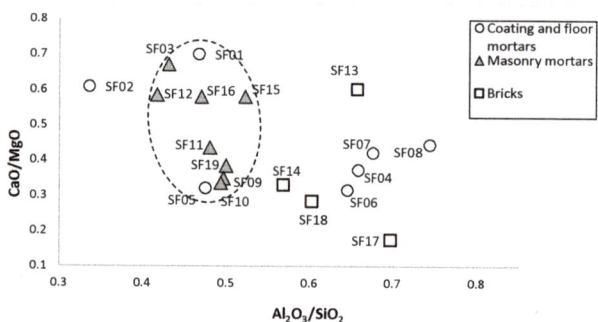

Figure 7. Mortars and bricks CaO/MgO vs. Al_2O_3/SiO_2 binary diagram.

The cluster analysis of the trace elements composition showed two main samples associations (Figure 8a): group i, including MM samples and CFM SF01, SF04, SF05 and SF07 samples, and group ii, with bricks and the remaining CFM SF02, SF06 and SF08 samples. The variable cluster analysis revealed that Ba, Cl and Zr concentration trends were distinctive from the others, forming a distinctive cluster (Figure 8b). Excluding some samples, in general, group i showed higher Ba (500–900 mg/kg), Sr (86–180 mg/kg, except in SF04 and SF07 samples, with 69.2 and 77.6 mg/kg, respectively) and Cl (160–610 mg/kg, excluding SF03 and SF09 both with 120 mg/kg). In group ii, higher Zr (370–590 mg/kg, excluding SF02 with 180 mg/kg) content was found. The CFM samples SF01, SF04 and SF07 in group i (excluding SF05), formed a distinct subcluster, mainly due to its higher Cl and lower Ba and Sr contents. In group ii, the CFM samples SF02, SF06 and SF08 also formed a distinctive cluster (except for SF18, included in CFM subcluster) mainly due to their higher Ba (400–490 mg/kg) and Cl (160–280 mg/kg) and lower Zr (370–480 mg/kg, excluding SF02 with 180 mg/kg) content. Bricks showed the lowest Sr content (35.6–46.1 mg/kg excluding SF17 with 89.8 mg/kg) among all samples.

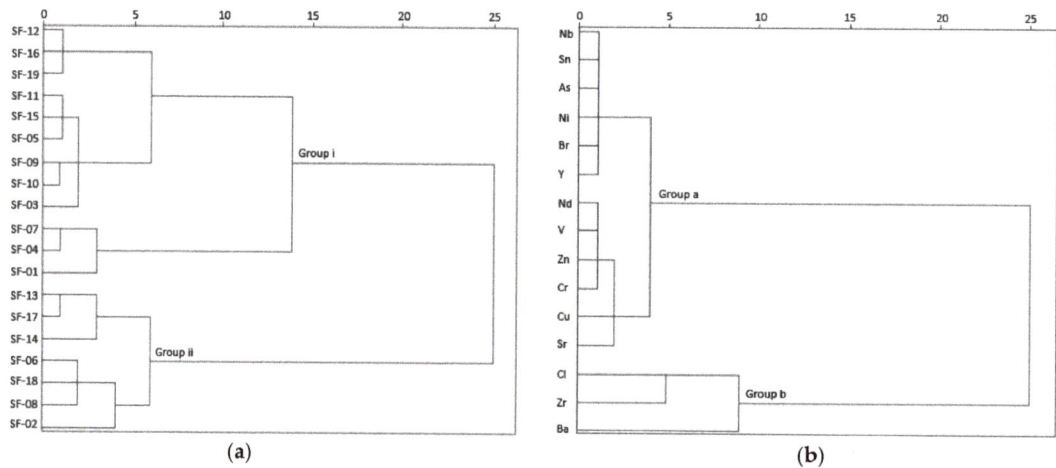

Figure 8. Cluster analysis of mortars and bricks samples trace element concentrations: (**a**) samples and (**b**) variables.

The individual trace elements cluster analysis of CFM, MM and bricks samples was performed. The CFM samples revealed two groups, a cluster with SF05 sample (V–VII centuries) revealing affinity to SF01 (IV–V centuries), SF04 and SF07 (V–VII centuries), with higher Ba (500–820 mg/kg) and Cl (470–610 mg/kg) contents than the other cluster formed with SF-08, SF-06 (V–VII centuries) and SF-02 (IV–V centuries) samples, with higher Zr (370–460 mg/kg, excluding SF02 with 180 mg/kg) content. In the analyses of the MM group, samples SF12, SF16 and SF19 (IV–V centuries) revealed a distinct composition from the other MM samples, showing higher Ba (800–900 mg/kg), Cl (320–370 mg/kg) and Sr (170–180 Mg/kg). The bricks group cluster analysis suggested that sample SF18 has a distinct composition from the remaining brick samples, with higher Cl (280 mg/kg) and V (67.6 mg/kg) content.

The Ba/Sr vs. Zr/Y binary diagram (Figure 9) revealed information on the employed raw materials. Lower Ba/Sr and higher Zr/Y ratios suggest the presence of natural rock fragments rather than ceramic materials (cocciopesto) in the aggregate [29]. Bricks and CFM samples presented higher Ba/Sr and lower Zr/Y ratios, which can be attributed to the presence of ceramic compounds. The CFM samples SF04, SF06, SF07 and SF08 (V–VIII centuries) showed lower Zr/Y and higher Ba/Sr ratios, with higher Al_2O_3, Fe_2O_3 and TiO_2 contents than the remaining CFM samples, which suggested their higher content in ceramic compounds. As expected, MM samples showed lower Ba/Sr and higher Zr/Y ratios. The SF13 and SF14 brick samples from the IV–V centuries showed similar Ba/Sr and Zr/Y ratios and suggested the use of the same source of raw materials. As a result, through the relation between Ba/Sr vs. Zr/Y, the compositional differences allowed to distinguish CFM (*opus signinum* from brick walls) from V–VII centuries and MM (from brick walls) from IV–V centuries.

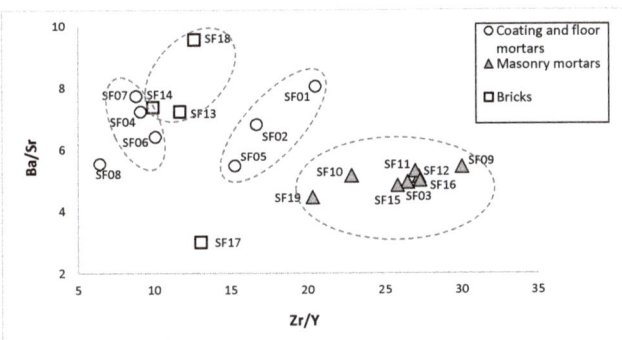

Figure 9. Mortars and bricks Ba/Sr vs. Zr/Y binary diagram.

5. Conclusions

The present study allowed us to confirm, in the Via XVII necropolis of the *Bracara Augusta* Roman city (Braga, Portugal), the distinctly chemical composition among coating and floor mortars (CFM), masonry mortars (MM) and bricks (B). The chemical affinity of each group (coating and floor mortars (CFM) from the V–VIII centuries, CFM from the IV–V centuries, masonry mortars (MM) from brick walls (IV–V centuries), MM from stone walls (V–VII centuries) and bricks ((B) from IV–V centuries) were confirmed by statistical analyses. Their composition was distinctly related to the use of different raw materials, according to their chronological context and, in mortars, according to their function in the structures and, in some samples, from contamination.

The mortar and brick low CaO content support the idea of using locally available poor-Ca raw materials. Additionally, a more careful selection of raw materials on bricks and CFM from V–VIII centuries was observed, using richer-Al kaolinitic clay, and the binder sources differed according to the different construction phases. A general higher compositional homogeneity was observed on MM compared to CFM. Some contamination was observed as a result of the proximity of some posterior funerary structures to the new building structure (made of Portland cement and concrete).

The characterization of the mortars and bricks from this funerary nucleus is the first step for the study of the provenance of the raw materials, the objective of future work. Moreover, the investigation on the compositional chemical data obtained for these original materials can be useful for the adequate reproduction of compatible mortars for conservation and restoration purposes, considering that this archaeological site is under musealization works.

Author Contributions: Conceptualization, A.F., J.R. and F.R.; methodology, A.F. and F.R.; statistical analysis, C.C.; validation, A.F. and F.R.; formal analyses, A.F., C.C., J.R., C.B., L.F., A.V. and F.R.; investigation, A.F. and J.R.; writing—original draft preparation, A.F., C.C. and J.R.; writing—review and editing, A.F., C.C., J.R. and F.R.; supervision, A.F. and F.R. All authors have read and agreed to the published version of the manuscript.

Funding: Ana Fragata acknowledges grant SFRH/BPD/101517/2014 from the Portuguese Foundation for Science and Technology (FCT). This research was supported by GeoBioTec Research Centre (UID/GEO/04035/2019 + UIDB/04035/2020), funded by FCT, FEDER funds through the Operational Program Competitiveness Factors COMPETE and by national funds (OE), through FCT, in the scope of the framework contract foreseen in the numbers 4, 5 and 6 of the article 23, of the Decree-Law 57/2016, of 29 August, changed by Law 57/2017, of 19 July.

Institutional Review Board Statement: Not applicable.

Informed Consent Statement: Not applicable.

Data Availability Statement: Not applicable.

Acknowledgments: The authors are grateful to Liberdade Street Fashion Shopping Centre—Cushman and Wakefield, and especially to its manager José Alberto Martins, for facilitating the access to the funerary nucleus, to the Archaeology Unit of University of Minho (UAUM), for the technical support, and to Cristina Sequeira for her assistance in mortars chemical analyses at Geosciences Department of University of Aveiro.

Conflicts of Interest: The authors declare no conflict of interest.

References

1. Fragata, A.; Ribeiro, J.; Candeias, C.; Velosa, A.; Rocha, F. Archaeological and Chemical Investigation on the High Imperial Mosaic Floor Mortars of the Domus Integrated in the Museum of Archaeology D. Diogo de Sousa, Braga, Portugal. *Appl. Sci.* **2021**, *11*, 8267. [CrossRef]
2. Silva, A.S.; Paiva, M.; Ricardo, J.; Salta, M.M.; Monteiro, A.M.; Candeias, A.E. Characterisation of Roman mortars from the archaeological site of Tróia (Portugal). *Mater. Sci. Forum* **2006**, *514*, 1643–1647. [CrossRef]
3. Cardoso, I.; Macedo, M.F.; Vermeulen, F.; Corsi, C.; Santos Silva, A.; Rosado, L.; Candeias, A.; Mirao, J. A multidisciplinary approach to the study of archaeological mortars from the town of Ammaia in the Roman province of Lusitania (Portugal). *Archaeometry* **2014**, *56*, 1–24. [CrossRef]
4. Borsoi, G.; Santos Silva, A.; Menezes, P.; Candeias, A.; Mirão, J. Analytical characterization of ancient mortars from the archaeological Roman site of Pisões (Beja, Portugal). *Constr. Build Mater.* **2019**, *204*, 597–608. [CrossRef]
5. Velosa, A.L.; Coroado, J.; Veiga, M.R.; Rocha, F. Characterisation of roman mortars from Conímbriga with respect to their repair. *Mater. Charact.* **2007**, *58*, 1208–1216. [CrossRef]
6. Ribeiro, J. *Roman Architecture in Bracara Augusta. Analysis of Building Techniques*; Afrontamento: Porto, Portugal, 2013. (In Portuguese)
7. Fernandes, F.M.; Lourenço, P.B.; Castro, F. Ancient Clay Bricks: Manufacture and Properties. In *Materials, Technologies and Practice in Historic Heritage Structures*; Dan, M.B., Přikryl, R., Török, Á., Eds.; Springer: Dordrecht, The Netherlands, 2010; pp. 29–48.
8. Prudêncio, M.I.; Braga, M.A.S.; Oliveira, F.; Dias, M.I.; Delgado, M.; Martins, M. Raw material sources for the Roman Bracarense ceramics (NW Iberian Peninsula). *Clays Clay Miner.* **2006**, *54*, 638–649. [CrossRef]
9. Al-Muheisen, Z. Archaeological Excavations at the Yasileh Site in Northern Jordan: The Necropolis. *Syr. Archéologie Art Et Hist.* **2018**, *85*, 315–337. [CrossRef]
10. Cochet, A.E.; Hansen, J. *Conduites et Objects de Plomb Gallo-Romaines de Vienne (Isére); Gallia, 46*; Centre National de la Recherche Scientifique: Paris, France, 1986. (In French)
11. Gillet, P.E.; Mahéo, N. Sarcophages en plomb gallo-romaines découverts à Amiens et dans ses environs (Somme). *Rev. Archéologique Picardie* **2000**, *3*, 77–118. (In French) [CrossRef]
12. Pereira, C. *The Roman Necropolis of Algarve (Portugal): About the Spaces of Death in the South of Lusitania*; Archaeopress Publishing Ltd.: Oxford, UK, 2015.

13. Di Benedetto, C.; Graziano, S.F.; Guarino, V.; Rispoli, C.; Munzi, P.; Morra, V.; Cappelletti, P. Romans' established skills: Mortars from D46b mausoleum, Porta Mediana Necropolis, Cuma (Naples). *Mediterr. Archaeol. Ar.* **2018**, *18*, 131–146.
14. Alberghina, M.F.; Germinario, C.; Bartolozzi, G.; Bracci, S.; Grifa, C.; Izzo, F.; La Russa, M.F.; Magrini, D.; Massa, E.; Mercurio, M.; et al. The Tomb of the Diver and the frescoed tombs in Paestum (southern Italy): New insights from a comparative archaeometric study. *PLoS ONE* **2020**, *15*, e0232375. [CrossRef]
15. Sanchez-Moral, S.; Cañaveras, J.C.; Benavente, D.; Fernandez-Cortes, A.; Cuezva, S.; Elez, J.; Jurado, V.; Rogerio-Candelera, M.A.; Saiz-Jimenez, C. A study on the state of conservation of the Roman Necropolis of Carmona (Sevilla, Spain). *J. Cult. Herit.* **2018**, *34*, 185–197. [CrossRef]
16. Martins, M.; Carvalho, H. Roman city of *Bracara Augusta* (*Hispania Citerior Tarraconensis*): Urbanism and territory occupation. *Agri Centuriate* **2017**, *14*, 79–98.
17. Braga, C.; Martins, M. The funerary topography of the Via XVII necropolis in Late Antiquity (Braga). *Agira* **2016**, *VIII*, 17–34.
18. Vaz, F.C.; Braga, C.; Tereso, J.P.; Oliveira, C.; Carretero, L.G.; Detry, C.; Marcos, B.; Fontes, L.; Martins, M. Food for the dead, fuel for the pyre: Symbolism and function of plant remains in provincial Roman cremation rituals in the necropolis of *Bracara Augusta* (NW Iberia). *Quatern. Int.* **2021**, *593*, 372–383. [CrossRef]
19. Braga, C. A new sector of Via XVII necropolis in *Bracara Augusta*: The High Empire phase. In Proceedings of the XVIII CIAC: Centro y Periferia en el Mundo Clásico, Mérida, Spain, 13–17 May 2013; Martínez, J.M., Trinidad, N.B., Llanza, I., Eds.; Museo Nacional de Arte Romano: Mérida, Spain, 2014; pp. 1253–1257.
20. Fontes, L.; Braga, C. Archaeological nucleus of Liberdade Street Fashion, Braga. *Forum* **2015**, *49–50*, 71–84. (In Portuguese)
21. Braga, C. Death, Memory and Identity: An analysis of *Bracara Augusta*'s Funerary Practices. Ph.D. Thesis, University of Minho, Braga, Portugal, 2018. (In Portuguese)
22. Özkaya, O.A.; Böke, H. Properties of Roman bricks and mortars used in Serapis temple in the city of Pergamon. *Mater. Charact.* **2009**, *60*, 995–1000. [CrossRef]
23. Ferreira, N. *Geological Chart of Portugal in Scale 1:50,000—Sheet 5-D, Braga*; Ministério da Economia, Instituto Geológico Mineiro, Departamento de Geologia: Lisboa, Portugal, 2000.
24. Lourenço, P.B.; Fernandes, F.M.; Castro, F. Handmade Clay Bricks: Chemical, Physical and Mechanical Properties. *Int. J. Archit. Herit.* **2010**, *4*, 38–58. [CrossRef]
25. Ontiveros-Ortega, E.; Rodríguez-Gutiérrez, O.; Navarro, A. Mineralogical and physical-chemical characterization of Roman mortars used for monumental substrates on the hill of San Antonio, in the Roman city of Italica (prov. *Baetica*, Santiponce, Seville, Spain). *J. Archaeol. Sci. Rep.* **2016**, *7*, 205–223.
26. Fragata, A.; Veiga, R.; Velosa, A. Substitution ventilated render systems for historic masonry: Salt crystallization tests evaluation. *Const. Build. Mater.* **2016**, *102*, 592–600. [CrossRef]
27. Ergenç, D.; Fort, R. Multi-technical characterization of Roman mortars from *Complutum*, Spain. *Measurement* **2019**, *147*, 106876. [CrossRef]
28. Baronio, G.; Binda, L. Study of the pozzolanicity of some bricks and clays. *Const. Build. Mater.* **1997**, *11*, 41–46. [CrossRef]
29. Miriello, D.; Bloise, A.; Crisci, G.M.; Apollaro, C.; La Marca, A. Characterisation of archaeological mortars and plasters from kyme (Turkey). *J. Archaeol. Sci.* **2011**, *38*, 794–804. [CrossRef]

Article
Characterization of In Situ Concrete of Existing RC Constructions

Marco Vona

School of Engineering, University of Basilicata, 85100 Potenza, Italy; marco.vona@unibas.it

Abstract: The strengths and mechanical characteristics of concrete play a key role in the safety levels for the recovery and reuse of existing RC buildings and civil engineering works. This is one of the main focuses of the current research trend. To this aim, the characteristics of concrete must be investigated: the characterization of the concrete and its in situ conditions play a key role. For these reasons, many studies on in situ and laboratory test methods and procedures have been carried out over the last two decades. In the past few years, non-destructive investigation methods have been considered reliable and used in many engineering applications, also for RC constructions. More recent codes and guidelines identify destructive test methods as a reference for practice application. However, non-destructive investigation methods can be used though exclusively in combination with destructive tests to support them. In this study, a significant database is considered to assess the reliability of the relationship between destructive and non-destructive methods for in situ concrete in existing RC constructions. The results of the analyses are used to verify the effectiveness of the methods and prediction models and suggest more effective test procedures. It can be stated that many of the existing empirical methods (based on pre-established correlations) are unable to provide a reliable evaluation of the compressive concrete strength and its variability. In practical applications, non-destructive methods often lead to unsatisfactory results for the existing reinforced concrete constructions. Finally, based on the results, some first operational indications are provided for practical investigations and future possible codes and guideline improvements.

Keywords: existing RC constructions; compressive concrete strength; in situ tests; NDT–DT correlations; code provision

1. Introduction

One of the main focuses of the current research trend is the recovery and reuse of existing RC buildings and civil engineering works. They are publicly and privately owned and are often severely degraded. Consequently, they are often underused and in many cases unused. The topic is particularly important in Europe, in which post-war reconstruction activities (after 1945) led to the construction of a very high number of RC buildings and civil engineering works. Currently, these buildings are very old and require assessment and retrofitting interventions.

For example, in Italy, the RC construction heritage is among the oldest in Europe and requires substantial and extraordinary maintenance as well as functional, structural, and seismic retrofitting. The criticalities are amplified by the recent seismic classifications. As an example, in Figure 1a, a concise representation of the distribution of residential buildings and civil engineering works is reported.

However, the seismic performance assessment of existing buildings and civil engineering works is a very complex and difficult task. In recent years, the research and professional activities and the effects of earthquakes [1] have highlighted strong differences between new and existing RC structures, especially the older ones. In particular, in Italy, there is a significant number of existing RC buildings and civil engineering works. In the last twenty years, the Italian code [2] has classified almost the entire Italian territory as seismic, highlighting the need to assess the performance of existing buildings through new and specific verification methods. Moreover, in Italy, about half of the existing buildings are made of reinforced concrete. In most cases, they are designed only for gravity loads, without specific seismic guidelines [3]. Similarly, civil engineering works (infrastructures) have, in many cases, been designed without seismic and modern standards and with less effective actions than those currently provided by the codes. The existing RC buildings and civil engineering works are increasingly under assessment.

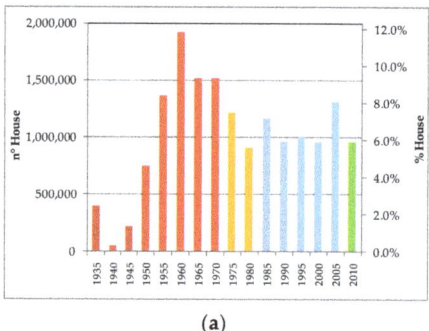 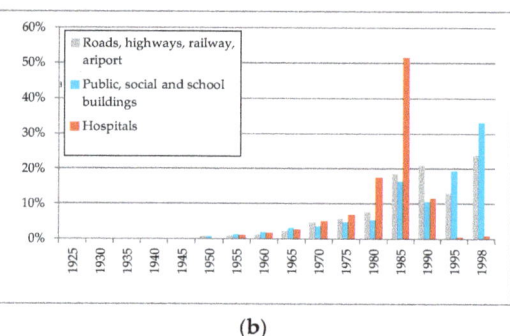

Figure 1. Residential buildings and main civil engineering works: (**a**) distribution according to age of the houses built; different colors refer to different Italian codes and/or new and significant seismic classification; (**b**) distribution of financial resources for engineering works that are totally or partially state-funded (source: www.istat.it (accessed on 24 January 2022)).

The increasing need for the use and rehabilitation of existing RC buildings and civil engineering works requires improving the assessment and investigation procedures, particularly for the execution and processing of the test results of the materials. The assessment and seismic retrofitting of the existing RC buildings has been dealt with in specific codes in many countries since the nineties [4] from which several guidelines [5], codes and recommendations [6], scientific publications [7] and consequent professional activities have derived to better explain the new trend in the codes. In Europe, the issue is considered in Eurocode 8, part 3, assessment and retrofitting of buildings [8]. The approaches of the European code [8] and those of the recent Italian code [2] are consistent with each other and were largely improved in the latest Italian code [9].

In the current Italian code, the part dedicated to the existing constructions is reported in chapter 8 which contains the general principles and the main issues of existing constructions; even though significant options and choices are left up to the professional's practice. More quantitative details are reported in the guidelines [10] which address the issues that support the professionals in acquiring the quantitative and qualitative information necessary for a broader knowledge of the existing characteristics of the building. The most significant differences that the recent codes have introduced for the assessment of existing constructions are in the: (i) geometry, (ii) construction details and (iii) materials. These parameters condition the assessment step and the subsequent intervention strategy. The third element (material characterization) plays a fundamental role, especially with regard to the concrete used in reinforced concrete structures. Strong differences and variability in the mechanical characteristics of materials, in particular with regard to concrete, were verified on many existing structures [11–13]. The concrete is often of a lower quality than expected. Under such conditions, the most widespread testing methodologies may not be effective due to several factors. First of all, the results of the tests depend on various factors such as specific weight, water/cement ratio, aggregate/cement ratio, type of aggregate, percentage of humidity, and presence of reinforcements. These factors are well known and extensively studied (for example, [14]). Moreover, other factors that are specific to old constructions can affect the results of the tests: the age of the concrete, the original construction procedures namely pouring, compacting, curing, quality controls (Figure 2), the period of service, the effects of the degradation induced by the environment and use, the effects of damage due to earthquakes and other actions and loads.

Figure 2. The old conventional concrete construction technology.

The latest Italian code [10] has introduced major innovations to define the level of knowledge on the basis of in situ and laboratory investigations. The amount of tests to be performed is linked to the level of knowledge by means of indications that are purely indicative. Therefore, the professional has a great responsibility for the design of the surveys, the choice and application of the investigation methods, and the elaboration of the results. The Italian code identifies tests (DTs) [15] to be the most effective method to directly estimate the compressive in situ strength of concrete. Other types of non-destructive tests (NDTs) [16] could be used, considering the relationships with the results of DTs [17,18]. However, non-destructive methods can be used only and exclusively in combination with DTs. Many design codes or guidelines provide indications for several procedures. Nevertheless, they are often not thorough due to the complexity of the problem. A typical problem for the existing RC constructions is where the concrete may be of poor quality and the operative applications of the non-destructive methods often lead to unsatisfactory results. Consequently, the practical efforts and required resources for NDTs do not seem to be consistent with the results obtained.

The purpose of this study is to identify the critical issues in current investigation procedures and possible improvements. Similar studies have been carried out on these topics in the past. However, no alternative and corrective methods useful for the applications were clearly identified.

In particular, in this study, the operational issues of DTs and NDTs for concrete in existing RC construction, as well as the reliable interpretation of the results, are investigated.

A contribution of the study is the database considered. The database was obtained from laboratory investigations and professional practice activities. Thanks to the database, the reliability of the relationship between destructive and non-destructive methods for in situ concrete in the existing RC constructions is considered and assessed.

As reported in the following sections, the investigation procedures for existing RC construction based on NDTs and the commonly used NDT–DT correlations highlighted significant criticalities, particularly in their ability to investigate and estimate the characteristics of the concrete. Such methods, which are often developed in laboratories and validated for new concrete, are often not suitable for investigating the existing RC constructions. Based on the reported experimental campaigns, some possible improvements are consequently defined from the obtained results.

2. Methods and Materials

To define the method followed in the study and before describing the materials, it is necessary to highlight the most recent and relevant literature, although a complete review is not among the objectives of the study. For the mechanical characterization of concrete (in short, the concrete strength), the codes and guidelines consider DTs and NDTs. For existing RC construction, NDTs [2,9–11,19,20] must always be correlated with DTs; they cannot be used alone. As a matter of fact, NDT methods are affected by many uncertainties [21–25]. Consequently, the relationship between DTs and NDTs is still the subject of many studies [26–36]. In the same way, the existing relationships between NDTs and concrete strength have been widely investigated [29,34–36] but should not be considered representative and simply applied for existing RC construction. They must be defined for the specific case under consideration.

In practical applications, one of the greatest difficulties is to identify the homogeneous areas of the structure in terms of concrete properties, and typically, NDT methods are used for this purpose [37]. However, each method (NDT) has variability that does not represent the mechanical characteristics of the concrete. Furthermore, the DT–NDT correlations have a high degree of uncertainty so the result is often unreliable. If this is the case, some NDT methods should also be excluded from surveys for the identification of concrete variability. Many interesting experiments were carried out [21–38] but the problem is still open, especially for the older RC structures. Uncertainties influence and can make DT–NDT correlations unusable.

The study of DT–NDT correlations (such as the SonReb method, SONic REBound) is still open and every possible contribution is still important. The SonReb method is undoubtedly the most common approach. SonReb procedures are based on rebound number (RN) and ultrasonic velocity (V). The approach was born in the 1970s and initially, its application was not based on direct correlation with destructive testing. In addition, with particular reference to the current application on existing structures, its use was significantly different. It was mainly dedicated to new concrete (for example, [39]). In the scientific literature, there are many studies about the SonReb method, with different empirical forms (i.e., based on linear, polynomial, power, exponential or logarithmic forms).

However, it must be emphasized that the most commonly used form in the last 30 years has been the exponential one where the relationship between the strength of the concrete and the NDTs is based on the form (1):

$$f_c = a \cdot RN^b \cdot V^c \qquad (1)$$

This form has been used since the 1970s but probably owes its greatest diffusion to Rilem [20]. The coefficients a, b and c are derived for each study with different experiments (see for example, [38]).

In this work, it was not considered necessary to report further details and indicate the various expressions and their comparisons. On the other hand, considering Italian data similar to what is contained herein, a SonReb reliability study [38] was recently carried out and an interesting review of the main forms and their comparison is reported.

More recently, some interesting applications have been made regarding a new emerging application area for artificial neural networks (ANNs) in civil engineering [40]. Some ANN-based techniques are used to study the strength of concrete by NDT [41]: these techniques are integrated with experimental results on cubic concrete specimens. The procedures have shown significant results and significant potential development by limiting negative effects of the natural dispersion of NDTs values [42].

It must be noted about the validation of the SonReb method that many forms were defined on laboratory-prepared samples, many even using cubic specimens [43–49]. In fact, much research can now be considered outdated.

To this aim, the time reference could be the European code [8] and therefore carried out well before the problem of the assessment of existing RC constructions became crucial and widespread. Only more recently, some empirical expressions have been defined considering cylindrical samples extracted from existing structures (core drilling) [50]. Several alternative procedures were proposed to obtain the relationship between in situ concrete strength and NDTs values, such as SonReb procedures, based on rebound number (RN) and ultrasonic velocity (V) for which specific coefficients are evaluated for investigated concrete [13,38,47,49].

Globally, the classic SonReb method seems inconsistent with the knowledge objectives of existing constructions. It is not consistent with the Italian code provision [10,11]. Furthermore, the overall costs associated with the investigation NDTs can still be considered high, compared with DTs for non-structural restoration elements, even more, when considered with the wide extension that is necessary for NDTs calibration, commonly considered as NDTs:DTs = 3:1.

For the above reasons, the evaluation of the mechanical characteristics of concrete (in particular, but not only, compressive strength) must be based mainly (or also exclusively) on DTs. NDTs must be considered only as a (non-essential) support for the evaluation of the homogeneity of the characteristics. Above all, the NDTs must never be used individually but only and always in conjunction with the DTs. In fact, NDTs are not satisfactory methods for the estimation of the mechanical characteristics and, in particular, the strength of concrete. These application restrictions are expressly provided for in the Italian code [10,11].

Consequently, the NDTs should not be used a priori for the rough determination of the characteristics of the concrete and neither for the a priori identification of homogeneous areas. Using them in this way could be misleading.

Nevertheless, DTs too present many uncertainties; they are linked to: (i) reliability and accuracy of the methods, (ii) characteristics and variability of the strength of the concrete in the structures, and (iii) execution of extraction tests. Conversely, some experimental programs on reinforced concrete elements showed the high within-member, within construction variability of in situ concrete strength [13,38,47,49] and the possible negative effects of core drilling on reinforced concrete columns [50], in particular for structural elements with low concrete strength. For the extracted concrete samples it must be taken into account that there are many differences between the resistance measured on the core samples and the actual insitu strength. Classically, the main considered factors [13] are the size and geometry of the cores, the coring direction, the presence of reinforcing bars or other inclusions, and the effect of drilling damage. However, age and design code of the construction, compressive stress, and management of the construction affect the results of the DTs but for these elements, there are still no reliable and univocal parameters and/or coefficients. The approach suggested in most codes based on the correlation of NDT and DT in situ results may not be reliable. Therefore, the implementation of the tests must be considered a very delicate issue.

In the practice application, design of investigation campaigns and results analyses play a key role in subsequent assessments. They are affected by the limited number of experimental observations and thus the obtained results (in terms of concrete characteristics) could not be statistically representative of the entire construction or its part. Consequently, many more studies (as this study) are needed for the validation of NDTs, and any new contribution could be useful.

Database

In the database, the samples are cylindrical samples extracted from existing RC structures by core drilling.

Moreover, some considerations regarding the usefulness of in situ NDTs are reported. The characteristics of the in situ tests are compared with the provisions of the codes in force at the time of the design and construction of the structures. It is to be noted that without this reference, the interpretation of the results of the tests would be ill-conditioned. In order to not make the study too long, further analyses of the database will be conducted in subsequent studies. As an example, the variability in the individual structures due to position, type of element, number of samples extracted, size of the structure, etc. has not been analyzed in depth. Similarly, no issues relating to the construction methods, management and degradation of the structures were developed, which also have a significant influence on the results of the tests.

The database is extremely interesting. It is derived from experimental and professional activities that were conducted or coordinated at the University of Basilicata and it considers activities carried out by other test centers and by professionals. This heterogeneous origin can be considered a valuable characteristic of the database which was created aftermany years of activities and research. The latest data were obtained in 2021. The database relates to hundreds of RC constructions (buildings and civil engineering works) located in the south-central Italian territory, over 7 regions. The RC buildings and civil engineering works were designed and used for public (infrastructures, schools, hospitals, barracks, offices, etc.) and private (mainly residential) use and were built from the 1950s. The database wascollected during several years of experimental programs and during several programs of seismic vulnerability assessments of public and private RC constructions. In these activities, the codes and guidelines for DTs and NDTs were used. Columns, walls and beams are the structural elements tested; their choice and the location of the tests were defined by the engineers, based on the construction characteristics considered.

The database contains DT and NDT results. In consideration of the main objective of this study, only the NDT cases were excluded, thus the following subsets were identified.

1. Subset DB_1: 2010 samples with compression tests ($f_{c,core}$) only (this is the main database).
2. Subset DB_2: 1039 samples with Specific Weight (SW).
3. Subset DB_3: 1175 samples with $f_{c,core}$, rebound hammer test (RN), and ultrasonic velocity test (V).
4. Subset DB_4: 202 samples with compression tests and ultrasonic velocity tests, $f_{c,core}$ and V.

The difference between DB_1 and DB_3 reflects the operational difficulty of carrying out NDTs in a widespread and reliable way on existing constructions. For example, in civil engineering works the execution of reliable NDTs (in particular ultrasound) is often prevented by the construction type, shape, size, and current state of the works under investigation, for example, see the work in Figure 3 where direct velocity values are not investigable (Figure 3a) and rebound hammer test is not possible or too expensive (Figure 3b). In ordinary buildings, the execution and diffusion of NDTs areoften hindered by functional issues, for example, the presence of non-structural elements and the state of conservation and degradation of buildings. Furthermore, in these cases, the overall cost of the single NDTs (including the costs of restoring the non-structural elements, Figure 4) is comparable with DTs.

Lastly, a limited subset of samples (DB_4.1) was considered based on the results of some laboratory and in situ investigations. DB_4.1 reports the data relating to the compression tests and direct velocity values, both performed in situ (on the structural elements) and in the laboratory (on the samples, before the compression tests). The database is constantly improved and updated and possesses the characteristics shown in the following figures and tables.

The database can be considered reliable in terms of credibility and adequacy. Major errors were removed (outliers and the like). Primarily, the quality of the database can be considered consistent with the main objectives of the study. The database is based on measurements that were conducted in certified laboratories considering the same standard code.

Figure 3. Investigation of the civil engineering works of a dam (**a**) and the current state of its concrete (**b**).

Figure 4. NDTs: preparation for Hammer Test and Ultrasonic velocity Test.

3. Results

Below, each database is investigated individually. Following the main objective of the study, the main results are reported but thanks to the significant value of the database, other studies and applications will be possible in the future beyond the objectives of this work.

3.1. Subset DB_1: Compression Test Analyses

Subset DB_1 considers all the results of the compression tests. It contains the data of the buildings and civil engineering works: 2010 samples in total. The results of the test show a significant variability of the compressive strength ($f_{c,core}$). The first analysis shows the distribution in compressive strength classes of different design and construction periods. The analysis of the $f_{c,core}$ values indicates the variability given by the period under consideration. The mean values show an anomaly in the age group <1961 (Figure 5a) as a result of the presence of a significant group of data obtained from civil engineering works. For this type of construction, the design value of the concrete resistance was decidedly higher and there was also greater care in the construction phases.

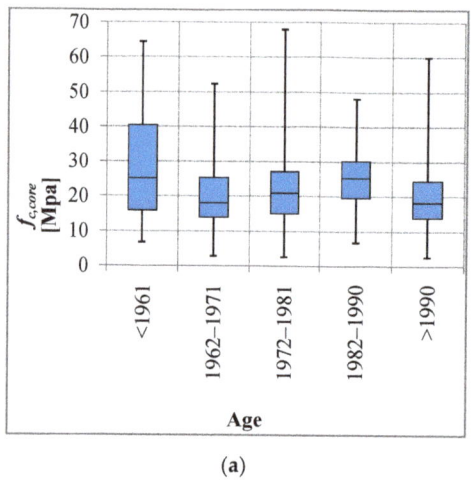
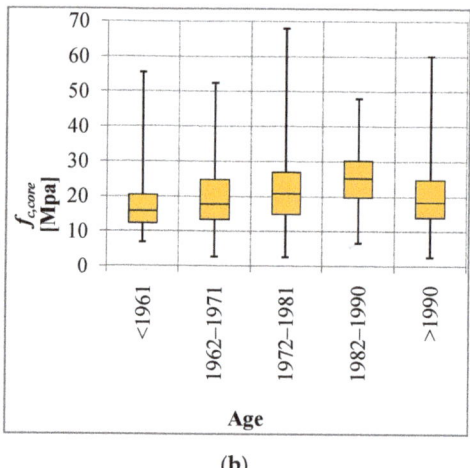

(a)   (b)

Figure 5. $f_{c,core}$—Age comparison with (**a**) or without (**b**) civil engineering works. Each box encloses 50% of the data with the median value of the maximum $f_{c,core}$ displayed as a thin line, within the box; the top and the bottom of the box, respectively, mark the 25% and 75% limits of the population; outside the box, the whiskers represent the maximum and minimum values of the population, respectively.

Historical analyses of the code provision support the investigation. Based on the historical analysis of the original design approach, it is possible to avoid incorrect interpretations of the strength data and structural capacities. The RC construction works before 1972 were designed and built in accordance with the original and earliest Italian code in force at the time: Royal Decree no. 2229 of 1939. In this code, the design strength for concrete was the average compressive strength, ranging from 120 to 160 Kg/cm^2. For structural elements subject only to axial load, the design strength was 180 Kg/cm^2. These values were commonly considered for buildings. For civil engineering works, the typical design strength was up to 225 Kg/cm^2. The above provision explained the high value of Figure 5a and then the value of the expected concrete compressive strength following the code in force at the time. In Figure 5b, the experimental data show a compressive strength consistent with the period. Since 1972, the design resistance was the characteristic strength of concrete R_{ck} which was not less than 150 Kg/cm^2; in design practice, the most common values were not lower than 250 Kg/cm^2. Thus the expected average values resulting from the tests should be higher than those of the previous period. Five periods were considered and summarized in Table 1, based on the design code and available data.

Table 1. DB_1: composition and main statistical values.

		$f_{c,core}$			
Construction Age	n° Samples	Mean Value [MPa]	Deviation Standard [MPa]	Min [MPa]	Max [MPa]
<1961	256	28.24	14.73	6.77	64.27
1962–1971	416	19.77	9.00	2.76	52.21
1972–1981	507	21.94	11.01	2.55	67.87
1982–1990	64	25.97	9.59	6.72	47.96
>1990	767	20.27	9.12	2.48	59.90
Total	2010				

3.2. Subset DB_2: Compression Test and Specific Weight

The analysis of the concrete specific weight (SW) is possible for a subset of 1039 samples. The specific weight depends on the water/cement ratio, aggregate/cement ratio, type of aggregate, original manufacturing procedures, and age. It can be considered to be representative of the global concrete quality: high porosity is generally associated with a lower strength of the structural elements. Therefore, SW could provide indications on performance and durability (which can affect the retrofitting strategies).In particular, the analyses look for the correlation between SW and compressive strength. Moreover, high porosity can affect the result of the compression test, as shown in Figure 6. Based on available data, four periods, summarized in Table 2, were considered, always based on the design code and available data.

(a) (b)

Figure 6. High porosity concrete in extracted samples: (**a**) Column, (**b**) Beam.

Table 2. DB_2: composition and main statistical values.

Construction Age	n° Samples	Percentage of Samples	Specific Weight Mean Value [kg/m³]	Deviation Standard [kg/m³]	Min [kg/m³]	Max [kg/m³]
<1962	35	3.37%	2277	34	2197	2345
1962–1971	248	23.87%	2260	79	2056	2444
1972–1981	265	25.51%	2194	110	1854	2455
>1981	491	47.26%	2257	112	1860	2619
	1039					

Furthermore, the SW—$f_{c,core}$ relationship is analyzed. Some SW values seem to be of critical value for compressive strength and, in some cases (on single buildings), it is possible to identify a clear correlation between SW and compressive strength. However, contrary to expectations, there is no clear and generalizable relationship between SW and compressive strength, as is reported in Figure 7 for different age of concrete based on the design code and available data.

3.3. Subset DB_3: Compression Test, Rebound Hammer Test, Ultrasonic Velocity Test

The DB_3 subset considers all cases (1175 samples, 80 buildings) in which $f_{c,core}$, RN (rebound index), V (direct ultrasonic velocity test) are simultaneously available. Cases where ultrasonic velocity was evaluated through surface measurements were excluded. To identify a potential relationship between DTs and NDTs, the analysis of the results of this subset plays a key role. However, the analyses generally show a poor or no correlation between the compressive strength and the considered NDTs. This result is particularly clear for the hammer tests (rebound number). The variability of the in situ compressive strength can be much greater than that shown by the NDTs (Figure 8).

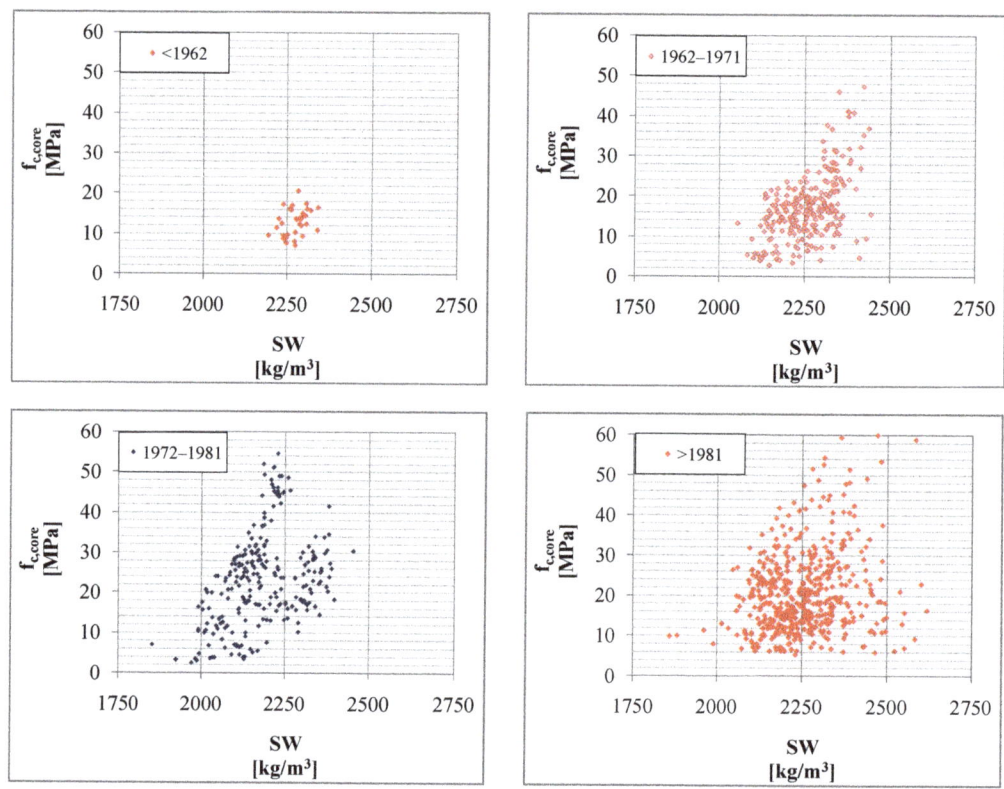

Figure 7. Correlation between Specific Weight (SW) and Compression Resistance of the samples, for different age of concrete based on the design code and available data.

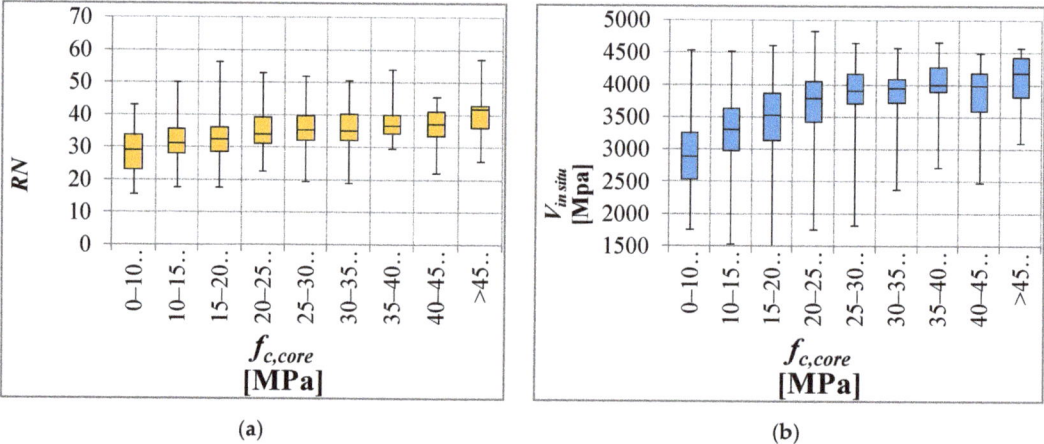

Figure 8. Box plot of NDT variability for several compressive strength classes: rebound number (**a**) and ultrasonic velocity tests (**b**).

Moreover, the relationship between NDT variability and DT variability for each construction was analyzed. A total of 80 buildings were analyzed and the results are shown in Figure 9. The variability of the DTs is not described by a similar variability of the NDTs, especially for the rebound number. In particular, the results of the rebound test show a poor correlation with concrete strength. Moreover, independently from the concrete strength, for the rebound test, the dispersion of the data is very high. The trend growth is significantly lower or negligible. NDTs show a limited variability which is intrinsic to the method and does not represent quality but an inability to represent the true characteristics of the concrete. In many cases, the coefficient of variation (CV) of the DTs is about 30% higher. However, this data is not indicative of problems within the data or that the investigation is not controlled but simply describes different concrete homogeneous areas. Therefore, in actual buildings, it seems unreliable to identify a relationship between NDTs and DTs. Looking at these results, identifying one or more concrete homogeneous areas based on the NDTs also appears to be difficult and unreliable.

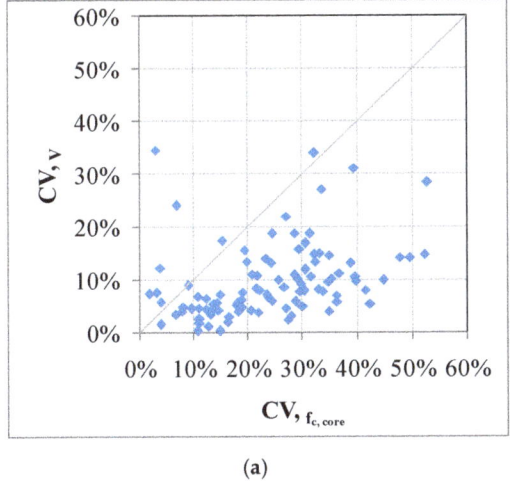
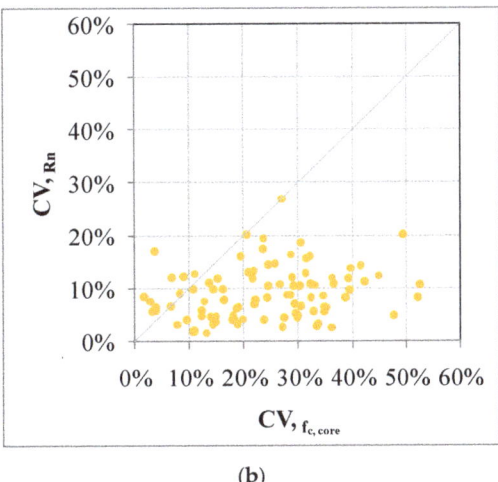

(a)   (b)

Figure 9. Relationship between DTs–NDTs within each building: rebound number (**a**) and ultrasonic velocity (**b**) tests.

For NDTs, considering the methods described in the previous paragraphs (in particular the combined SonReb method), it is believed that it is more reliable to use only the results of the ultrasonic tests to build a calibrated relationship between DTs and NDTs. This result is consistent with other recent studies (for example [51]) which, however, are based only on an experimental laboratory program (samples made in the laboratory). Derived from the exponential form SonReb, the calibrated expression could therefore be:

$$f_c = a \cdot V^b \quad (2)$$

where a and b are defined based on the DT and NDT results. In this way, it is also possible to consider a lower number of in situ tests, increasing the reliability of the tests and thereby reducing costs. The method can be effectively applied with the correction reported in the next paragraph.

3.4. Subset DB_4: Compression Test and Ultrasonic Velocity Test

Subset DB_4 reports the ultrasonic velocity as NDTs. DB_4 was defined considering only a subset of samples with a direct velocity measurement both on the structural elements (in situ) and on the single extracted samples (in the same points) before the compression test (Figure 10). In total, 45 data triples ($V_{c,core}$, $V_{c,insitu}$, $f_{c,core}$) were obtained from an experiment conducted in the laboratory of the University of Basilicata [48] on fourbeams extracted from an existing and demolished RC school building. In addition, 202 homogeneous triples were also considered, divided into five homogeneous groups relating to 18 school buildings subjected to seismic assessment. The grouping considered in the following was defined by considering that the buildings belong to school complexes as well as their homogeneity (age and design code); the structural elements considered are mainly columns. Another 202 data triples ($V_{c,core}$, $V_{c,insitu}$, $f_{c,core}$) were obtained. All these values are subsequently compared and Figure 11 reports the result of the first comparison.

Figure 10. Direct measurement of ultrasonic velocity on single concrete samples.

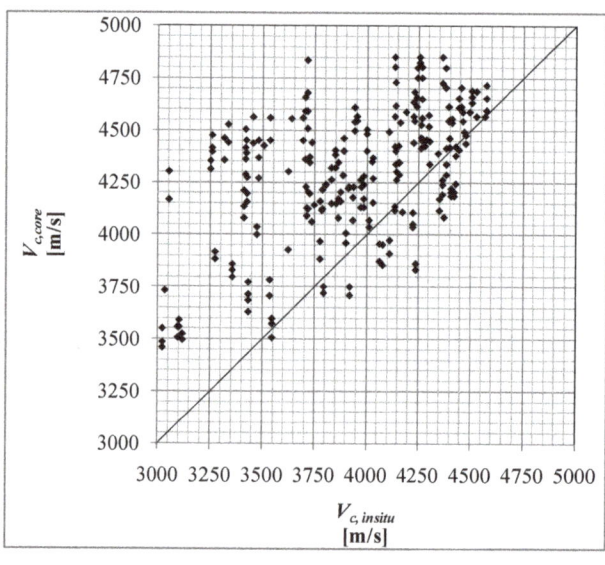

Figure 11. In situ direct ultrasonic velocity ($V_{c,insitu}$) test vs. ultrasonic velocity test samples ($V_{c,core}$).

As expected, in Figure 11, the in situ ultrasonic velocities significantly underestimate the "true velocity" values obtained from the samples. This difference could be influenced by the surface degradation of the in situ concrete.

In Figure 12, the results of the ultrasonic velocity NDTs are compared with the result of the compression tests. In Figure 12, 45 data triples of the University of Basilicata experiment [48] are reported, considering the relationships between $V_{c,core}$-$f_{c,core}$ and $V_{c,insitu}$-$f_{c,core}$ in Figure 12a whereas in Figure 12b; the best fit of the data is reported.

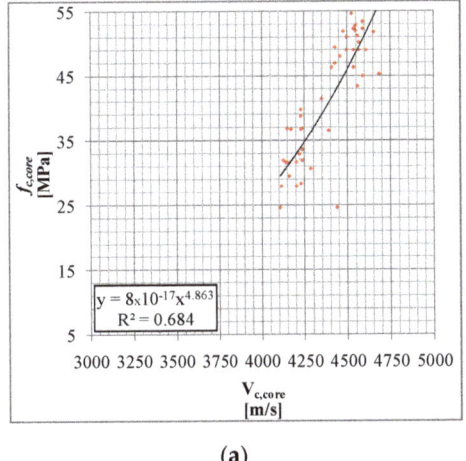 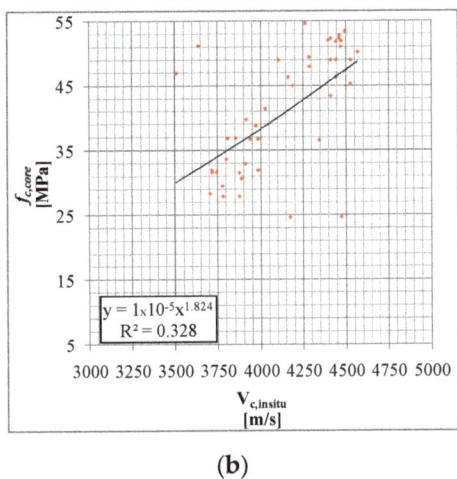

(a) (b)

Figure 12. Comparison of direct ultrasonic velocity samples vs. compressive concrete strength (**a**) and in situ ultrasonic velocity vs. compressive concrete strength (**b**).

In Figure 13a, for those buildings under investigation, the relationships between $V_{c,core}$-$f_{c,core}$ are reported; whereas in Figure 13b, for those buildings under investigation, the relationships $V_{c,insitu}$-$f_{c,core}$ are reported; the best fit of the data is reported. In the first case, the internal variability of the ultrasonic method is very limited.

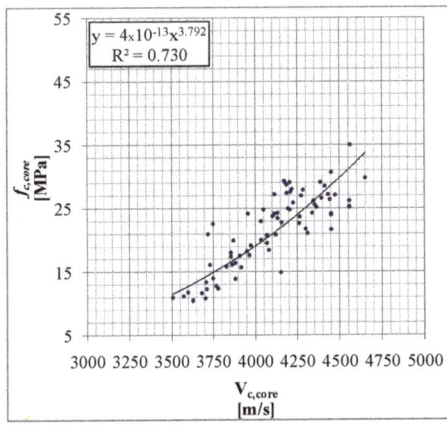 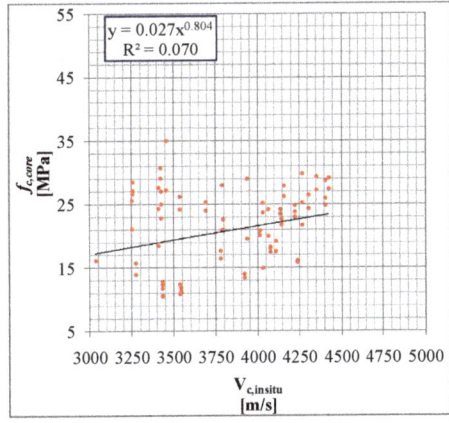

Figure 13. *Cont.*

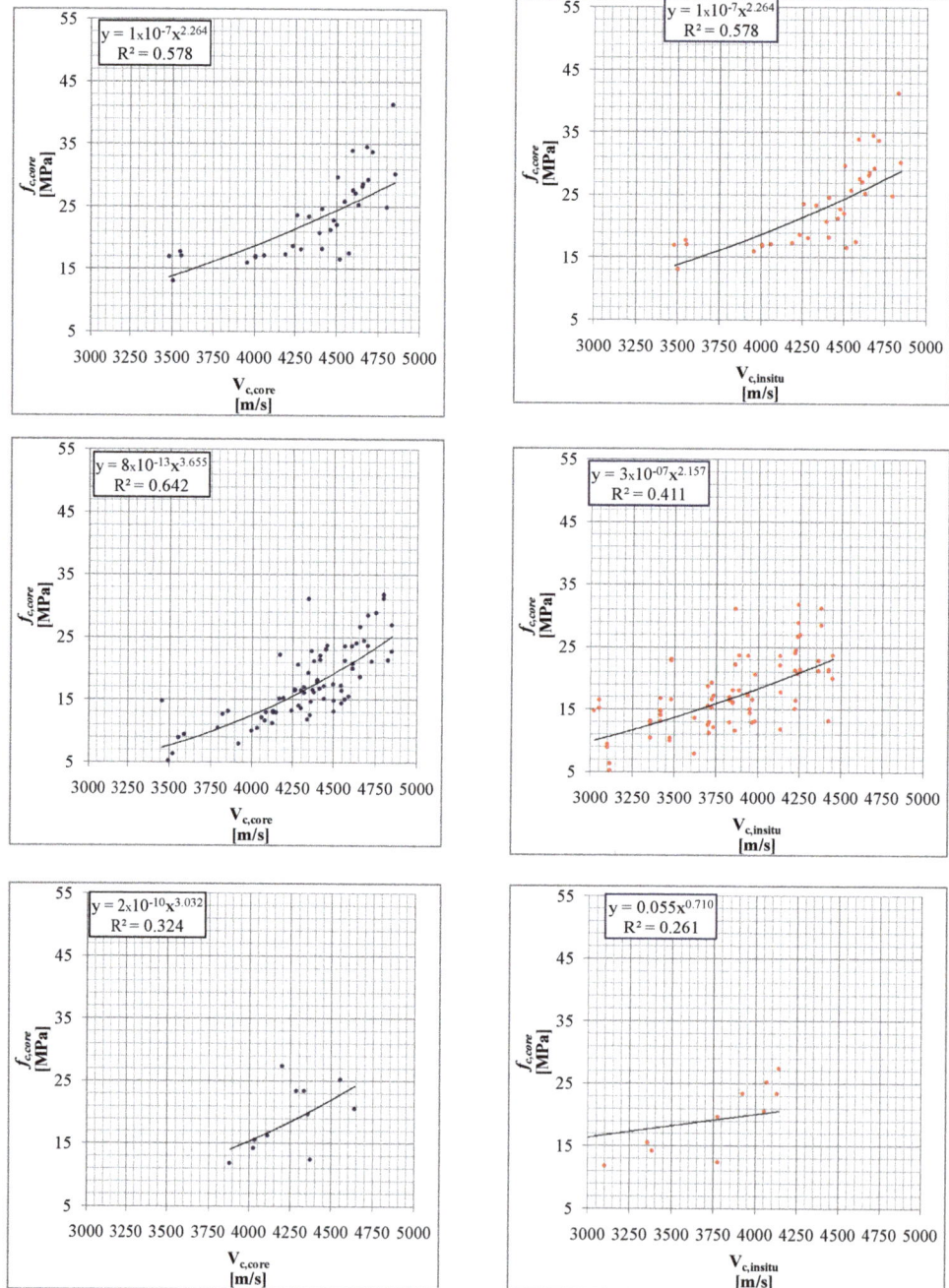

Figure 13. Cont.

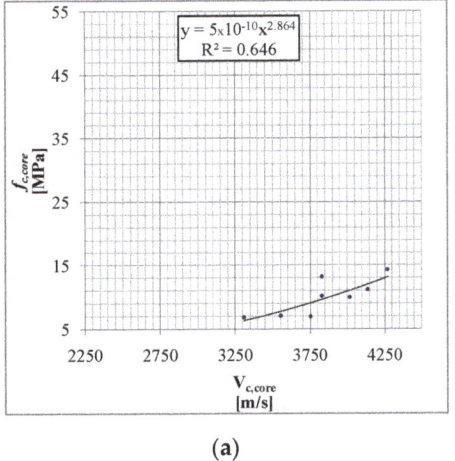 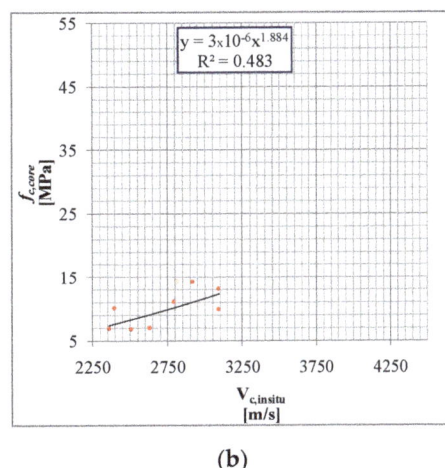

Figure 13. Ultrasonic velocity samples vs. compressive concrete strength (**a**) and in situ ultrasonic velocity vs. compressive concrete strength (**b**). The comparisons are related to the various samples taken in situ at different constructions and sites.

It is heavily influenced by the material and by the context [48] and it seems obvious that for in situ tests the measurement errors are systematic, and difficult to either estimate or eliminate. On the other hand, even in cases of high variability in $f_{c,core}$, $V_{c,core}$ seems able to correctly estimate the $f_{c,core}$ values and follow its evolution. The effects of age and compressive stress, also referred to in several studies, seem less evident in the $V_{c,core}$-$f_{c,core}$ relationship whereas for the $V_{c,insitu}$-$f_{c,core}$ relationship, age also has a clear influence on the deterioration of the concrete and subsequent reduction of the reliability of the correlation.

The above-illustrated results show the ability of the ultrasonic tests to evaluate the concrete strength if the ultrasonic velocity measurement is carried out directly on the sample. Therefore, in situ ultrasonic tests could be used to estimate the compressive concrete strength by applying the expression reported in the previous paragraph, in which, however, the in situ ultrasonic velocity values are corrected with those measured on the extracted sample by adopting, for example, an average correction value given by the ratio between $V_{c,core}$ and $V_{c,insitu}$. This proposal is also consistent with other studies (e.g., [12,15]).

3.5. Discussion

In the literature and previous studies, methods based on the NDT–DT relationship are considered reliable and are considered for applications also on existing RC constructions. As a matter of fact, as reported in the previous sections, in the last few years, for existing RC construction the investigation procedures based on NDTs and the commonly used NDT–DT correlations have shown significant critical issues, particularly in their ability to investigate and estimate the characteristics of the concrete. Such methods, which are often developed in laboratories and validated for new concrete, are often not suitable for investigating the existing RC constructions.

On the contrary, this study shows a significant degree of uncertainty, operational difficulties and potential errors in interpreting the results; they are based on experimental experiences on existing RC constructions. Based on the reported experimental campaigns, the critical issues in current NDT–DT relationships are clearly shown and some possible improvements are consequently defined from the obtained results (Sections 3.3. and 3). For this reason, the data shown below can be considered an excellent reference to estimate the reliability of the strength values of concrete, based on past experience or the existence of similar structures. The availability of new additional data and the criticalities in practical professional applications allow for updating the procedures and evaluating their reliability.

4. Conclusions

In the last few years, the characteristics of concrete have been investigated using empirical expressions in order to evaluate the in situ compressive strength using NDTs, namely the rebound hammer test and ultrasonic pulse velocity test. In other studies, the variability of the most common NDT (rebound hammer test and ultrasonic pulse velocity test) was analyzed [12,13,52,53]. Generally, the variability within these methods is very small.

The present study reported data and procedures, based on a database of real data. A new contribution from the study is really the database. The database was obtained from laboratory investigations and professional practice activities. On the contrary, several interesting and recent studies are developed in laboratories and validated only for new concrete where the results do not take into account the quality of the concrete in existing reinforced concrete constructions and its low strength and degradation problems.

However, based on the results of this study, it can be stated that many of the existing empirical methods (based on pre-established correlations) are unable to provide a reliable evaluation of the compressive concrete strength and its variability.

In particular, the results obtained show variability in compressive concrete strength but highlighted the non-variability of the Rebound Number (RN) and its non-correlation with concrete compressive strength. Ultrasonic velocity shows an extremely limited variability although it can be increased by relating the in situ ultrasonic velocity and the ultrasonic velocity samples.

In conclusion, based on this work, it can be highlighted that:

1. The rebound hammer test is not representative of the compressive concrete strength and is also misleading. It must be excluded. The classic SonReb method should not be used;
2. Ultrasonic velocity tests can be used only if suitably calibrated with Ultrasonic velocity tests on the extracted concrete before the compression tests;
3. Using current methods and procedures, NDTs cannot be used a priori to identify homogeneous areas.

Lastly, this study contains the experimental data useful as a reference both for practical professional activities and for the comparison and validation results used in research activities [54].

This study allows for further developments due to:

- An increase in the amount of experimental data available;
- The improvement of data collection procedures and standardization of analysis procedures;
- The updated indication in the codes and guidelines.

In the next step of the study, new procedures and methods will be developed based on the reported results, in particular, following the results of Section 3.4; the new research trend based on ANNs can also be applied using the real database, suitably treated and analyzed. The strengths and mechanical characteristics of concrete play a key role in the safety levels of construction. These issues were often underestimated but represent one of the most important issues to define optimal retrofitting strategies [55].

Funding: This research was partially funded by the 2020 MIUR PON R&I 2014-2020 Program (project MITIGO, ARS01_00964).

Institutional Review Board Statement: Not applicable.

Informed Consent Statement: Not applicable.

Data Availability Statement: Not applicable.

Conflicts of Interest: The author declares no conflict of interest.

References

1. Dolce, M.; Di Bucci, D. Comparing recent Italian earthquakes. *Bull. Earthq. Eng.* **2015**, *15*, 497–533. [CrossRef]
2. OPCM 3274, Ordinanza del Presidente del Consiglio dei Ministri nr. 3274 del 20 Marzo 2003 "Primi Elementi in Materia di Criteri Generali per la Classificazione Sismica del Territorio Nazionale e di Normative Tecniche per le Costruzioni in Zona Sismica"; Presidente del Consiglio dei Ministri: Roma, Italy, 2003. (In Italian)
3. Masi, A.; Vona, M. Vulnerability assessment of gravity-load designed RC buildings, evaluation of seismic capacity through non linear dynamic analyses. *Eng. Struct.* **2012**, *45*, 257–269. [CrossRef]
4. Applied Technology Council. *ATC-40—Seismic Evaluation and Retrofit of Concrete Buildings, Was Developed under a Contract from the California Seismic Safety Commission*; Report No. SSC 96-01; Applied Technology Council: Redwood City, CA, USA, 1996.
5. Federal Emergency Management Agency (FEMA). *FEMA 356—Prestandard and Commentary for the Seismic Rehabilitation of Buildings*; Federal Emergency Management Agency (FEMA): Washington, DC, USA, 2000.
6. New Zealand Society for Earthquake Engineering (NZSEE). *Assessment and Improvement of the Structural Performance of Buildings in Earthquakes, Recommendations of a NZSEE Study Group on Earthquake Risk Buildings*; New Zealand Society for Earthquake Engineering (NZSEE): Wellingto, New Zealand, 2006.
7. Manfredi, G.; Masi, A.; Pinho, R.; Verderame, G.; Vona, M. *Valutazione di Edifici Esistenti in C.A.*; IUSS Press: Pavia, Italy, 2007.
8. *European Standard EN 1998-3-2005*; Eurocode 8: Design of Structures for Earthquake Resistance—Part 3: Assessment and Retrofitting of Buildings. Comité Européen de Normalisation (CEN): Brussels, Belgium; European Committee for Standardization: Brussels, Belgium, 2005.
9. NTC 2008, Decreto del Ministro delle Infrastrutture. 2008 Norme Tecniche per le Costruzioni, Suppl. Or. G.U., n.29 del 04 febbraio 2008—Serie Generale. (In Italian)
10. NTC 2018, Decreto del Ministro delle Infrastrutture e dei Trasporti del 17 Gennaio 2018, Nuove Norme Tecniche per le Costruzioni. Suppl. Or. G.U., n. 42 del 20 Febbraio 2018—Serie Generale. (In Italian)
11. Ministero delle Infrastrutture e dei Trasporti. *Circolare 21 Gennaio 2019, n. 7 C.S.LL.PP. Istruzioni per L'applicazione dell'«Aggiornamento delle "Norme Tecniche per le Costruzioni"» di cui al Decreto Ministeriale 17 Gennaio 2018. Circolare n. 7 C.S.LL.PP. Suppl. Or. G.U., n. 35 del 11 Febbraio 2019—Serie Generale*; Ministero delle Infrastrutture e dei Trasporti: Roma, Italy, 2019. (In Italian)
12. Vona, M. A review of experimental results about in situ concrete strength. International Conference on Materials for Renewable Energy and Environment, MREE 2013, 15–16 May 2013. *Adv. Mater. Res.* **2013**, *773*, 278–282. [CrossRef]
13. Masi, A.; Vona, M. *La Stima della Resistenza del Calcestruzzo In-Situ: Impostazione delle Indagini ed Elaborazione dei Risultati, Progettazione Sismica*; No. 1/2009; IUSS Press: Pavia, Italy, 2009, ISSN 1973-7432. (In Italian)
14. Masi, A.; Vona, M. Prove distruttive e non distruttive su materiali ed elementi strutturali di edifici esistenti in cemento armato. In Proceedings of the Conferenza Nazionale sulle NDT, Monitoraggio e Diagnostica, 12° Congresso Nazionale dell'AiNDT, Biennale NDT-MD, Milano, Italy, 11–13 October 2007. (In Italian).
15. Sabbağ, N.; Uyanık, O. Prediction of reinforced concrete strength by ultrasonic velocities. *J. Appl. Geophys.* **2017**, *141*, 13–23. [CrossRef]
16. *UNI EN 12504-2*; Testing Concrete in Structures—Non-Destructive Testing—Determination of Rebound Number. UNI Ente Italiano di Normazione: Milano, Italy, 2001. (In Italian)
17. *UNI EN 12504-1*; Testing Concrete in Structures—Cored Specimens—Taking, Examining and Testing in Compression. UNI Ente Italiano di Normazione: Milano, Italy, 2002. (In Italian)
18. *UNI EN 12504-4*; Testing concrete—Part 4: Determination of ultrasonic pulse velocity. UNI Ente Italiano di Normazione: Milano, Italy, 2005.
19. Malhotra, V.M. Contact strength requirements cores versus in-situ evaluation. *ACI J. Proc.* **1977**, *74*, 163–172.
20. *RILEM NDT 4*; Recommendations for In Situ Concrete Strength Determination by Combined Non-Destructive Methods, Compendium of RILEM Technical Recommendations. E&FN Spon: London, UK, 1993.
21. Breysse, D. Nondestructive evaluation of concrete strength: An historical review and a new perspective by combining NDT methods, *Constr. Build. Mater.* **2012**, *33*, 139–163. [CrossRef]
22. De Stefano, M.; Tanganelli, M.; Viti, S. Variability in concrete mechanical properties as a source of in-plan irregularity for existing RC framed structures. *Eng. Struct.* **2014**, *59*, 161–172. [CrossRef]
23. Pereira, N.; Romão, X. Assessing concrete strength variability in existing structures based on the results of NDTs. *Constr. Build. Mater.* **2018**, *173*, 786–800. [CrossRef]
24. Reysse, D.; Balayssac, J.P.; Biondi, S.; Borosnyói, A.; Candigliota, E.; Chiauzzi, L.; Garnier, V.; Grantham, M.; Gunes, O.; Luprano, V.; et al. Non destructive assessment of in situ concrete strength: Comparison of approaches through an international benchmark. *Mater. Structutres* **2017**, *50*, 133. [CrossRef]
25. Breccolotti, M.; Bonfigli, M.F.; Materazzi, A.L. SonReb concrete assessment for spatially correlated NDT data. *Constr. Build. Mater.* **2015**, *192*, 391–402. [CrossRef]
26. Verderame, G.M.; Manfredi, G.; Frunzio, G. Le proprietà meccaniche dei calcestruzzi impiegati nelle strutture in cemento armato realizzate negli anni '60. In Proceedings of the X Convegno Nazionale "L'Ingegneria Sismica in Italia", Potenza-Matera, Italy, 9–13 September 2001. (In Italian).
27. Pucinotti, R. Assessment of in situ characteristic concrete strength. *Constr. Build. Mater.* **2013**, *44*, 63–73. [CrossRef]

28. Nguyen, N.T.; Sbartai, Z.M.; Lataste, J.F.; Breysse, D.; Bos, F. Assessing the spatial variability of concrete structures using NDT techniques-laboratory tests and case study. *Constr. Build. Mater.* **2013**, *49*, 240–250. [CrossRef]
29. Breysse, D.; Martinez-Fernandez, J. Assessing concrete strength with rebound hammer: Review of key issues and ideas for more reliable conclusions. *Mater. Struct.* **2013**, *47*, 1589–1604. [CrossRef]
30. Pucinotti, R. Reinforced concrete structure. Non destructive in situ strength assessment of concrete. *Constr. Build. Mater.* **2015**, *75*, 331–341. [CrossRef]
31. Samarin, A. Combined methods. In *Handbook on Nondestructive Testing of Concrete*, 2nd ed.; Malhotra, V.M., Carino, N.J., Eds.; CRC Press: Boca Raton, FL, USA, 2004.
32. Quagliarini, E.; Clementi, F.; Maracchini, G.; Monni, F. Experimental assessment of concrete compressive strength in old existing RC buildings: A possible way to reduce the dispersion of DT results. *J. Build. Eng.* **2016**, *8*, 162–171. [CrossRef]
33. Alwash, M.; Sbartaï, Z.M.; Breysse, D. Non-destructive assessment of both mean strength and variability of concrete: A new bi-objective approach. *Constr. Build. Mater* **2016**, *113*, 880–889. [CrossRef]
34. Ali-Benyahia, K.; Sbartaï, Z.M.; Breysse, D.; Kenai, S.; Ghrici, M. Analysis of the single and combined non-destructive test approaches for on-site concrete strength assessment: General statements based on a real case-study. *Case Stud. Constr. Mater.* **2017**, *6*, 109–119. [CrossRef]
35. Alwash, M.; Breysse, D.; Sbartaï, Z.M. Non-destructive strength evaluation of concrete: Analysis of some key factors using synthetic simulations. *Constr. Build. Mater.* **2015**, *99*, 235–245. [CrossRef]
36. Alwash, M.; Breysse, D.; Sbartaı, Z.M. Using Monte-Carlo simulations to evaluate the efficiency of different strategies for nondestructive assessment of concrete strength. *Mater. Struct.* **2017**, *50*, 90. [CrossRef]
37. Masi, A.; Chiauzzi, L.; Manfredi, V. Criteria for identifying concrete homogeneous areas for the estimation of in-situ strength in RC buildings. *Constr. Build. Mater.* **2016**, *121*, 576–587. [CrossRef]
38. Cristofaro, M.T.; Viti, S.; Tanganelli, M. New predictive models to evaluate concrete compressive strength using the Sonreb method. *J. Build. Eng.* **2020**, *27*, 100962. [CrossRef]
39. Ferreira, R.M.; Jalali, S. NDT measurements for the prediction of 28-day compressive strength. *NDTE Int.* **2010**, *43*, 55–61. [CrossRef]
40. Khashman, A.; Akpinar, P. Non-Destructive Prediction of Concrete Compressive Strength Using Neural Networks. *Procedia Comput. Sci.* **2017**, *108*, 2358–2362. [CrossRef]
41. Asteris, P.G.; Mokos, V.G. Concrete compressive strength using artificial neural networks. *Neural Comput. Appl.* **2020**, *32*, 11807–11826. [CrossRef]
42. Asteris, P.G.; Skentou, A.D.; Bardhan, A.; Samui, P.; Lourenço, P.B. Soft computing techniques for the prediction of concrete compressive strength using Non-Destructive tests. *Constr. Build. Mater.* **2021**, *303*, 124450. [CrossRef]
43. Bellander, U. NDT testing methods for estimating compressive strength in finished structures—Evaluation of accuracy and testing system. In *RILEM Symposium Proceedings on Quality Control of Concrete Structures*; Swedish Concrete Research Institute: Stockholm, Sweden, 1979; Volume 1, Session 2.1; pp. 37–45.
44. Bocca, P.; Cianfrone, F. Le Prove Non Distruttive Sulle Costruzioni: Una Metodologia combinata. *L'Industria Ital. Cem.* **1983**, *6*, 429–436. (In Italian)
45. Kheder, G.F. A two stage procedure for assessment of in-situ concrete strength using combined non-destructive testing. *Mater. Struct.* **1999**, *32*, 410–417. [CrossRef]
46. Caiaro, R.; De Paola, S.; Porco, G. Indagini non distruttive per il controllo dei calcestruzzi di media ed alta resistenza. In *Atti del 10° Congresso Nazionale dell'AIPnD*; AIPnD: Ravenna, Italy, 2003; pp. 360–371. (In Italian)
47. Faella, G.; Guadagnuolo, M.; Donadio, A.; Ferri, L. Calibrazione sperimentale del metodo Sonreb per costruzioni della Provincia di Caserta degli anni '60 '80. In Proceedings of the Atti del XIV Convegno Nazionale L'Ingegneria Sismica in Italia, Potenza, Italy, 18–22 September 2011. (In Italian).
48. Masi, A.; Chiauzzi, L. An experimental study on the within-member variability of in situ concrete strength in RC building structures. *Constr. Build. Mater.* **2013**, *47*, 951–961. [CrossRef]
49. Masi, A.; Dolce, M.; Vona, M.; Nigro, D.; Pace, G.; Ferrini, M. Indagini sperimentali su elementi strutturali estratti da una scuola esistente in c.a. In Proceedings of the Atti del XII Convegno Nazionale L'Ingegneria Sismica in Italia, Pisa, Italy, 10–14 June 2007.
50. Vona, M.; Nigro, D. Evaluation of the predictive ability of the in situ concrete strength through core drilling and its effects on the capacity of the RC columns. *Mater. Struct./Mater. Constr.* **2015**, *48*, 1043–1059. [CrossRef]
51. Poorarbabi, A.; Ghasemi, M.; Moghaddam, M.A. Concrete compressive strength prediction using non-destructive tests through response surface methodology. *Ain Shams Eng. J.* **2020**, *11*, 939–949. [CrossRef]
52. Pessiki, S.P. *In-Place Methods to Estimate Concrete Strengths*; ACI 228.1R-03 Report; American Concrete Institute: Indianapolis, IN, USA, 2003.
53. Brozovsky, J. Evaluation of calculation correlation efficiency as mentioned in EN13791 in order to determination concrete compression strength by non destructive testing. In Proceedings of the 10th International Conference of the Slovenian Society for NDT, Ljubljana, Slovenia, 1–3 September 2009; pp. 221–231.

54. Croce, P.; Marsili, F.; Klawonn, F.; Formichi, P.; Landi, F. Evaluation of statistical parameters of concrete strength from secondary experimental test data. *Constr. Build. Mater.* **2018**, *163*, 343–359. [CrossRef]
55. Vona, M.; Anelli, A.; Mastroberti, M.; Murgante, B.; Santa-Cruz, S. Prioritization strategies to reduce the seismic risk of the public and strategic buildings. *Disaster Adv.* **2017**, *10*, 1–15.

Article

The Cultural Heritage of "Black Stones" (*Lapis Aequipondus/Martyrum*) of Leopardi's Child Home (Recanati, Italy)

Patrizia Santi [1], Stefano Pagnotta [2], Vincenzo Palleschi [3], Maria Perla Colombini [4] and Alberto Renzulli [1,*]

[1] Dipartimento di Scienze Pure e Applicate, Università degli Studi di Urbino Carlo Bo, Campus Scientifico "Enrico Mattei", Via Cà Le Suore 2, 61029 Urbino, Italy; patrizia.santi@uniurb.it
[2] Dipartimento di Scienze della Terra, Università di Pisa, Via Santa Maria 53, 56126 Pisa, Italy; stefano.pagnotta@unipi.it
[3] Applied Laser Spectroscopy Laboratory, ICCOM-CNR U.O.S., Pisa, Via G. Moruzzi 1, 56124 Pisa, Italy; vincenzo.palleschi@cnr.it
[4] Dipartimento di Chimica e Chimica Industriale, Università di Pisa, Via G. Moruzzi 13, 56124 Pisa, Italy; maria.perla.colombini@unipi.it
* Correspondence: alberto.renzulli@uniurb.it

Abstract: A macroscopic lithological study and physical (hardness, size, weight) investigations, coupled with laser-induced breakdown spectroscopy (LIBS) and X-ray fluorescence (XRF) chemical analyses of three egg- and one pear-shaped polished black stones, exposed in the library of the child home of the famous poet Giacomo Leopardi, at Recanati (Italy), were carried out. They are characterized by different sizes: two with the same weight of 16.9 kg and the two smaller ones of 5.6 kg each, corresponding to multiples of standard roman weights (*drachma* and *scrupulum*). These features and the presence of some grooves on the rock artefacts, probably for grappling hooks, suggest an original use as counterweight for the four black stones herein classified as amphibole-bearing serpentinites whose lithologies are far away from Recanati (probably coming from geological outcrops in Tuscany). The four serpentinite stones closely match with the so-called *Lapis Aequipondus* used in antiquity by the Romans as counterweights. Due to the presence of lead rings or iron hooks in these stones, *Lapis Aequipondus* were also used for martyrdoms during the persecution of Christians in the Roman period, attached to the necks of martyrs that were then thrown in the wells or attached to the ankles of hanging bodies. This is the reason why these stones are also known as *Lapis Martyrum*, venerated with the relative martyrs, in several churches of Rome. The four black stones investigated probably arrived at Recanati from Rome after the middle of the 19th century. In the past, Christians also called *Lapis Martyrum* the "devil's stones" (*Lapis Diaboli*). This could also be the reason for the popular belief that black stones cannot be touched by people, except those of the Leopardi dynasty. This work contributes to the cultural heritage of Leopardi's child home, as the four black stones had never been investigated.

Keywords: ultramafic metamorphic rock; stone artefact; *Lapis Aequipondus/Martyrum*; cultural heritage; counterweight; LIBS; XRF

Citation: Santi, P.; Pagnotta, S.; Palleschi, V.; Colombini, M.P.; Renzulli, A. The Cultural Heritage of "Black Stones" (*Lapis Aequipondus/Martyrum*) of Leopardi's Child Home (Recanati, Italy). *Materials* **2022**, *15*, 3828. https://doi.org/10.3390/ma15113828

Academic Editors: Žiga Šmit and Eva Menart

Received: 11 April 2022
Accepted: 17 May 2022
Published: 27 May 2022

Publisher's Note: MDPI stays neutral with regard to jurisdictional claims in published maps and institutional affiliations.

Copyright: © 2022 by the authors. Licensee MDPI, Basel, Switzerland. This article is an open access article distributed under the terms and conditions of the Creative Commons Attribution (CC BY) license (https://creativecommons.org/licenses/by/4.0/).

1. Introduction

In the child home of the famous Italian poet Giacomo Leopardi (born in Recanati on 29 June 1799 and died in Naples on 14 June 1837), some black stones are exposed on the table of one of the library's rooms. Recanati is a small village (at present, of about twenty thousand inhabitants) in central Italy (Marche Region) just 12 km away from the Adriatic Sea.

Historical data on the provenance of these stones and the way they arrived to Recanati are not available. Most probably, they reached Recanati after the middle of the 19th century, since, according to Count Vanni Leopardi, one of the last members of the Leopardi family at the time of this study (he later died in 2019), they are not recorded in the detailed catalog of rare objects written by Monaldo Leopardi (1776–1847), father of the famous poet. During the shooting of the movie *Il giovane favoloso* (Italy, 2014; director, Mario Martone), dedicated to the life of Giacomo Leopardi, the actor Massimo Popolizio (Monaldo Leopardi in the movie) was impressed by the fact that the director said to him that, according to a popular belief, "only people of the Leopardi family could touch the black stones" [1], as other people touching them would be hit by a curse.

That probably may derive from the mysterious atmosphere around these stones because of their exotic appearance and the black color, which is very different from the chromatic features (pale-yellow to whitish) of the sedimentary rocks in the surroundings of Recanati [2]. Some historical reasons at the base of the popular idiom should, however, exist, and they could be unraveled through the present work on these stone artefacts based on a petrographic macroscopic observation (naked eye and hand lens), morphological and physical (size, weights, hardness) studies and a chemical survey, conducted through laser-induced breakdown spectroscopy (LIBS) and X-ray fluorescence (XRF). As a matter of fact, this is the first petrographic investigation on the four black stones exposed in the library of the child home of the famous poet Giacomo Leopardi. Qualitative results of the analyses (LIBS and XRF spectra) will be discussed and compared in order (i) to confirm that the combination of these two techniques is useful for chemical investigations in the field of cultural heritage and (ii) to determine the petrographic classification of the stones. In addition, petrographic and physical comparisons of the four black stones with similar lithologies known and used in antiquity will be carried out and addressed in an archaeometric framework. In particular, of paramount importance will be a comparison, by lithology, of physical properties and/or shape with (a) similar black stones, which are present nowadays in several churches in Rome [3–7], (b) the available geological information of similar rocks, which are present in limited outcrops in Tuscany and used as building or ornamental stones in this Italian region [8–10] and (c) photographic collections of similar black stones [11,12]. Unfortunately, no analyses with the LIBS and XRF were possible to carry out in similar black stones found in the Roman churches (no permission was given). However, comparisons based on robust mineralogic, chemical and petrographic methods enabled an archaeometric study of the four black stones, leading to reasonable conclusions concerning their significance in the framework of the cultural heritage of Leopardi's child home.

2. Materials and Methods

The investigated stones consist of three roughly egg-shaped and one pear-shaped polished rock artefacts (Figure 1). With permission of Count Vanni Leopardi, a series of physical measurements (size, weight) and macroscopic observation of physical properties (color, shining, hardness) of the rock type were carried out. No sampling was allowed, but in situ chemical analyses of the four stones were, nevertheless, permitted. The analysis was performed using mobile dual-pulse instrument for laser-induced breakdown spectroscopy (Modì), a LIBS system made by Marwan Technology (Pisa) and Elio, a portable X-ray fluorescence (XRF) instrument by Bruker Co. These techniques require no sample pre-treatment, a considerable advantage with respect to traditional spectroscopic destructive techniques, which need sample mineralization by acid attack. LIBS and XRF are somewhat complementary elementary techniques. The XRF technique is very versatile, but it cannot detect elements lighter than Al (atomic weight 13); it provides volume-integrated analysis and has a lateral spatial resolution, typically of the order of 1 mm. On the other hand, the LIBS technique can detect light elements, such as Na (atomic weight 11) or Mg (atomic weight 12), has a lateral spatial resolution of the order of a few tens of micrometers and provides an in-depth compositional analysis [13]. The identification and the assignment of emission lines relevant to single atomic species allows us to determine the sample elemental composition. The calibration-free LIBS method was used for the quantitative analysis of the

LIBS spectra (CF-LIBS) [14]. In fact, CF-LIBS does not require calibration standards, since all the information needed for the determination of the sample composition is extracted from the LIBS spectrum itself. The main drawback of LIBS, on the other hand, is its micro-destructivity. The joint use of XRF and LIBS thus allows us to exploit the benefits of both techniques, mitigating their drawbacks (for example, reducing the number of LIBS points of analysis by performing a preliminary compositional screening using the non-destructive XRF technique). These techniques were additionally coupled with fundamental lithological analyses and physical properties characterization. In addition, the comparison with compatible stone artefacts (by lithology) used in antiquity (throughout Italy) and their relative use/s was performed, thus allowing us to focus the present study on the four black stones in the framework of a contribution to the cultural heritage of Leopardi's child home as well. Table 1 synthesizes the methods and rationale adopted in the present study.

Figure 1. The four black stones. Two with the same weight of 16.9 kg (**1,4**) and two smaller ones (**2,3**) of 5.6 kg each. The red arrows indicate the traces of partially smoothed grooves originally used to fix a harness or a ring and the relative hook (or a grappling hook). The graphic scale is 10 cm (the small black and white squares in the right part are 1 cm^2 each). ©Famiglia Leopardi Recanati. Courtesy of Leopardi family, any reproduction is forbidden.

Table 1. Methods and rationale of the present study (in four phases) aimed at unraveling the origin of the investigated black stones of Leopardi's child home.

1	Analysis of the four black stones of the Leopardi's child home (LIBS, XRF, structure, lithology, colour, shining, hardness)	⇨	Petrographic classification
2	Comparison with the stone artifacts used in antiquity having the same petrographic classification, physical properties, shape, size, weight, and surface features	⇨	Matching with *Lapis Aequipondus*
3	Deepening the alternative use of *Lapis Aequipondus*	⇨	*Lapis Martyrum*
4	Petrographic comparison with compatible (by classification and colour) black serpentinites of Italy	⇨	Outlining the most likely provenance area of the *Lapis Aequipondus/Martyrum* and the four black stones of Leopardi's child home themselves

3. Results

3.1. Physical Properties, Lithology, Mineralogy and Petrography

The four black stones are similar from a macroscopic point of view and can thus be all classified as belonging to the same lithology, a fine- to medium-grained metamorphic rock consisting of mafic minerals. No orientation of minerals occurs (isotropic structure). The rocks cannot be scratched with the fingernail and hardly scratched with a sharpener or a coin. In this way, they are characterized by a medium-high hardness, between 5 and 7 in the Mohs scale. In addition, they are partially greasy to the touch.

On the rock surfaces, traces of partially smoothed and circular grooves (up to 1 cm wide) are present (in the upper portions of stone 1, 2, 4 of Figure 1). The grooves may have been originally used to fix a harness or a ring and the relative hook (grappling hooks?). Two out of four rock artefacts are partially damaged in the upper portion, probably due to their use, where the mechanical stress of the harness (or ring) produced a rock rupture (upper portions of stone 3 and 4; Figure 1). Although the black color prevails in the well-polished surfaces, having a metallic shine, millimetric to centimetric pale- to olive-green spots are also present, mostly where some roughness occurs or along the grooves. As shown in Figure 1, the stones have different sizes: two (1 and 4) with the same weight of 16.9 kg and two smaller ones (2 and 3) of 5.6 kg each. These weights roughly correspond to five Roman *drachmae* (i.e., 17.04 kg; 1 Roman *drachma* = 3.408 kg) and five Roman *scrupuli* (i.e., 5.68 kg; 1 Roman *scrupulum* = 1.136 kg). The two largest (16.9 kg) egg-shaped stones have the two axes of the rotation ellipsoid constituting a solid ovoidal form of ca. 23.4 × 36.4 and 24.4 × 41.3 cm, respectively, whereas the smaller egg-shaped one has the two axes ca. 18.7 × 26.7 cm. The pear-shaped stone has a maximum width of ca. 36.3 cm.

The main mafic minerals recognized macroscopically (naked eye and hand lens) are serpentine and amphibole. The LIBS and XRF spectra (Figure 2) are very similar for each of the four samples and reveal a chemical qualitative composition, which is referrable to a serpentinite. In fact, from the LIBS spectrum, a stoichiometric ratio Mg/Si between 1.4 and 1.6 is evaluated using the calibration-free LIBS (CF-LIBS) approach developed by ALS Lab [14], which agrees with the serpentine formula $Mg_6[(OH)_8Si_4O_{10}]$. The presence of clear peaks of Ca and Fe (both in LIBS and XRF spectra) indicates the presence of amphibole of the tremolite-actinolite species $Ca_2(Mg,Fe)_5[OH,F(Si_4O_{11})]_2$. In addition, the presence of peaks of Na and Al (LIBS) suggests the possible presence of low amount of jadeite (a pyroxene, $NaAlSi_2O_6$). As the four black stones are ultramafic metamorphic rocks of the greenschist facies [15], deriving from the serpentinization of mantle peridotites, it is not surprising to also see in the spectra Ni (XRF) and Cr (both LIBS and XRF) peaks, which are probably from relict minerals of olivine and Cr-spinel. The sulphur peak (XRF) agrees with the common presence of sulphides in serpentinites as accessory minerals [16], possibly pyrite (FeS_2) or chalcopyrite ($CuFeS_2$), as Cu also appears in both LIBS and XRF spectra. Traces of sphalerite [$(Zn,Fe)S$] can also be inferred due to Zn peak in the XRF spectrum. As the XRF analysis reported in Figure 2 was performed close to the smoothed grooves originally used to fix a harness or a ring and the relative hook, the detected traces of Pb could be referrable to grappling hooks sealed with lead (plumbing). Weak peaks of strontium, visible in both XRF and LIBS spectra, could be associated with Ca having a chemical affinity (charge, ionic radius) and usually present in accessory minerals, such as calcite ($CaCO_3$), in this kind of serpentinite lithology. The XRF spectrum shows the characteristic fluorescence emission of the X-ray tube anode (Rh), while the emission lines from ambient air (N, O, H) are detectable in the LIBS spectrum.

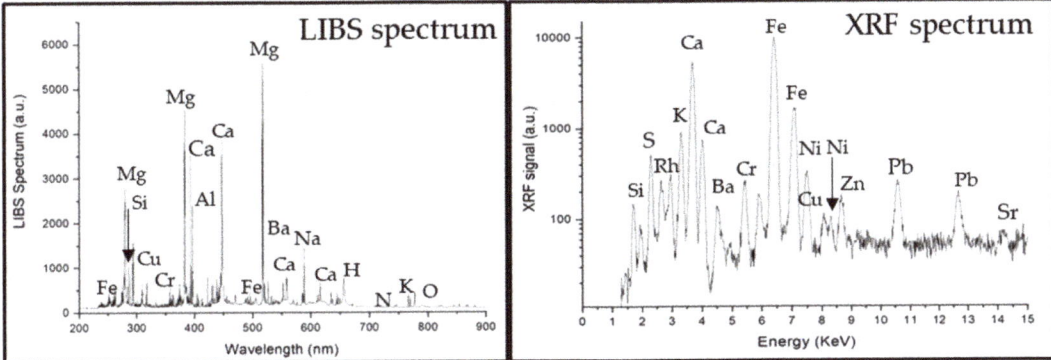

Figure 2. Representative analyses of the four investigated black stones by laser-induced breakdown spectroscopy (LIBS) and X-ray fluorescence (XRF). Spectra (both LIBS and XRF) are very similar for each of the four black stones of the Casa Leopardi's child home.

3.2. Comparative Results with Black Serpentinites Used in Antiquity

In the four black stones investigated, the correspondence of the weights to multiples of some standard Roman weights and the presence of grooves to hook a harness or a ring clearly address their origin as counterweights. The lithological similarity to what is already known as *Lapis Aequipondus* or *Lapis Martyrum* [4,5,17] is straightforward (Figure 3). The Romans utilized this stone to make scale weights. There were weights of different sizes, and for this reason, lead rings or iron hooks were attached to them. According to Corsi [17], *Lapis Aequipondus* or *Lapis Nephriticus* has a relatively high hardness, pale to olive green color, belonging to the jade group of rocks. Jade is used as a synonym of Na-pyroxenite, having Na-pyroxene (jadeite, Fe-jadeite, Mg- and Fe-omphacite or a mix of them) as the most abundant phase. It corresponds only in part to the gemological term «jade» [18,19], in which nephrite is also included [20]. The term *"Nephriticus"* clearly refers to the nephrite, which is an amphibole of the tremolite-actinolite series, $Ca_2(Mg, Fe)_5Si_8O_{22}(OH)_2$ whose name was, however, abolished by the International Mineralogical Association (IMA). In addition, a misunderstanding may exist concerning the *Lapis Aequipondus*, as the ancient Roman stone makers also gave the same name to a dark green to black serpentinite, partially greasy to the touch, that can be well polished and again used for counterweights [17]. The four black stones of Leopardi's child home thus clearly refer to this latter variety of *Lapis Aequipondus*.

In the framework of mafic and ultramafic metamorphic rocks of the greenschist facies used in antiquity and greasy to the touch, the gray to pale green soapstones (talc- and magnesite-bearing chlorite schist) recognized as a subgroup of the so called pietra ollare [21–24] can be ruled out (for different modal mineralogy, color and hardness) as the lithology of the four stones of Leopardi's child home.

The rings or hooks of the *Lapis Aequipondus* also allowed the use of these stones as martyrdom instruments during the Christian persecution [3,5,6,17], mostly attaching these stones to the neck of the martyrs who were then thrown in wells. *Lapis Aequipondus* is, therefore, generally known as *Lapis Martyrum* and also as the "devil's stone" (*Lapis Diaboli*) [6]. The use of *Lapis Martyrum* is well represented in some frescos of the Basilica of Santo Stefano Rotondo (Rome; Figure 4), with the stone (a spherical or flattened ball shape) attached to the ankles of hanging bodies to make the tortures more painful. This is the reason why these stones started to be highly venerated by the Christians as *Lapis Martyrum*, and several ovoidal to disc-shaped or truncated cone samples (e.g., with a shape of a circular loaf) are present nowadays in several Roman churches, such as Santa Maria in Cosmedin, Santa Maria in Trastevere, Santa Sabina and San Lorenzo Fuori le Mura (Figure 5).

Figure 3. Macroscopic comparison among: (**a**) one of the black stones (1) from Leopardi's child home; (**b**) *Lapis Martyrum* located at Santa Maria in Trastevere church (Rome); (**c**) particular of *Lapis Aequipondus* with permission of G. Giardini who took the photo [11] at the Lateran Museum (Rome); (**d**) *Lapis Martyrum* 4th century AD (private collection [12]).

Figure 4. Indoor frescos inside the Basilica of Santo Stefano Rotondo, Rome (Antonio Tempesta, Pomarancio and Matteo da Siena, 1586) representing the martyrdom of: *San Primus* (**a**); *San Felicianus* (**b**); a nameless martyr (**c**). *Primus* and *Felicianus* brothers suffered martyrdom around the year 297 AD, during the Diocletian persecution. The relatively bad quality of photos (**b**,**c**) is because the frescos are not well preserved.

Examples of *Lapis Martyrum* are reported in the literature [5,7] in other Roman churches, such as S. Clemente, S. Prassede, S. Pudenziana, S. Paolo alle Tre Fontane, S. Nicola al Carcere Tulliano, SS. Cosma and Damiano, museums (Musei Capitolini and Museo Lateranense of Rome) and S. Angelo church of Perugia. Finally, a series of weight units for measurement are exposed at the Museum of Terme di Diocleziano (Rome).

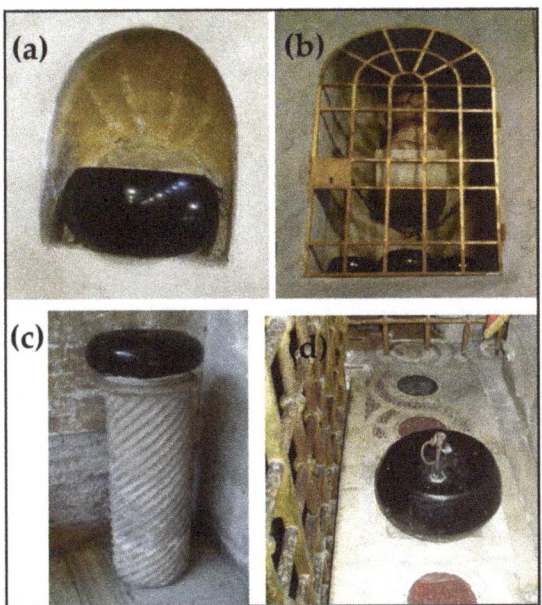

Figure 5. Some *Lapis Martyrum* hosted in different churches of Rome; (**a**) Santa Maria in Cosmedin; (**b**) Santa Maria in Trastevere; (**c**) Santa Sabina; (**d**) San Lorenzo Fuori le Mura (presently missing).

4. Discussion and Conclusions

The black stones found at Giacomo Leopardi's child home had never been classified according to mineralogy and petrography. As no destructive analyses were permitted on the artefacts' material, LIBS and portable XRF chemical surveys therefore represented fundamental analytical techniques, which allowed, along with macroscopic structure investigation and physical properties analyses, to define (for the first time) an appropriate and rigorous petrographic classification of the four black stones that arouse so much curiosity at Leopardi's child home: amphibole-bearing serpentinites. Comparisons with stone artefacts possessing very similar lithologies and physical features also lead to a recognition of the four black stones as counterweights, known by the Romans as *Lapis Aequipondus* and also used for martyrdoms (*Lapis Martyrum*) during the persecutions against Christians.

However, such metamorphic rocks are not present in the surroundings of Recanati, nor within the radius of hundreds of kilometers from the hometown of the famous poet, an area only characterized by sedimentary rocks. As a matter of fact, serpentinites (greenschist facies) have their origin in rocks belonging to ophiolite sequences and, according to the knowledge on the regional petrology and geological maps of Italy, the best candidate areas for their provenance can be represented by Liguria and Tuscany regions in the Northern Apennines [25,26]. Of course, ophiolite sequences are also widespread in the Alps and throughout the Mediterranean area, but the presence of some old quarries of green to dark green to black serpentinites exploited in ultramafic metamorphic rocks of the ophiolite sequences of Tuscany and Liguria addresses our lithological comparisons to central Italy. As a matter of fact, various lithotypes of serpentinites from the Northern Apennines were largely used in the historical architecture of the above two regions [27]. The area of Pian di Gello near Monte Ferrato, north of Prato, was recognized by Del Riccio and Rodolico [3,10] as a source area of black serpentinites used for building stones. In particular, in his "History of the Stones", the monk Agostino Del Riccio [3] describes a "Paragone" Stone near Prato (Sacca and Sant'Anna), which can be considered another variety of dark green to black serpentinite. He also reports the term "Frombole di Mare" for some of these black ("Paragone") polished stones with rounded shape, which were present in Rome and used for Christian persecutions during the Roman period [3].

A dark green to black serpentinite quarried in the surroundings of Florence (near Antella) was used to pave the floor of Santa Maria del Fiore Cathedral in Florence [8]. By contrast, from the Sacca di Prato quarry come the black "Paragone" stone and the green "Green of Prato" stone used for the external walls of the Cathedral itself [8]. Finally, it is worth noting that Giamello et al. [9] indicate dark green to black serpentinite quarried in the area of Casciano of Murlo/Vallerano (Siena) as the source area for many lithotypes used on the floor of the Siena Cathedral.

Clues therefore exist to locate the most probable area of provenance of the four black stones of Leopardi's child home (and thus of *Lapis Aequipondus* or *Lapis Martyrum*), namely, the quarrying sites of Tuscany. Although Monaldo Leopardi (the father of the famous poet) used to record every kind of object present in the home, nevertheless, there is not a trace of them in the Leopardi archive. It can thus be hypothesized that the four black stones probably arrived in Recanati, from Rome, after the death of Monaldo (after the mid-19th century). As the Leopardi family was deeply Catholic, the bad reputation of the four black stones and the legend about their curse on all people touching them (except those of the Leopardi dynasty) could probably derive from the use of this kind of stone as *Lapis Martyrum*. In any case, the overall study of these stones, from petrographic classification to comparison with other artefacts and their use in the past, sheds light on a matter stimulating curiosity to all the visitors of Leopardi's child home, where the four black stones are exposed in a room of the library.

Author Contributions: Conceptualization, P.S. and A.R.; methodology, P.S., A.R., V.P., S.P. and M.P.C.; formal analysis, V.P., S.P. and M.P.C.; investigation, P.S., A.R., V.P., S.P. and M.P.C.; writing—original draft preparation, P.S. and A.R.; writing—review and editing, P.S., A.R., V.P., S.P. and M.P.C.; visualization, P.S., A.R., V.P., S.P. and M.P.C.; supervision, P.S. and A.R. All authors have read and agreed to the published version of the manuscript.

Funding: This research was financially supported by University of Urbino Carlo Bo (Dipartimento di Scienze Pure e Applicate, DISPEA_SANTI_AFFIDAMENTI_2019).

Institutional Review Board Statement: This study does not require ethical approval.

Informed Consent Statement: Not applicable.

Data Availability Statement: Not applicable.

Acknowledgments: We are very grateful to the late Conte Vanni Leopardi who authorized this research with his curiosity-driven interest and enthusiasm. Carmela Magri and Sara Eugeni are also acknowledged for their useful information during our survey of the library of Leopardi's child home. We are also grateful to Giorgio Giardini for the copyright permission to reproduce his photo (Figure 3c in this work) present in Giardini and Colasante [11]. This work benefited from comments and suggestions of two anonymous reviewers.

Conflicts of Interest: The authors declare no conflict of interest.

References

1. Available online: https://sentierofrancescano.wordpress.com/2014/10/16/il-giovane-favoloso-giacomo-leopardi/ (accessed on 30 March 2022).
2. Sarti, M.; Coltorti, M. *Note Illustrative della Carta Geologica d'Italia Alla Scala 1:50.000, Foglio 239 Osimo*; ISPRA: Roma, Italy, 2011; 139p.
3. Del Riccio, A. *Istorie Delle Pietre*, reprinted in 1979 by Bazzocchi P.; S.P.E.S.: Firenze, Italy, 1597.
4. Borghini, G. *Marmi Antichi*, 5th ed.; De Luca Ed. d'Arte s.r.l.: Roma, Italy, 2004; 342p.
5. Pullen, H.W. *Manuale dei Marmi Romani Antichi*; A cura di F. Crocenzi; Gangemi Editore: Roma, Italy, 2016; 252p.
6. De Matthaeis, N. *Andar per Miracoli. Guida All'affascinante Mondo Delle Reliquie Romane*; Intra Moenia: Oria, Italy, 2013.
7. Barbiero, A.S. *Paolo e le Tre Fontane—XXII Secoli di Storia Messi in luce da un Monaco Cisterciense (Trappista)*; Abbazia "Nullius": Roma, Italy, 1938; 143p.
8. Bastogi, M.; Fratini, F. Geologia, litologia, cave e deterioramento delle pietre fiorentine. *Mem. Descr. Carta Geol. d'It.* **2004**, *LXVI*, 27–42.

9. Giamello, M.; Droghini, F.; Mugnaini, S.; Guasparri, G.; Sabatini, G.; Scala, A.; Morandini, M. Il pavimento marmoreo del Duomo di Siena. Caratterizzazione dei materiali e dello stato di conservazione. In *Studi Interdisciplinari sul Pavimento del Duomo di Siena. Iconografia, Stile, Indagini Scientifiche*; Edizioni Cantagalli: Siena, Italy, 2005; pp. 173–197.
10. Rodolico, F. *Le pietre delle città d'Italia*; Le Monnier: Firenze, Italy, 1965.
11. Giardini, G.; Colasante, S. Pietre decorative antiche. Collezioni "Federico Pescetto" e "Pio De Santis". Ist.Poligrafico e Zecca dello Stato. *Mem. Carta Geol. d'It.* **2002**, *XV*, 232.
12. Available online: https://www.invaluable.com/auction-lot/lapis-aequipondus-martyrum-ou-pierre-des-martyrs-130-c-4bd4399b16# (accessed on 24 March 2022).
13. Palleschi, V. Laser-induced breakdown spectroscopy: Principles of the technique and future trends. *ChemTexts* **2020**, *6*, 18. [CrossRef]
14. Ciucci, A.; Corsi, M.; Palleschi, V.; Rastelli, S.; Salvetti, A.; Tognoni, E. New procedure for quantitative elemental analysis by laser-induced plasma spectroscopy. *App. Spectr.* **1999**, *53*, 960–964. [CrossRef]
15. Bucher, K.; Frey, M. *Petrogenesis of Metamorphic Rocks*; Springer: Berlin/Heidelberg, Germany, 2002; 341p.
16. Park, A.F. Serpentinite. In *The Encyclopedia of Igneous and Metamorphic Petrology*; Bowes, D.R., Ed.; Van Nostrand Reinhold: New York, NY, USA, 1990; p. 525.
17. Corsi, F. *Delle Pietre Antiche con Notabile Aggiunta al III Libro*; Generic: Roma, Italy, 1845.
18. Webster, R. Gems. Butterworth &, Co.: London, UK, 1983; p. 652.
19. Hurlbut, C.S.; Kammerling, R.C. *Gemmology*; Wiley & Sons: New York, NY, USA, 1991.
20. D'Amico, C.; Starnini, E.; Gasparotto, G.; Ghedini, M. Eclogites, jades and other HP-metaophiolites employed for prehistoric polished stone implements in Italy and Europe. Special Issue 3: A showcase of the Italian research in applied petrology. *Per. Min.* **2003**, *73*, 17–42.
21. Mannoni, T.; Messiga, B. La produzione e la diffusione dei recipienti di pietra ollare nell'Alto Medioevo. In *6° Congresso Internazionale di Studio Sull'alto Medioevo: Longobardi e Lombardia:Aspetti di Civiltà Longobarda (Milano, 21–25 Ottobre 1978)*; Centro di studi sull'alto Medioevo: Spoleto, Italy, 1980; pp. 501–522.
22. Mannoni, T.; Pfeifer, H.R.; Serneels, V. Giacimenti e cave di pietra ollare nelle Alpi, La pietra ollare dalla preistoria all'età moderna. In Proceedings of the Atti del Convegno, Como, Italy, 16–17 October 1982; New Press: Como, Italy, 1987; pp. 7–45.
23. Santi, P.; Renzulli, A.; Antonelli, F.; Alberti, A. Classification and provenance of soapstones and garnet chlorite schist artifacts from Medieval sites of Tuscany (Central Italy): Insights into the Tyrrhenian and Adriatic trade. *J. Archaeol. Sci.* **2009**, *36*, 2493–2501. [CrossRef]
24. Mini, F.M.; Santi, P.; Renzulli, A.; Riccardi, M.P.; Antonelli, F.; Alberti, A. Representative archaeological finds of pietra ollare from Comacchio (Italy): Identifying provenance and high-T mineral breakdown reactions hindering lithotype classification. *Archaeol. Anthropol. Sci.* **2016**, *8*, 135–148. [CrossRef]
25. Schwarzenbach, E.M.; Vogel, M.; Früh-Green, G.L.; Boschi, C. Serpentinization, carbonation and metasomatism of ultramafic sequences in the Northern Apennine ophiolite (NW Italy). *J. Geophys. Res. Solid Earth* **2021**, *126*, e2020JB020619. [CrossRef]
26. Conti, P.; Cornamusini, G.; Carmignani, L. An outline of the geology of the Northern Apennines (Italy), with geological map at 1:250,000 scale. *Ital. J. Geosci.* **2020**, *139*, 149–194. [CrossRef]
27. Fratini, F.; Rescic, S.; Pittaluga, D. Serpentinite and ophicalcite in the architecture of eastern Liguria and as decoration of Tuscan religious buildings. *Res. Policy* **2022**, *75*, 102505. [CrossRef]

Article

Degradation of CdS Yellow and Orange Pigments: A Preventive Characterization of the Process through Pump–Probe, Reflectance, X-ray Diffraction, and Raman Spectroscopy

Francesca Assunta Pisu *, Pier Carlo Ricci, Stefania Porcu, Carlo Maria Carbonaro and Daniele Chiriu *

Department of Physics, University of Cagliari, Cittadella Universitaria, 09042 Cagliari, Italy
* Correspondence: francescaassunta.pisu@dsf.unica.it (F.A.P.); daniele.chiriu@dsf.unica.it (D.C.)

Abstract: Cadmium yellow degradation afflicts numerous paintings realized between the XIXth and XXth centuries. The degradation process and its kinetics is not completely understood. It consists of chalking, lightening, flaking, spalling, and, in its most deteriorated cases, the formation of a crust over the original yellow paint. In order to improve the comprehension of the process, mock-up samples of CdS in yellow and orange tonalities were studied by means of structural analysis and optical characterization, with the principal techniques used in the field of cultural heritage. Mock ups were artificially degraded with heat treatment and UV exposure. Relevant colorimetric variation appears in CIE Lab coordinates from reflectance spectra. XRD, SEM-EDS, and Raman spectroscopy revealed the formation of cadmium sulfate, whilst time-resolved photoluminescence and pump–probe transient absorption spectroscopy suggest the formation of a defective phase, compatible with Cd vacancies and the formation of both CdO and $CdSO_4$ superficial clusters.

Keywords: cadmium sulfide; degradation; Raman spectroscopy; transient absorption spectroscopy; reflectance; photoluminescence

1. Introduction

During the nineteenth century, new inorganic pigments were synthesized and used extensively by coeval artists [1,2] for their characteristics, such as high color intensity, low cost, and covering power; they substituted the well-known ancient pigments in numerous Impressionist and early Modernist paintings.

All the pigments, organic and inorganic, synthetic and mineral, can be subjected to time corruption. Degradation can arise from environmental conditions, light exposure, bacterial attacks, or the use of erroneous solvents during the restoration process.

As is known, the most famous pigments used in painting are usually semiconductors. Depending on the surrounding environment, these materials are potentially not stable and degradation processes can take place through reactions with the atmosphere, especially with molecules, such as carbon dioxide, sulfur, oxygen, water, etc. [3,4]. The result implies the formation of patinas of new compounds due to carbonation or oxidation processes. Sometimes, the environmental conditions cause a phase transformation in the mineral that can present a different color. The reaction depends on many important parameters, such as temperature, light, relative humidity, pristine defect density in the materials, points of nucleation, time, etc.

We decided to concentrate our attention on nucleation processes due to an initial presence of defects in the structure. Those point defects could be related to different phenomena, such as, for example, light exposure. Once the degradation process is observed, one can trace back the numerous steps that led to the deteriorated materials, down to the presence of nucleation areas where the process started. Sometimes, these nucleation areas can be readily observed on the surface of the applied pigment. The questions we would like to answer are the following: Are we able to detect a primordial state of nucleation area where only starting point defects are present? Which non-destructive diagnostic tools can be promoted to detect those defects and to prevent degradation? The answer can arrive from the combined use of some optical technique, such as Raman spectroscopy, reflectance, and pump–probe spectroscopy [5–10]. An additional value can be associated to pump–probe spectroscopy, not conventionally used in the field of Cultural Heritage. Here, the electronic properties can be studied by analyzing all the optical transitions that involve not only the natural energy level positions, but even the presence of a distribution of some electronic levels due to color centers. These centers are associated with point defects, from which the mentioned nucleation process can start. In this work, we explored the combination of some conventional techniques and the pump probe one to provide an answer to the degradation of yellow cadmium pigment.

1.1. Yellow and Orange Cadmium Pigments

Yellow cadmium pigment is constituted by a cadmium sulfide (CdS) semiconductor, presenting both crystalline and amorphous forms. The hexagonal phase (α-CdS) is found in nature as the mineral greenockite, with space group P6$_3$mc, the cubic structure belongs to the space group F$\bar{4}$3m (β-CdS), known as hawleyite mineral [11], while the amorphous form is a chemically synthesized product.

The synthesis procedure can be divided into wet and dry processes. The latter could have, as starting materials, cadmium oxide, cadmium metal, or cadmium carbonate, each of which can be mixed with sulfur and heated to 300°–500° in the absence of air. The wet process consists in a solution with soluble sulfide (such as hydrogen sulfide, sodium sulfide, or barium sulfide) and soluble cadmium salts, such as chloride, cadmium nitrate, cadmium sulfate, or cadmium iodide, with a final precipitation of CdS [12,13].

During the 20th century, to produce different hues from the light yellow typical of CdS, its synthesis started to be changed by inserting zinc to lighten the color and selenium to increase the red hue. The use of selenium during the production allows one to obtain a range of colors from pale yellow to red, depending on the percentage of selenium, and these pigments are known as cadmium sulfoselenide compounds. Around 1920, to obtain another lighter shade, the use of additive barium sulfate was included in the synthesis procedure. These pigments can be prepared by calcining the hexagonal CdS plus selenium again or calcinating the precipitate resulting from the treatment of a cadmium salt with an alkaline selenide and sulfide [14].

1.2. CdS Degradation Phenomena

The degradation process leads to various alteration degrees in the original yellow paint till the formation of an overlaying crust in the worst cases [15]. Degradation of cadmium yellow was observed in works of many famous artists, such as Pablo Picasso, Vincent Van Gogh, Georges Seurat, Henri Matisse, Ferdinand Leger, Edvard Munch, and James Ensor [13,15–18].

Many studies were performed on Impressionist paints to understand the degradation process and identify which degradation products arise. In degraded areas of James Ensor's paint, Van der Snickt et al. [13] found the formation of disfiguring white crystals that are present on the paint surface in approximately 50 µm sized globular agglomerations. They noted that the degradation penetrated below the paint surface about ca. 1–2 µm, suggesting that the degradation process appears to have affected only the material that is in direct contact with moisture and (UV) light. The degradation process is due to oxidation in the CdS pigment, with the final formation of cadmium sulfate as a degradation product. They found traces of $CdCO_3$ and $PbSO_4$. The former was attributed to a reaction from the starting material; the presence of the latter was not clarified. In the work of Van Der Snickt et al. [17], the authors suggested the presence of cadmium carbonate as a secondary degradation product following the primary photo-degradation of CdS, perhaps by the capture of atmospheric CO_2, or a tertiary process involving a further breakdown in cadmium oxalate into cadmium carbonate. Further, Pouyet and coworkers [16] reported that the degraded white crust of the paints showed a high presence of $CdCO_3$ and traces of $CdSO_4 \cdot nH_2O$ and CdC_2O_4. The presence of cadmium carbonate was explained both as a starting reagent of the wet process and as second degradation product of photo-alteration.

Other degradation products individuated in degraded paints are cadmium oxalate, cadmium sulphate, and cadmium oxide. All of them can be derived from starting materials or filler products in paints but, as asserted by Mass et al. [15], their presence on the surface and not in the depth of the paint layer, is the proof of their degradation character. The cadmium oxalates are concentrated near the surface of the paint layer, demonstrating that this is a photo-oxidation product rather than a paint filler in the cadmium yellow paints. Cadmium sulfates have high solubility, so their presence in traces throughout the paint layer can be explained as mobile photo-oxidation products or, again, as starting reagents.

In addition to paintings, hexagonal CdS is one of the most important semiconductors for high-tech applications, for solar energy harvesting, among others, because of its bandgap around 2.4 eV [19] and calcination studies of synthetic CdS and consequent photo-oxidation by visible light exposure were carried out for this research field [20]. CdO was detected as a photo-degradation product and it was also found that the calcination treatment decreases the trap states. In addition, by increasing the calcination time, the color of the sample changes from yellow to pale yellow, with variations in the absorption curve that become sharper.

Therefore, the different degradation pathways of CdS pigment proposed in the literature can be summarized. The action of light with energy equal to or higher than CdS band gap leads to the formation of electron-hole pairs:

$$CdS + h\nu \rightarrow e^- + h^+ \tag{1}$$

The holes oxidize the cadmium compound with:

$$CdS + 2h^+ \rightarrow Cd^{2+} + S. \tag{2}$$

At this point, the contact with air oxygen leads to the oxidation of sulfur to sulfate:

$$CdS + 2O_2 \rightarrow Cd^{2+} + SO_4^{2-} \tag{3}$$

More generally, the paint's exposure to moisture and air can lead to a fading in yellow with the formation of cadmium sulfate hydrate, as confirmed by XRD measurements in [13]. Therefore, the chemical reaction is:

$$CdS + 2O_2 + H_2O \rightarrow CdSO_4 \cdot H_2O \tag{4}$$

L. Monico et al. proposed another mechanism of sulfate formation considering the action of surface holes [21]:

$$CdS + 4h^+_{surf} + 2H_2O + O_2 \rightarrow CdSO_4 + 4H^+ \tag{5}$$

The presence of lead carbonate (not clear yet) with the formation of cadmium sulfate could cause the presence of lead sulfate as another degradation product:

$$PbCO_3 + SO_4^{2-} \rightarrow PbSO_4 + CO_3^{2-} \tag{6}$$

The degradation phenomena of CdS pigments are clearly related to changes in CdS optical proprieties. Comelli et al. [18], by studying the degraded paint of Picasso's Femme, demonstrated that PL emissions from trap states (TS) are highly favored in relation to the near band edge (NBE) recombination, indicating a higher surface reactivity. This TS emission occurs at wavelengths shorter than those of not degraded CdS yellow. This behavior could be related to environment reactive nanometric-size grains of yellow pigment, resulting after the synthesis procedure without further calcination. In addition, a few computation studies in the literature [22–24] tried to explain the CdS degradation process and its origin.

In this work, to understand the degradation process of CdS pigment, the effect of calcination, the role of inner defects, and to clarify which degradation products could be originated, we simulated several artificial aging processes on two kinds of CdS commercial samples: the yellow and orange ones. The pigments were artificially degraded through different accelerated ageing processes: heat treatment and UV light exposure.

We studied the degradation phenomenon with various optical techniques in order to select which guarantees the fast results useful to determine the conservation state. In the past, artificial aging was already realized [12,21,25], aiming at studying the complex system made by the pigment with its binder (oil). In this study, we want to focus on the role of the pigment alone in the degradation process of CdS paints, after two different simulated artificial degradation processes, to understand which final degradation products are formed after them. In future work, the study will be extended to other components in the complex system of paint to include both the binder and the support.

2. Materials and Methods

2.1. Mock-Up Samples

Cd pigments used for accelerated degradation studies were bought from Kremer pigments. We used pigments number 21,040 (white yellow) and 21,080 (orange) and called them C-A and S-A respectively. The artificially aged samples were named with the initial C or S to indicate the yellow or orange paints, followed by a number indicating the temperature of heat treatment (300–400–500) and duration of heating (1 h, 2 h, 6 h). With the "UV" term we expressed the UV exposure realized by Hg lamp, with wavelength at 365 nm, followed by the relative exposure time (in hours) up to 56 h with a density power of 7 mW/ cm^2.

The thermal treatment is obtained using a temperature range between 300 and 500 °C for different heating times (1 h, 2 h, 4 h, 6 h). Their chromatic and structural variation were measured with micro-Raman spectroscopy, reflectance, TR-PL spectroscopy, X-ray diffraction, and transient absorption. Then these samples were stored at room temperature, 20% relative humidity (RH) in contact with air, illuminated by ambient light from Compact Fluorescent Lamps (CFL) for 4 h/days for 6 months.

To perform UV exposure the acquired powder was mixed with distilled water and dispersed on a slide. The sample was exposed with Hg-lamp combined with a filter to remove the visible components and leave mainly the 365 nm component. The exposure was made at room temperature and with an RH value of 20%. The reflectance spectra were collected by step of 8 h a day and the other measurements are made only on the raw samples and at the end of the process for a total period of three weeks of air exposure.

2.2. Reflectance Measurements

Reflectance measurements were performed by means of a Laser-driven Xenon lamp (EQ-99X) with wide emission spectrum from 200 to 2000 nm and average optical power of 1 mW/nm. The source was coupled with an optical fiber to an integrating sphere and to an Advantes Sensiline Avaspec-ULS-TEC Spectrometer (spectral range 250–900 nm). The measurements were acquired in 10° reflection mode, with a BaSO$_4$ plate as reference. The results were related to the D65 illuminant and the CIELab standard colorimetric observables. The CIE coordinates were obtained using the ColorConvert v. 7.77 software.

2.3. Raman Measurements

High-resolution micro-Raman scattering measurements were obtained in back-scattering geometry through the confocal system SOL Confotec MR750 equipped with a Nikon Eclipse Ni microscope. Raman spectra were gathered by using, as excitation wavelength, the 532 nm line of a solid-state laser (DPSS LASOS Instruments, Jena, Germany). A grating with 1800 grooves/mm was used to obtain a resolution of 0.2 cm^{-1}. Near-infrared micro-Raman scattering measurements were carried out in back-scattering geometry with a 1064 nm line of an Nd:YAG laser. Measurements were performed with a compact spectrometer B&WTEK (Newark, NJ, USA) i-Raman Ex integrated system with a spectral resolution of 8 cm^{-1}. For each experimental setup, all the spectra were collected with an acquisition time of about 60 s (five replicas) and power excitation between 5 and 10 mW concentrated in a spot of 0.3 mm^2 on the surface through a Raman Video MicroSampling System (Nikon, Tokyo, Japan) Eclipse for high-resolution and BAC151B in the other case equipped with a 20× Olympus objective to select the area on the samples. Each measurement area is represented by 5 measures over the surface of 1 cm^2 and the average value of all the spectra is proposed in the experimental results. A total variation of less than 2% assures homogeneity in the surface degradation.

2.4. Time-Resolved Photoluminescence (TR-PL)

TR-PL measurements were recorded by using different excitation systems:

- Excitation with 100 fs long pulses delivered by an optical parametric amplifier (Light Conversion TOPAS-800-fs-UV-1) pumped by a regenerative Ti:sapphire amplifier (Coherent Libra-F-1K-HE230). The repetition frequency was 1 kHz.
- Excitation with 100 fs long pulses from Ti:sapphire oscillator Coherent Chameleon Ultra II having a repetition rate of 80 MHz.

PL signal was recovered by a streak camera (Hamamatsu C10910, Hamamatsu, Photonics, Herrsching am Ammersee, Germany) equipped with a grating spectrometer (Princeton Instruments, Trenton, NJ, USA, Acton Spectra Pro SP-2300). All of the measurements were collected in the front-face configuration to reduce inner-filter effects. Proper optical filters were applied to remove the reflected contribution of the excitation light.

2.5. Pump and Probe Transient Absorption Spectroscopy (TAS)

Transient absorption measurements were performed with a pump–probe differential spectrometer (Ultrafast Systems HELIOS-EOS), with both pump and probe wavelengths generated by a Ti:Sapphire regenerative amplifier (Coherent Libra-F-1K-HE-230), which delivers 100 fs long pulses at 800 nm with 1 KHz repetition rate. The main emission from the regenerative amplifier was split into two branches: one sent to an optical parametric amplifier (Light Conversion TOPAS C), in order to generate the pump wavelengths, and the other sent to the sapphire plate of the HELIOS spectrometer, where multi-color probe beam was generated by means of white light supercontinuum generation. The probe pulses were time delayed with respect to the pump pulses, by passing through a variable digitally controlled optical delay line. The pump and probe beams were then non-collinearly focused and overlapped on the sample surface, with the pump being chopped at 500 Hz, so that half of the transmission spectra were recorded with the pump on and half with pump off. The transmission spectra from the probe beam were recorded as a function of the relative delay time, by means of CCD spectrometers

2.6. XRD Measurements

X-ray patterns were collected at room temperature by using a Bruker D8 Advance diffractometer operating at 30 kV and 20 mA, equipped with a Cu tube ($\lambda = 1.5418$ Å), a Vantec-1 PSD detector, and an Anton Parr HTK2000 high-temperature furnace. Powder patterns were recorded in the $15° \leq 2\theta \leq 85°$ range.

2.7. SEM-EDS

SEM images were gathered by a scanning electron microscope ESEM: FEI Quanta 200 under low-vacuum conditions. EDS semiquantitative analyses were obtained with the help of Thermo Scientific (Waltham, MA, USA) EDS UltraDry INTX-10P-A system equipped with Pathfinder. Each point of analysis was collected with an acceleration voltage of 20 kV and live time of 30 s.

3. Results and Discussion

3.1. Heating Process

The starting pigments were characterized first by XRD to obtain information about the phase composition and additive compounds. At a later time, a second analysis with reflectance and Raman spectroscopy was conducted to have more detail on structural modifications during the aging process. XRD analyses revealed the presence of barium sulfate (see Figure 1a), in a percentage of about 7(1)% with respect to the remaining $Cd_{1-x}Zn_xS$ in yellow cadmium (x = 0.19). The orange one revealed the presence of hexagonal $CdS_{1-x}Se_x$ and CdS (Figure 1b). The phase identification confirms the mixture of CdZnS and $BaSO_4$ for yellow C-samples and CdSeS for orange S-samples. Thermal treatments on the two mentioned samples, C-A and S-A, were made at different temperatures (300 °C, 400 °C, 500 °C) for 1 h and, for the 500 °C temperature, we performed the calcination also for 2 h, 4 h, and 6 h until obtaining a notable chromatic change. The XRD measurements performed for the samples heated at 300–400–500 °C for 1 h are reported in Figure 1a,b. As displayed in the patterns, a perceptible difference is shown only in the C-500-1 h curve, in which a broadening of the region between 24 and 34° is present. This broadening could be associated with an increase in structural disorder related only to the CdZnS compound. Actually, studies in the literature on thermal stability on $BaSO_4$ demonstrate that its XRD pattern did not change for this compound [26]. Contrary to what was discussed in the introduction, no evidence can be reported in this analysis about the formation of other compounds, such as $CdSO_4$ and CdO. Only in the case of heat treatment at 500 °C for 6 h, the formation of these compounds is detected, justifying the observed change in color. Actually, by increasing the heating time until 6 h at 500 °C, the XRD measurements were able to detect the degradation products. As can be seen from Figure 1a,b, the C-500-6 h patterns show peaks of $CdSO_4 \cdot 2CdO$ and in the S-500-6 h sample, also the presence of a pure Cd-sulfate phase.

The chromatic variation among the samples was measured from the reflectance spectra. The first derivative spectra for the yellow cadmium are shown in Figure 1c and the strongest variations recorded at 500 °C after 6 h are summed up in the CIE color chart in Figure 1d. The calculated CIE coordinates, relative simulated colors, and the maximum value of the first derivative (X_0) are listed in Table 1.

Table 1. Chromatic variation in C-A and S-A samples after heating process obtained by reflectance spectra.

Sample	L	a	b	X_0 [nm]	Simulated Color	Sample	L	a	b	X_0 [nm]	Simulated Color
C-A	94.1	−11.1	101	496.9		C-A	94.1	−11.1	101	496.9	
C-300 °C-1 h	103	−17.3	101	494.2		C-500 °C-2 h	100	−5.91	84	506.4	
C-400 °C-1 h	93.7	−8.89	101	497.7		C-500 °C-4 h	104	−5.66	99.3	508	
C-500 °C-1 h	85.8	−5.08	75.6	502.3		C-500 °C-6 h	91	−0.81	86.9	511	
S-A	78.2	35.3	92.6	514.5, 556.0		S-A	78.2	35.3	92.6	514.5, 556.0	
S-300 °C-1 h	92.2	42.4	107	512.5		S-500 °C-2 h	89,7	37.1	95	514.0, 553.0	
S-400 °C-1 h	76.8	28.4	85.3	514.0, 552.8		S-500 °C-4 h	91,5	43.2	105	513.0, 555.0	
S-500 °C-1 h	76.4	29.9	89.7	514.3, 552.9		S-500 °C-6 h	70.4	32.1	62.6	515.0, 556.0	

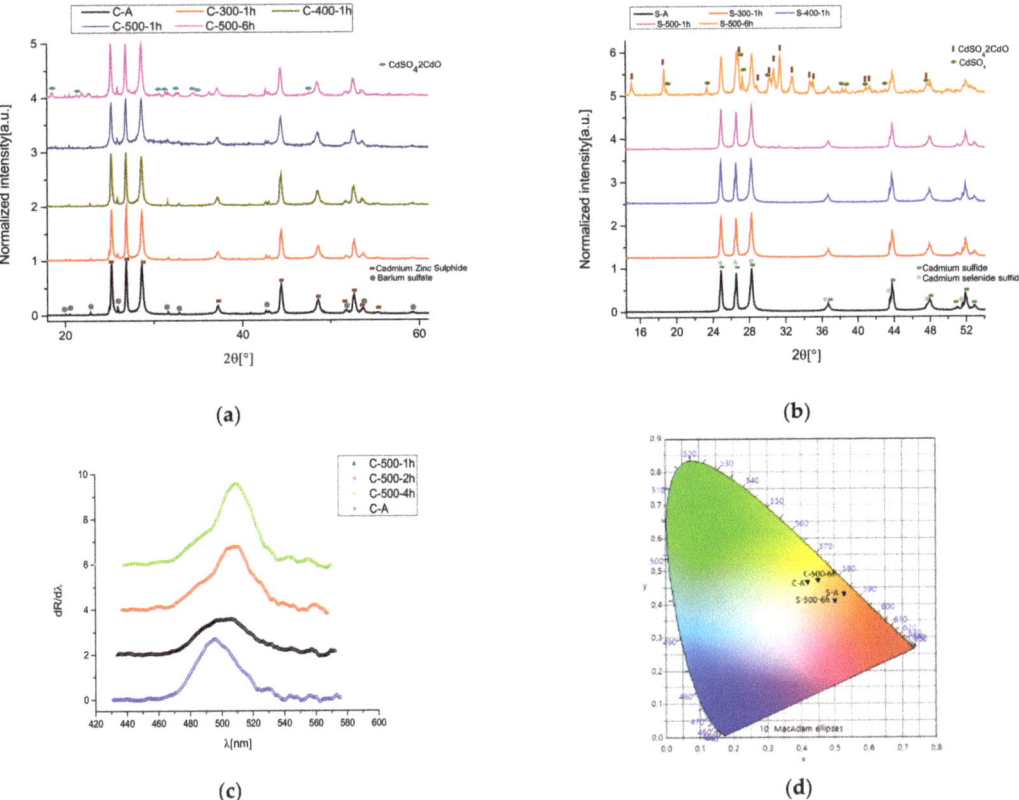

Figure 1. (a) XRD patterns of C-A heated and no-aged samples; (b) XRD patterns of S-A heated and no-aged samples; (c) first derivative reflectance spectra for heated C-A samples; and (d) the color chart of C-A, S-A, C-500-6 h, and S-500-6 h samples.

By heating the C-A sample, a notable shift in X_0 towards high wavelengths is produced. The shift in the S-sample is not remarkable, as can be seen from the X_0 values in Table 1, where the peaks of the first derivative range from 514.5 and 556 (for the no-treated S-sample) to 515 and 556 (for S-500-6 h). As is known from the literature, the X_0-value is linked to the direct bandgap value [27–29], so the shift in sample C is compatible with a decrease in the direct bandgap and it can be explained by the formation of the CdO compound after a thermal treatment, as proved by [20,30]. On the contrary, reflectance measurements suggest another change: heated samples reveal an increment in the luminosity L parameter in CIE Lab coordinates. This change could be associated with the formation of whitish compounds, probably $CdSO_4$ [31].

As the calcination time increases, we generally register growth in the L parameter (Table 1) for both S-A and C-A samples. We can explain this behaviour considering that the colorimetric coordinates can be influenced by the combination of grain dimensions with the formation of mentioned compounds, CdO and $CdSO_4$. If the temperature increases, the grain size in the sample is larger (see Comelli et al.) and this means a decrease in luminosity but also a major formation of cadmium sulfate and the presence of a darker compound, such as CdO. Therefore, the effect of the co-presence of a whitish and a darker compound, in addition to the variation in grain sizes of the sample, can influence the variability in the L parameter, which does not follow a linear increase as a function of the heat treatment. Actually, luminosity L decreases again after 6 h at 500 °C, during which, as hypothesized before, a predominant brownish compound is forming (CdO), inducing a reduction in this parameter.

To confirm this assumption, we performed TR-PL measurements and the results are included in Figure 2a: we reported the emission of C-samples before and after the thermal treatment, where, again, a visible red shift is present. Actually, the peak moves from 480 nm in sample C-A to 510 nm in the sample heated at 500 °C for 6 h. In Figure 2b, the TR-PL kinetics for the S-samples that do not present substantial variation as a function of the temperature are shown.

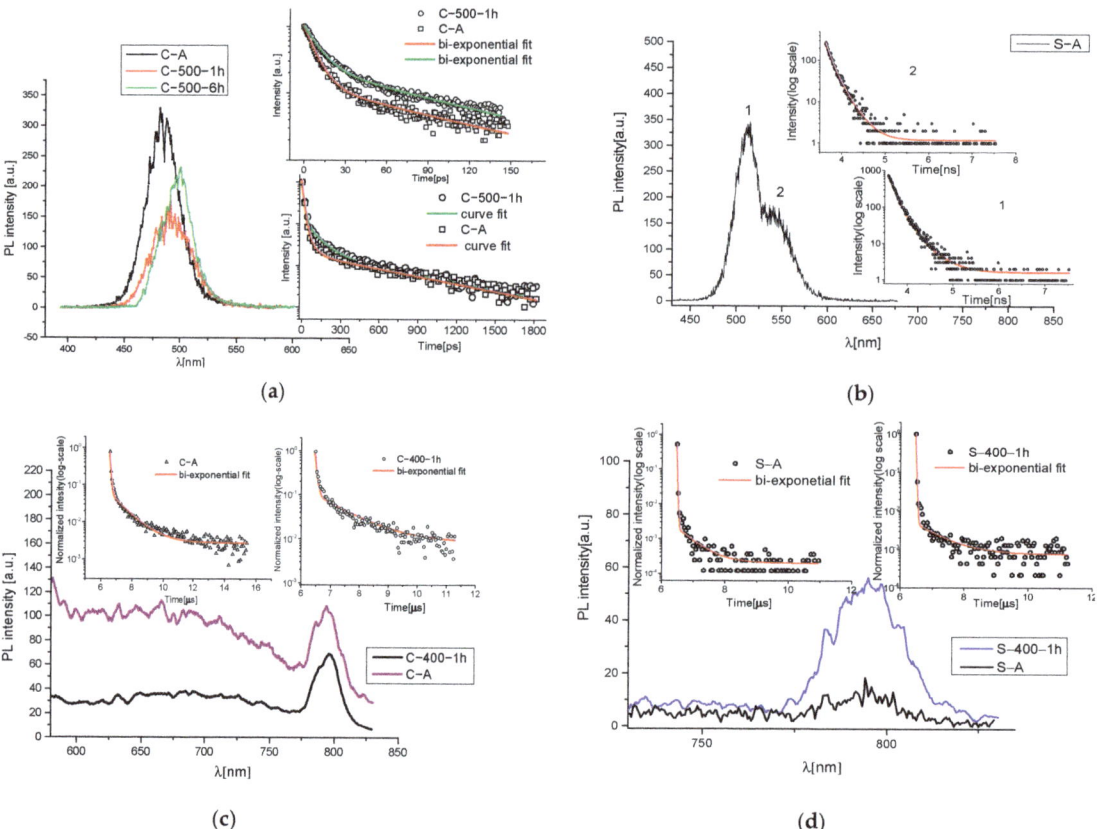

Figure 2. (**a**) Comparison between the PL emission of C-A sample and the C-samples heated at 500 °C for 1 h and for 6 h and respective decay fits for C-A and C-500-1 h samples; (**b**) TRPL spectra of S-A sample; in inset, the time-resolved spectra of the two bands of emission (one related to CdS and the second to -Se inclusion) with relative bi-exponential decay fit; (**c**) TRPL microsecond-scale analysis for C-A and C-400-1 h samples; in inset, time-resolved fit of 700–800 nm emission band for both the samples; (**d**) TRPL microsecond-scale analysis for S-A and S-400-1 h samples; in inset, time-resolved fit of 700–800 nm emission band, exciting wavelength 430 nm, P = 50 uW.

A Gaussian deconvolution of the spectra for different temperatures was calculated for sample C and the respective band positions in eV are represented in Table 2. TR-PL analysis was made on the time scale, ranging from a picoseconds regime to the nanosecond scale, revealing the presence of three decay times derived by three emission channels (τ_1, τ_2 and τ_3). In the case of the CA sample, as indicated by previous studies in the literature [32], the shorter time t_1 around 8–13 ps is attributable to the bandgap emission, while $t_2 < 100$ ps and t_3 of 730 ps are associated with superficial and intermediate structural defects, respectively. While t_1 and t_3 times remain constants as a function of the temperature, the intermediate time t_2 changes assuming a maximum value of 200 ps at 400 °C. This behavior seems to be compatible with the formation of CdO nuclei. Actually, in the literature, pure CdO presents two decay times, one in a range of 100–500 ps and one in a range of 1–3 ns. [33] This assumption is not exhaustive for the presence of this compound and a detailed study is necessary to corroborate this hypothesis.

Table 2. Value of emission channels for exciting wavelength λ = 450 nm, P = 235 μW.

Sample	E (eV)	τ_1 [ps]	τ_2 [ps]	τ_3 [ps]
C-A	2.56	8–13	50	730
C-300-1 h	2.55	8–13	150	727
C-400-1 h	2.56	8–13	200	730
C-500-1 h	2.51	8–13	100	728
C-500-6 h	2.48	8–13	50	707

TRPL was performed also in the microseconds range, confirming the presence of deep trap states (TS) emissions [32,34], with a slow broad emission in the spectral region 650–750 nm and a strong sharp peak at 790 nm due to crystal defects (see Figure 2c,d). Even using this technique, no evidence of time decays ascribable to $CdSO_4$ can be reported, leaving unresolved the variation in the L paramater discussed before, associated with the formation of a withish compound.

To shed light on this behavior, a complete Raman characterization of the samples, at different wavelengths, was essential to better understand which probable compounds are formed by heating. For confirming the CdO formation hypothesis, the Raman spectra could not be helpful. In the literature, there are different and contrasting Raman spectra associated to cadmium oxide [35–37]; otherwise, according to some authors, cadmium oxide should not be Raman active [38]. Therefore, the individuation of this secondary compound cannot be undoubtedly approved from our Raman spectra.

However, the other discussed possible compound was detected. Actually, the Raman spectrum collected with 1064 nm shows the presence of a big amount of sulfate compound (Figure 3a), as demonstrated by the presence of a strong peak around 1000 cm^{-1}, typical for SO_4 vibration modes. The existence of this band is revealed also in the samples treated at lower temperatures than 500 °C. To define, with greater precision, the position of the sulfate vibration and, therefore, to be able to identify the compound, the Raman spectra were acquired with a better resolution (1 cm^{-1}), with a 532 nm source. The Raman spectra with high resolution presented a shift of about 6 cm^{-1} between the peaks due to the heat treatment, visible also with the help of a deconvolution procedure by Lorentzian curves around 990–1000 cm^{-1} (SO_4 vibration), see Figure 3b, confirming the whitish $CdSO_4$ formation.

As previously discussed, no peaks related to CdO and $CdSO_4$ compounds at lower temperature than 500 °C are noticeable by XRD, suggesting the possible formation of a thin superficial layer of whitish $CdSO_4$ crust, undetectable with the XRD technique because of the detection limit threshold. Only a broadening in the region between 24° and 34° was revealed, suggesting a progressive structural disorder in the phase CdZnS. This hypothesis can be confirmed with a detailed analysis of some vibrational modes of the Raman spectra. As reported in the literature [39,40] the ratio between the 215 cm^{-1} (TO multi-phonon process) and 300 cm^{-1} (LO) peaks of Cd-pigment Raman spectrum can provide information about the structural disorder, comparing the spectra of the natural and heated samples. In our case, the calculated ratio is drawn in Figure 3c. As reported in [40], the increase in structural disorder and zinc content leads to a decrease of 215 cm^{-1} band for the TO mode. In our yellow samples, the heat treatment produces a decrement in

TO-peak intensity, as can been seen directly by the spectra reported in Figure 3c, confirming the hypothesis.

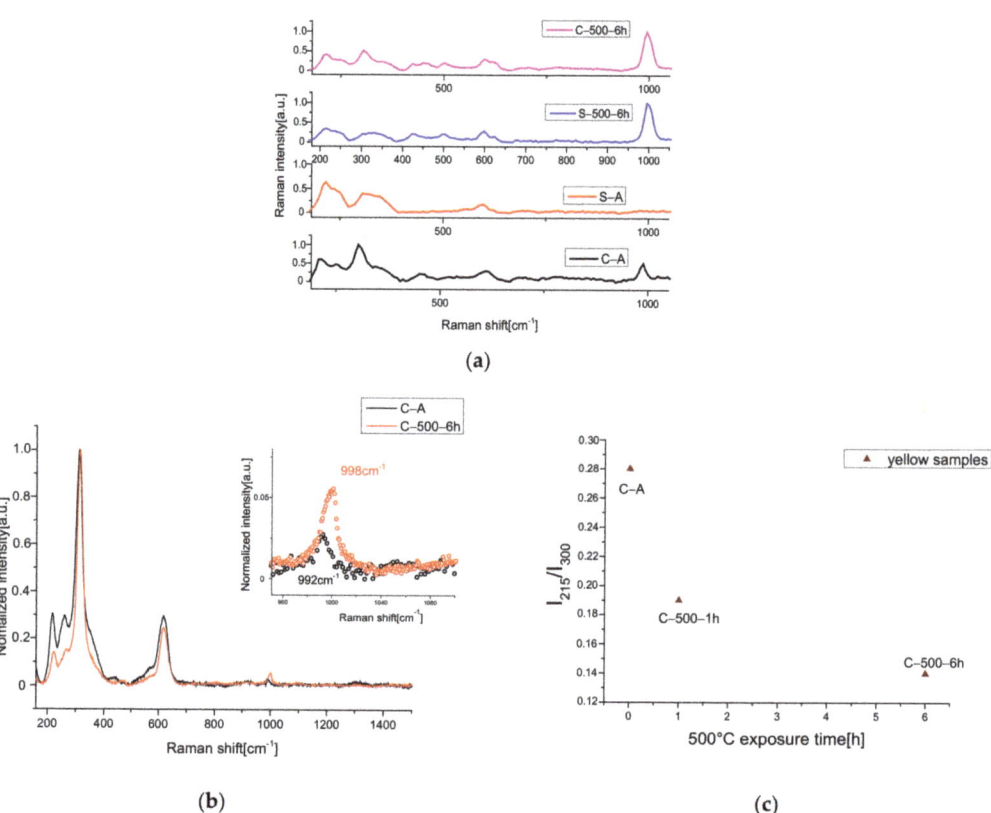

Figure 3. (a) 1064 nm excited Raman spectra before and after 500 °C for 6 h; (b) 532 nm-Raman spectra of C-A and C-500-6 h; in inset, the R.O.I with the value of the peaks obtained by the deconvolution process. (c) Variation in TO/LO bands in high-resolution 532 nm excited Raman spectra of yellow samples.

Finally, to obtain further confirmation and information about the superficial effect of heating treatment, SEM-EDS measurements were also performed on these two samples. In Table 3, stoichiometric calculations on the element percentages to reach Cd saturation suggest, again, the presence of hydrate sulfate compounds and almost a double amount of Cd-oxide.

3.2. UV Exposure Process

In order to establish the light stability in Cd pigments used in paints, a UV treatment at 365 nm, with different exposure times, was conducted. To characterize the pigment variations, after this degradation process, we adopted Raman spectroscopy, reflectance, luminescence, and transient absorption. As found in the literature [21], the UV action in the presence of oxygen can produce the formation of sulfate compounds.

In particular, Raman analyses made on the yellow sample confirmed a broad band in the region of sulfate vibrations between 990 cm^{-1} and 1008 cm^{-1} (see Figure 4a).

Table 3. Calculated Cd- compounds percentage by saturation of Cd- amount for S-500-6 h and C-500-6 h.

Calculation of Cadmium Saturation for S-500-6 h Sample		
	Point 1 [%]	Point 5 [%]
Cd (Se,S)	26.3	27.6
CdSO$_4$(8/3 H$_2$O)	24.4	26.3
CdO	49.3	46.1
	Point 2 [%]	Point 3 [%]
CdZnS	32.7	38.7
CdSO$_4$(8/3 H$_2$O)	32.2	23.7
CdO	35.1	37.6

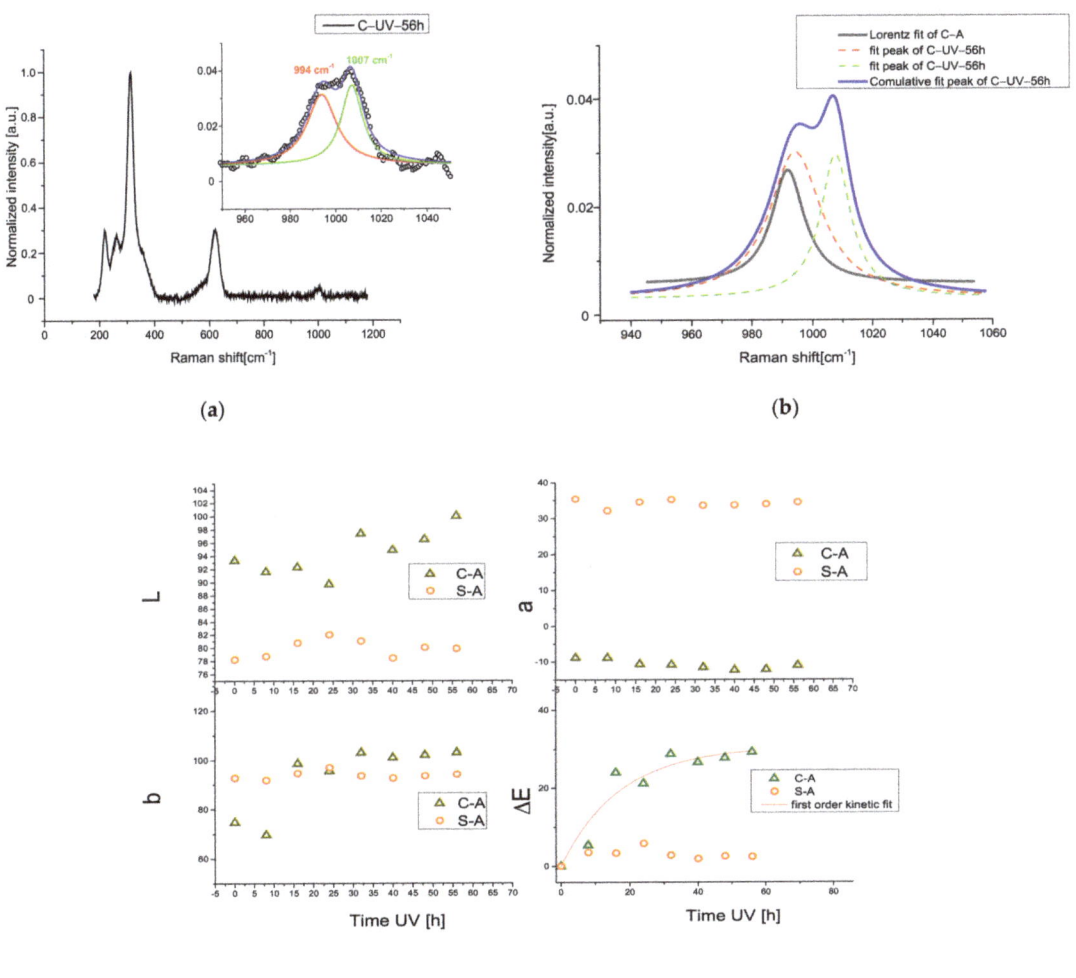

Figure 4. (a) Raman spectrum of C-UV-56 h sample with 532 nm excitation wavelength; in inset, the deconvolution of the region of sulfate; (b) deconvolution of the region 900–1060 cm^{-1} of C-A Raman spectrum; (c) CIE coordinate trend for C-A and S-A samples exposed to UV radiation. The ΔE for the yellow sample shows exponential growth.

To establish if the new sulfate compound is derived from CdS degradation or if the starting barium sulfate converted to another form, in Figure 4b, the deconvolutions (Lorentzian functions) for the artificially and no-degraded samples are presented in the region of interest (R.O.I). The C-UV-56 h signal consists of two peaks, one located at 994 cm^{-1} with an area of 0.89 and the other one at 1007 cm^{-1} with an area of 0.5. The former is very similar to the one of the C-A band (grey line) located at 992 cm^{-1} (attributed to BaSO$_4$). The deconvolution procedure presents a slight red shifting and about a doubling of the relative area, suggesting a conversion of barium sulfate to another form. The latter can be associated with CdS degradation in cadmium sulfate. In fact, the peak at 1007 cm^{-1} is usually associated with cadmium sulfate compounds bound with the $x \cdot H_2O$ molecule [41]. To confirm the formation of this species, a compositional analysis by means of SEM-EDS was made and reported point by point in the Supplementary Materials. For the C-UV-56 h sample, the results summarized in Table 4 confirm the presence of Cadmium hydrate sulfate and a possible excess of -S^{2-}, suggesting the formation of a notable amount of Cd vacancies inside the CdS crystal after light exposure. This is another known cause for the color change in the Cd pigment, as previously reported in the literature [22]. Actually, the Cd vacancies led to the formation of an intra-gap level with NIR emission and time decays of some microseconds, as already discussed before in the case of thermally treated samples.

Table 4. Calculated Cd- compound percentage by saturation of Cd- amount.

C-UV-56 h	Point 1 [%]	Point 2 [%]
CdZnS	13.3	64
CdSO$_4$ (8/3 H$_2$O)	86.7	36
CdO	-	-
-S^{2-} in excess	1.3	1.7

With the intention of determining the real effect of color change, a detailed colorimetric analysis was performed. Reflectance spectra were collected with 8 h steps of UV exposure and the related CIE Lab coordinates were calculated. The results are summarized in the Supplementary Materials (Figure S1). The L parameter tends to increase, mainly for the C-A sample with light exposure and the shape of the first derivative (Figure S1A,B), after 56 h, widens, showing a blue shift, typical of the addition of light and white shades in yellow pigments, as studied by Gueli et al. [31]. The total variation in CIE coordinates is represented in CIE space. The S-A sample did not show a remarkable difference in CIE coordinates after 56 h of UV exposure as visible from the color space diagram (see Figure S1C).

To speed up the reading of chromatic variations, a graph with the relative value for each coordinate is shown in Figure 4c, where we tried to express the amount of the total color variation by the ΔE value through a first kinetic model (exponential fit) for the C-A sample:

$$\Delta E = A(1 - e^{\frac{-t}{\tau}}) \tag{7}$$

where A is the asymptotic value of the ΔE curve (the final step of conversion of CdS to CdSO$_4$, involving the chromatic change from yellow colors to white ones) and tau is the characteristic time of reaction. After a time of about 40 h, the conversion is completed.

To understand what UV light accomplishes in the process, we monitored the stability of the studied pigments, kept in standard environmental conditions, and we observed that, effectively, the degradation also started slowly in a natural way. After a deposition above a slide, the samples, heated and no-heated (Figure 5), were kept at room temperature with 20 RH%, illuminated by artificial light (Compact Fluorescence Lamps) for 4 h/days for 6 months. NIR-Raman spectra were acquired showing the development of a new shoulder in the sulphate region. In Figure 5, the deconvolution of this region is represented. Even in this case, the formation of sulphate compounds was recorded, revealing that the action of light and mainly of the oxygen leads to degradation in Cd pigments.

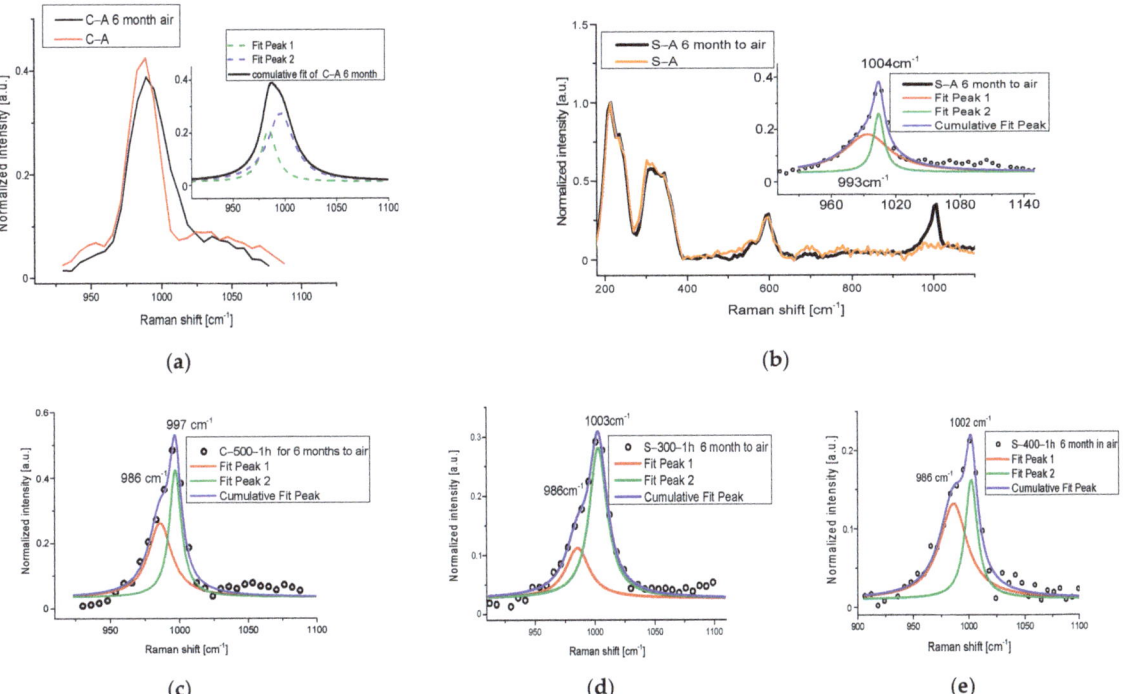

Figure 5. (**a**) Raman spectrum of C-A sample exposed to 6 months in air in a range between 900 and 1110 cm^{-1}; in inset, the band deconvolutions; (**b**) Raman spectra of S-A sample and S-A after 6 months in air; in inset, the deconvolutions of the R.O.I (sulphate region) for the sample exposed to air; (**c**) deconvolutions of the R.O.I for the C-500-1 h sample exposed to 6 months to air; (**d**) deconvolutions of the R.O.I for the S-300-1 h sample exposed for 6 months to air; (**e**) deconvolutions of the R.O.I for the S-400-1 h sample exposed for 6 months to air.

The optical variations registered on cadmium yellow can be explained in detail with an in-depth study of the electronic properties of this pigment and its behavior after the accelerated degradation process. For this reason, we characterized, with pump–probe spectroscopy, the optical differences induced in our samples after UV exposure and heat treatment. To clarify the used nomenclature, we will describe, as a short-lived signal, those that last some picoseconds up to tens of picoseconds and long-lived signals, those ranging from hundreds of ps to ns. As reported in Figure 6a, the C-A sample shows a broad positive signal (excited state absorption—ESA), centered at 477 nm, with a duration of 300 ps. In addition, a shorter negative broad signal in the region between 650 nm and up to 800 nm is presented.

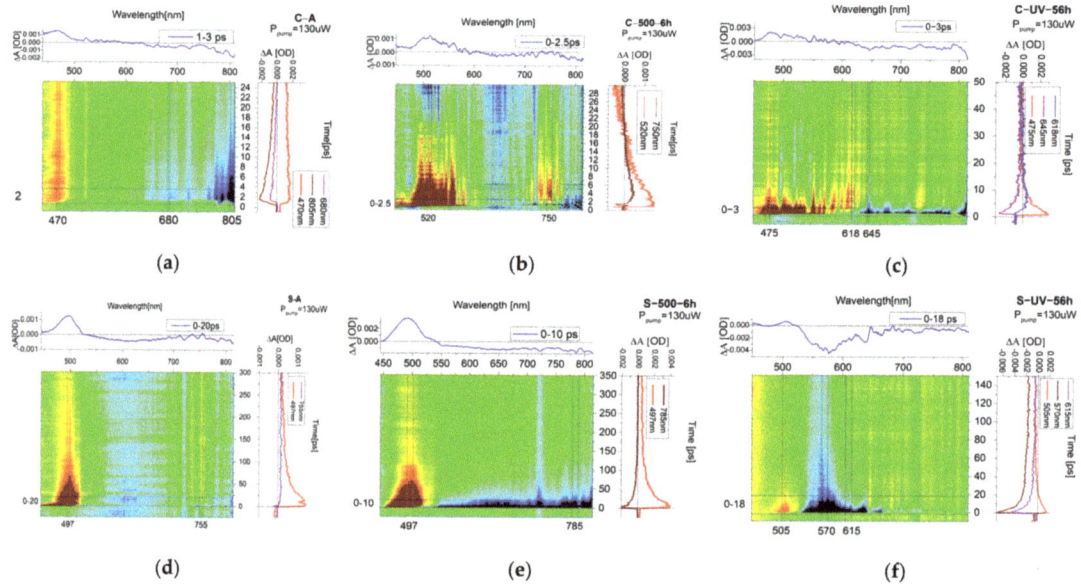

Figure 6. Transient absorption map of (**a**) C-A sample, (**b**) C-500-6 h sample, (**c**) C-UV-56 h sample, (**d**) S-A sample, (**e**) S-500-6 h sample, and (**f**) S-UV-56 h sample.

As a comparison, thermal treatment at 500 °C for 6 h drastically changed the previous ESA signal (Figure 6b). In this case, the spectrum is composed mainly of a broad short-lived signal, having different peaks in the region between 475 and 580 nm. Before 475 nm, the same ESA observed for the C-A sample is observed, but with a duration of only 5 ps. The broad short-lived signal after 20 ps converted into a negative signal, probably stimulated emission (SE). Around 750 nm, we have another positive signal with the same duration (about 20 ps) but with a longer rise time. The ground state depletion (GSD) signal in the near infrared is located only in the region 775–815 nm and with a very short time of about 5 ps. The disappearing of the broad trap state band (negative signal between 600 and 750) after the thermal treatment agrees with the TR-PL measurements in the micro-second scale mentioned before (Figure 2c).

The UV exposure led a further broadening of the ESA signals (Figure 6c) with respect to the C-A sample until 630 nm, after which a broad short-lived negative signal started from 640 nm. Additional positive bands start a few picoseconds after the pump absorption, suggesting a non-radiative relaxation to a lower level, from which starts the probe absorption.

Concerning the orange sample, the no-treated powder is composed of a long-lived ESA signal centered at around 495 nm. No negative signals were detected for this sample (Figure 6d). The heated sample (Figure 6e) showed the same ESA signal and, in addition, a new very broad long-lived negative signal in a region between 550 nm and 815 nm (2.25–1.52 eV) attributed to a GSD. This result is compatible with TR-PL measurements where the new trap states around 750–820 nm were obtained after the heating relax after microseconds (Figure 2d).

The UV-exposed sample showed variations in the kinetics of ESA signal, which became shorter (duration of about 40 ps) and the formation of a new broad negative signal centered at 570 nm. The latter can be broken up into two regions with different kinetics, as can be clearly seen from Figure 6f, associable to a long-lived GSD signal at 570 nm and short-lived GSD at 650 nm, both towards trap states.

To produce a first interpretation of these results, we can hypothesize that the negative signal in the NIR around 780 nm, found in both normal and UV-exposed samples, could be linked to Cd vacancies, as confirmed by SEM-EDS calculation and by previous authors [22]. Furthermore, the substantial difference in UV exposure for the C sample resides in the broadening of absorption signals, attributable to a change inside the conduction band structure, new levels due to the formation of defects, and changes in electronic transfer, as suggested by different kinetics observed in some positive signals. The thermal treatment and UV exposure, mainly for orange samples but even, to a lesser degree, in white yellow, led the formation of new trap states inside the bandgap, defects responsible for the darkening effect in aged samples.

4. Conclusions

In summary, in this study, we thoroughly analyzed the degradation of CdS yellow pigment used in paintings, with the purpose of shedding light about the possible causes and providing useful information for the field of Cultural Heritage. Light-yellow and orange pigments from Kremer were artificially degraded through different accelerated ageing: heat treatment and UV light exposure. Whereas the orange pigment seems more stable, the yellow one degrades more markedly. Reflectance spectra and chromatic coordinates in the CIELab space revealed that the heating treatment, executed in a range of 100–500 °C until 6 h, causes a prominent color variation for the light-yellow pigment in terms of bleaching. XRD and Raman spectroscopy suggest that the cause is attributable to the formation of sulfate compounds and possible cadmium oxide, detected in the XRD pattern of the sample heated at 500 °C for 6 h, after the reaction with the oxygen present in the environment atmosphere. This interpretation is also confirmed by the punctual SEM/EDS analysis. The formation of deep trap states and oxide products after thermal treatment was also confirmed by TRPL measurement, performed in the ps and micro-second scale. The same conclusion for the formation of whitish compounds can be adopted for UV treatment, which allows one to demonstrate, by means of Raman spectroscopy and SEM-EDX, the formation of a superficial sulfate phase. In addition, the action of UV light for the yellow sample, and both UV and thermal exposure for the orange one, seems to produce a defective phase where intra-gap energy levels are generated. Actually, with the help of pump–probe measurements, GSD and long ESA signals due to the formation of trap levels are evidenced in the visible and near-infrared region, both for light-yellow and orange pigments. If structural defectivities are activated by light exposure, the reaction with atmosphere seems to also produce darker compounds, such as cadmium oxide, and whitish compounds, such as cadmium sulfate, as previously reported in the literature.

Supplementary Materials: The following supporting information can be downloaded at: https://www.mdpi.com/article/10.3390/ma15165533/s1, Figure S1: First derivative reflectance spectra for C-A samples (a) and S-A sample (b) before and after 56 h of UV exposure, (c) representation of CIE value of these samples in the CIE color space. Tables S1 and S2: elemental compositions of C-A sample. Table S3: Semi-quantitative analysis based on SEM EDX results for C-A and S-A. Table S4: SEM-EDX quantitative elementary analysis for sample C-UV-56 h. Table S5: SEM-EDX quantitative elementary analysis for sample C-500-6h. Table S6: SEM-EDX quantitative elementary analysis for sample S-500-6h.

Author Contributions: Conceptualization, D.C. and F.A.P.; methodology, D.C.; validation, D.C., P.C.R. and C.M.C.; formal analysis, F.A.P.; investigation, F.A.P. and S.P.; data curation, F.A.P. and D.C.; writing—original draft preparation, F.A.P.; writing—review and editing, D.C. and C.M.C.; supervision, D.C. All authors have read and agreed to the published version of the manuscript.

Funding: This research received no external funding.

Data Availability Statement: Not applicable.

Conflicts of Interest: The authors declare no conflict of interest.

References

1. Shank, J.W.; Feller, R.L. Artists' Pigments. A Handbook of Their History and Characteristics, Vol. 1. *Leonardo* **1989**, *22*, 267–268. [CrossRef]
2. Barnett, J.R.; Miller, S.; Pearce, E. Colour and art: A brief history of pigments. *Opt. Laser Technol.* **2006**, *38*, 445–453. [CrossRef]
3. Anaf, W.; Schalm, O.; Janssens, K.; De Wael, K. Understanding the (in) stability of semiconductor pigments by a thermodynamic approach. *Dye. Pigment.* **2015**, *113*, 409–415. [CrossRef]
4. Anaf, W.; Trashin, S.; Schalm, O.; Van Dorp, D.; Janssens, K.; De Wael, K. Electrochemical photodegradation study of semiconductor pigments: Influence of environmental parameters. *Anal. Chem.* **2014**, *86*, 9742–9748. [CrossRef] [PubMed]
5. Chiriu, D.; Pisu, F.A.; Ricci, P.C.; Carbonaro, C.M. Application of raman spectroscopy to ancient materials: Models and results from archaeometric analyses. *Materials* **2020**, *13*, 2456. [CrossRef] [PubMed]
6. Chiriu, D.; Pala, M.; Pisu, F.A.; Cappellini, G.; Ricci, P.C.; Carbonaro, C.M. Time through colors: A kinetic model of red vermilion darkening from Raman spectra. *Dye. Pigment.* **2021**, *184*, 108866. [CrossRef]
7. Chiriu, D.; Ricci, P.C.; Carbonaro, C.M.; Nadali, D.; Polcaro, A.; Collins, P. Raman identification of cuneiform tablet pigments: Emphasis and colour technology in ancient Mesopotamian mid-third millennium. *Heliyon* **2017**, *3*, e00272. [CrossRef]
8. Pisu, F.A.; Carbonaro, C.M.; Corpino, R.; Ricci, P.C.; Chiriu, D. Fresco paintings: Development of an aging model from 1064 nm excited raman spectra. *Crystals* **2021**, *11*, 257. [CrossRef]
9. Yu, J.; Warren, W.S.; Fischer, M.C. Visualization of vermilion degradation using pump-probe microscopy. *Sci. Adv.* **2019**, *5*, eaaw3136. [CrossRef]
10. Villafaña, T.E.; Samineni, P.; Warren, W.S.; Fischer, M.C. Historical pigments revealed by pump-probe microscopy. In Proceedings of the Laser Science, Rochester, NY, USA, 14–18 October 2012; LS2012.
11. Traill, R.J.; Boyle, R.W. Hawelite, Isometric Cadmium Sulphide, A New Mineral. *Am. Mineral. J. Earth Planet. Mater.* **1955**, *40*, 555–559.
12. Leone, B.; Burnstock, A.; Jones, C.; Hallebeek, P.; Boon, J.J.; Keune, K. The deterioration of cadmium sulphide yellow artists' pigments. In Proceedings of the 14th Triennial Meeting, ICOM Committee for Conservation, The Hague, The Netherlands, 12–16 September 2005.
13. Van Der Snickt, G.; Dik, J.; Cotte, M.; Janssens, K.; Jaroszewicz, J.; De Nolf, W.; Groenewegen, J.; Van Der Loeff, L. Characterization of a degraded cadmium yellow (CdS) pigment in an oil painting by means of synchrotron radiation based X-ray techniques. *Anal. Chem.* **2009**, *81*, 2600–2610. [CrossRef]
14. Kulkarni, V.G.; Garn, P.D. Study of the formation of cadmium sulfoselenide. *Thermochim. Acta* **1986**, *99*, 33–36. [CrossRef]
15. Mass, J.; Sedlmair, J.; Patterson, C.S.; Carson, D.; Buckley, B.; Hirschmugl, C. SR-FTIR imaging of the altered cadmium sulfide yellow paints in Henri Matisse's le Bonheur de vivre (1905-6)-examination of visually distinct degradation regions. *Analyst* **2013**, *138*, 6032–6043. [CrossRef]
16. Pouyet, E.; Cotte, M.; Fayard, B.; Salomé, M.; Meirer, F.; Mehta, A.; Uffelman, E.S.; Hull, A.; Vanmeert, F.; Kieffer, J.; et al. 2D X-ray and FTIR micro-analysis of the degradation of cadmium yellow pigment in paintings of Henri Matisse. *Appl. Phys. A Mater. Sci. Process.* **2015**, *121*, 967–980. [CrossRef]
17. Van Der Snickt, G.; Janssens, K.; Dik, J.; De Nolf, W.; Vanmeert, F.; Jaroszewicz, J.; Cotte, M.; Falkenberg, G.; Van Der Loeff, L. Combined use of Synchrotron Radiation Based Micro-X-ray Fluorescence, Micro-X-ray Diffraction, Micro-X-ray Absorption Near-Edge, and Micro-Fourier Transform Infrared Spectroscopies for Revealing an Alternative Degradation Pathway of the Pigment Cadmium Ye. *Anal. Chem.* **2012**, *84*, 10221–10228. [CrossRef]
18. Comelli, D.; Maclennan, D.; Ghirardello, M.; Phenix, A.; Schmidt Patterson, C.; Khanjian, H.; Gross, M.; Valentini, G.; Trentelman, K.; Nevin, A. Degradation of Cadmium Yellow Paint: New Evidence from Photoluminescence Studies of Trap States in Picasso's Femme (époque des "demoiselles d'Avignon"). *Anal. Chem.* **2019**, *91*, 3421–3428. [CrossRef]
19. Armani, N.; Salviati, G.; Nasi, L.; Bosio, A.; Mazzamuto, S.; Romeo, N. Role of thermal treatment on the luminescence properties of CdTe thin films for photovoltaic applications. *Thin Solid Film.* **2007**, *515*, 6184–6187. [CrossRef]
20. Fan, Y.; Deng, M.; Chen, G.; Zhang, Q.; Luo, Y.; Li, D.; Meng, Q. Effect of calcination on the photocatalytic performance of CdS under visible light irradiation. *J. Alloys Compd.* **2011**, *509*, 1477–1481. [CrossRef]
21. Monico, L.; Chieli, A.; De Meyer, S.; Cotte, M.; de Nolf, W.; Falkenberg, G.; Janssens, K.; Romani, A.; Miliani, C. Frontispiece: Role of the Relative Humidity and the Cd/Zn Stoichiometry in the Photooxidation Process of Cadmium Yellows (CdS/Cd1−xZnxS) in Oil Paintings. *Chem.–A Eur. J.* **2018**, *24*, 11584–11593. [CrossRef]
22. Giacopetti, L.; Nevin, A.; Comelli, D.; Valentini, G.; Nardelli, M.B.; Satta, A. First principles study of the optical emission of cadmium yellow: Role of cadmium vacancies. *AIP Adv.* **2018**, *8*, 065202. [CrossRef]
23. Giacopetti, L.; Satta, A. Degradation of Cd-yellow paints: Ab initio study of native defects in {10.0} surface CdS. *Microchem. J.* **2016**, *126*, 214–219. [CrossRef]
24. Giacopetti, L.; Satta, A. Reactivity of Cd-yellow pigments: Role of surface defects. *Microchem. J.* **2018**, *137*, 502–508. [CrossRef]
25. Paulus, J.; Knuutinen, U. Cadmium colours: Composition and properties. *Appl. Phys. A Mater. Sci. Process.* **2004**, *79*, 397–400. [CrossRef]
26. Mikhailov, M.M.; Yuryev, S.A.; Lapin, A.N.; Lovitskiy, A.A. The effects of heating on BaSO4 powders' diffuse reflectance spectra and radiation stability. *Dye. Pigment.* **2019**, *163*, 420–424. [CrossRef]

27. Riveros, G.; Gómez, H.; Henríquez, R.; Schrebler, R.; Córdova, R.; Marotti, R.E.; Dalchiele, E.A. Electrodeposition and characterization of ZnX (X=Se, Te) semiconductor thin films. *J. Chil. Chem. Soc.* **2002**, *47*, 411–429. [CrossRef]
28. Marotti, R.E.; Guerra, D.N.; Bello, C.; Machado, G.; Dalchiele, E.A. Bandgap energy tuning of electrochemically grown ZnO thin films by thickness and electrodeposition potential. *Sol. Energy Mater. Sol. Cells* **2004**, *82*, 85–103. [CrossRef]
29. Henríquez, R.; Grez, P.; Muñoz, E.; Gómez, H.; Badán, J.A.; Marotti, R.E.; Dalchiele, E.A. Optical properties of CdSe and CdO thin films electrochemically prepared. *Thin Solid Film.* **2010**, *518*, 1774–1778. [CrossRef]
30. Nair, P.K.; Daza, O.G.; Reádigos, A.A.C.; Campos, J.; Nair, M.T.S. Formation of conductive CdO layer on CdS thin films during air heating. *Semicond. Sci. Technol.* **2001**, *16*, 651. [CrossRef]
31. Gueli, A.M.; Gallo, S.; Pasquale, S. Optical and colorimetric characterization on binary mixtures prepared with coloured and white historical pigments. *Dye. Pigment.* **2018**, *157*, 342–350. [CrossRef]
32. Cesaratto, A.; D'Andrea, C.; Nevin, A.; Valentini, G.; Tassone, F.; Alberti, R.; Frizzi, T.; Comelli, D. Analysis of cadmium-based pigments with time-resolved photoluminescence. *Anal. Methods* **2014**, *6*, 130–138. [CrossRef]
33. Zou, B.S.; Volkov, V.V.; Wang, Z.L. Optical properties of amorphous ZnO, CdO, and PbO nanoclusters in solution. *Chem. Mater.* **1999**, *11*, 3037–3043. [CrossRef]
34. Nevin, A.; Cesaratto, A.; Bellei, S.; D'Andrea, C.; Toniolo, L.; Valentini, G.; Comelli, D. Time-resolved photoluminescence spectroscopy and imaging: New approaches to the analysis of cultural heritage and its degradation. *Sensors* **2014**, *14*, 6338–6355. [CrossRef] [PubMed]
35. Srihari, V.; Sridharan, V.; Ravindran, T.R.; Chandra, S.; Arora, A.K.; Sastry, V.S.; Sundar, C.S. Raman scattering of cadmium oxide: In B1 phase. In Proceedings of the AIP Conference, Manipal, India, 26–30 December 2010.
36. Kumar, S.; Ojha, A.K. Synthesis, characterizations and antimicrobial activities of well dispersed ultra-long CdO nanowires. *AIP Adv.* **2013**, *3*, 052109. [CrossRef]
37. Ganesh, V.; Manthrammel, M.A.; Shkir, M.; AlFaify, S. Investigation on physical properties of CdO thin films affected by Tb doping for optoelectronics. *Appl. Phys. A Mater. Sci. Process.* **2019**, *125*, 249. [CrossRef]
38. Popović, Z.V.; Stanišić, G.; Stojanović, D.; Kostić, R. Infrared and Raman Spectra of CdO. *Phys. Status Solidi B* **1991**, *165*, K109–K112. [CrossRef]
39. Chi, T.T.K.; Gouadec, G.; Colomban, P.; Wang, G.; Mazerolles, L.; Liem, N.Q. Off-resonance Raman analysis of wurtzite CdS ground to the nanoscale: Structural and size-related effects. *J. Raman Spectrosc.* **2011**, *42*, 1007–1015. [CrossRef]
40. Rosi, F.; Grazia, C.; Gabrieli, F.; Romani, A.; Paolantoni, M.; Vivani, R.; Brunetti, B.G.; Colomban, P.; Miliani, C. UV–Vis-NIR and micro Raman spectroscopies for the non destructive identification of Cd1−xZnxS solid solutions in cadmium yellow pigments. *Microchem. J.* **2016**, *124*, 856–867. [CrossRef]
41. Falgayrac, G.; Sobanska, S.; Brémard, C. Heterogeneous microchemistry between $CdSO_4$ and $CaCO_3$ particles under humidity and liquid water. *J. Hazard. Mater.* **2013**, *248*, 415–423. [CrossRef]

Article

X-ray Imaging and Computed Tomography for the Identification of Geometry and Construction Elements in the Structure of Old Violins

Mariana Domnica Stanciu [1,2,*], Mircea Mihălcică [1,*], Florin Dinulică [3], Alina Maria Nauncef [4], Robert Purdoiu [5], Radu Lăcătuș [5] and Ghiorghe Vasile Gliga [6,7]

1. Faculty of Mechanical Engineering, Transilvania University of Brasov, B-dul Eroilor 29, 500360 Brasov, Romania
2. Russian Academy of Natural Sciences Sivtsev Vrazhek, 29/16, 119002 Moscow, Russia
3. Department of Forest Engineering, Forest Management Planning and Terrestrial Measurements, Transilvania University of Brașov, 500123 Brașov, Romania; dinulica@unitbv.ro
4. Faculty of Music, Transilvania University of Brașov, 500360 Brașov, Romania; a_nauncef@unitbv.ro
5. Faculty of Veterinary Medicine Cluj Napoca, University of Agricultural Sciences and Veterinary Medicine Cluj-Napoca, Calea Mănăștur 3-5, 400374 Cluj-Napoca, Romania; robert.purdoiu@usamvcluj.ro (R.P.); radu.lacatus@usamvcluj.ro (R.L.)
6. Faculty of Furniture Design and Wood Engineering, Transilvania University of Brasov, B-dul Eroilor 29, 500360 Brasov, Romania; vasile.gliga@unitbv.ro
7. S.C. Gliga Musical Instruments S.A., str. Pandurilor 120, 545430 Mureș, Romania
* Correspondence: mariana.stanciu@unitbv.ro (M.D.S.); mircea.mihalcica@unitbv.ro (M.M.)

Citation: Stanciu, M.D.; Mihălcică, M.; Dinulică, F.; Nauncef, A.M.; Purdoiu, R.; Lăcătuș, R.; Gliga, G.V. X-ray Imaging and Computed Tomography for the Identification of Geometry and Construction Elements in the Structure of Old Violins. *Materials* **2021**, *14*, 5926. https://doi.org/10.3390/ma14205926

Academic Editors: Žiga Šmit and Eva Menart

Received: 1 September 2021
Accepted: 7 October 2021
Published: 9 October 2021

Publisher's Note: MDPI stays neutral with regard to jurisdictional claims in published maps and institutional affiliations.

Copyright: © 2021 by the authors. Licensee MDPI, Basel, Switzerland. This article is an open access article distributed under the terms and conditions of the Creative Commons Attribution (CC BY) license (https://creativecommcns.org/licenses/by/4.0/).

Abstract: Numerous studies on heritage violins have shown that there are a number of factors that contribute to the acoustic quality of old violins. Among them are the geometric shape of the violin, the thickness of the tiles, the arching of the tiles, the dimensions and position of the bass bar, the size and position of the acoustic holes. Thus, the paper aims to compare the structural and constructive elements of old violins made in various famous violin workshops (Stainer, Klotz, Leeb, Babos Bela), using nondestructive and noncontact techniques based on image analysis. The violins that were studied date from 1716 to 1920, being in good condition, most of them being used by artists from the Brașov Philharmonic of Romania. In the first stage of the study, the violins were optically analyzed and scanned to identify the structure of the resonant wood, using the WinDENDRO Density 2007 program. X-ray imaging and computed tomography (CT) were also used. Combining the types of analyses, capitalizing on the expertise of violin producers and the knowledge of researchers in the field, valuable data on the geometric and constructive characteristics of old violins were extracted.

Keywords: old violin; X-ray imaging; computed tomography; resonance wood; constructive elements

1. Introduction

It is unanimously recognized that the queen of stringed musical instruments is the violin, an instrument whose shape, size and materials have reached the highest performance of musical sounds, through the ancient and established luthiers Andrea Amati (1505–1578), Giuseppe Guarneri del Gesù (1698–1744) and Antonio Stradivari (1644–1737). The current shape of the violin was established by Andrea Amati (1505–1578); over time, violin makers brought only small changes into the constructive elements, almost imperceptible to an ordinary visual analysis. Many of these musical instruments are rare examples of high artistic mastery and are still used as a reference in the contemporary manufacture of violins [1–3]. In addition to the consecrated violin makers mentioned above, violin making workshops have been developed in other regions of Europe, through violin makers who completed their apprenticeship in Italian violin makers' workshops and who imprinted the specifics of the area on their violin models.

Therefore, Jacobus Stainer (1619–1683) was the most famous luthier of the Austrian-German schools, being born in Absam (Tyrol). He is supposed to have been the disciple of Nicolò Amati of Cremona, although the manuscripts and historical evidence are not complete enough to justify this assumption. In any case, his work, the oldest of which dates from the 1630s, bears a strong resemblance to that of Amati. Stainer eventually settled in his hometown of Absam in 1656, where he began producing some of his best instruments, which appear to be inspired by Amati's models. During this period, Stainer created his own style, producing exceptional instruments that rivaled or even surpassed the works of his Cremonese contemporaries of the seventeenth century. As specific constructive features, we can notice that the arching of the front plates is higher than that of the rear plates; the growth is maintained up to half the length; and the finish used is yellow, with a shade of pale rose [4–6]. The resonant bar is terminated at its two ends by bevels that extend on the sound plate to which it is glued, being placed under the G string of the instrument.

Joseph Thomas Klotz (1743–1819) was the son of Sebastian Klotz, one of Stainer's best disciples and successors, and he had his workshop in Mittenwald, Germany. Historians believe that this artist built the violins according to his father's system, but he knew the qualities of wood better; therefore, his instruments were superior in tone, but inferior in the finish (in their initial, original state).

Johann Georg (II) Leeb (1740–1813) was the son of Johann Georg Leeb (I), both of whom marked the Hungarian violin school. Johann Georg Leeb worked in Presburg (now Bratislava). Leeb violin designs are quite distinct, with flat arches with a certain influence of the violinist Carlo Bergonzi (1683–1747). Johann G. (II) Leeb was a prolific manufacturer and produced violins of different quality classes, using different materials. He was succeeded by his son Johann Georg (III), born in 1779.

Babos Bela was a representative of the Hungarian violin school from the beginning of the 20th century [5,6]. In Romania, in the well-known city of violins, Reghin, the art of violin making developed to a level of qualitative and aesthetic perfection, with numerous violin workshops, due to the high quality of wood existing in the area known in antiquity as the Italian valley. At first glance, the constructive form of the violin and the wood in its structure have been preserved over the centuries, but there are constructive details that differentiate the style of the violinists and even the acoustics of the instrument. The constructive elements of a violin have both a functional and an aesthetic role. Thus, the body of the violin, composed of the upper plate, the back plate, straps and counter-straps, has the acoustic role of amplifying the musical sounds emitted during the movement of the bow over the strings. For reasons of mechanical strength, the body of the violin also contains constructive elements that fix the two plates (by means of straps, counter-straps, hubs and corners) and elements that support and fix the neck of the violin (Figure 1). The plates have a spatially curved shape in both the longitudinal and transverse directions. Their thickness varies from the center (the area between the holes f) to the edges. From the wood species point of view, a selected (for the structural characteristics) softwood (spruce—Picea Abies L. Karst) is used for the top plates of the violins (as well as of all stringed instruments) and curly maple wood (Acer pseudoplatanus L.) is used for the back plates. Previous research has shown that old violins emit much clearer, brighter, louder sounds than new violins. Over time, the determining factors for this aspect have been analyzed, starting from the structural quality of the wood, moisture content, wood ageing, slab geometry (thickness/arching), finishes, constructive elements of the violins (stern, position and shape of acoustic holes, sounding bar, gag) and string quality [7–11]. It has not yet been possible to detect the predominant factor, thus the research remains open. The non-invasive structural analysis of historical musical instruments is a fundamental tool for defining restoration and conservation protocols, as well as for the study of ancient manufacturing techniques and acoustic analysis related to this class of cultural objects. The importance and value of typical bowed string instruments, on the other hand, requires a non-destructive approach with strict environmental control, fast acquisition times and high spatial resolution [12–16].

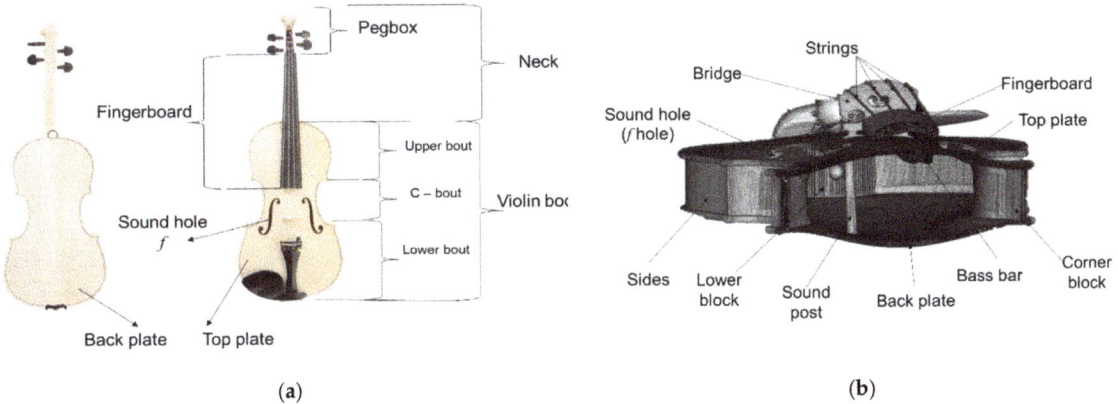

Figure 1. The constructive elements of the violin: (**a**) the front view of the violin; (**b**) cross section through the violin.

Through various non-invasive and non-destructive modern methods and techniques, numerous researchers have performed structural assessments of musical instruments, highlighting the richness of details, characterizing their internal structure, identifying defects, assessing the thickness of structural elements of wood and its density and conducting a dendrochronological investigation of historical violins [16–19].

From a constructive and technological point of view, the top plates of a violin are obtained from two halves, cut radially longitudinally from the logs (as in Figure 2), which, after their natural drying, are conditioned in a drying chamber up to a moisture content of 6–8%. Then, the pairs boards are glued lengthwise, obtaining a plate with an anatomical structure of the wood symmetrical to the median longitudinal axis.

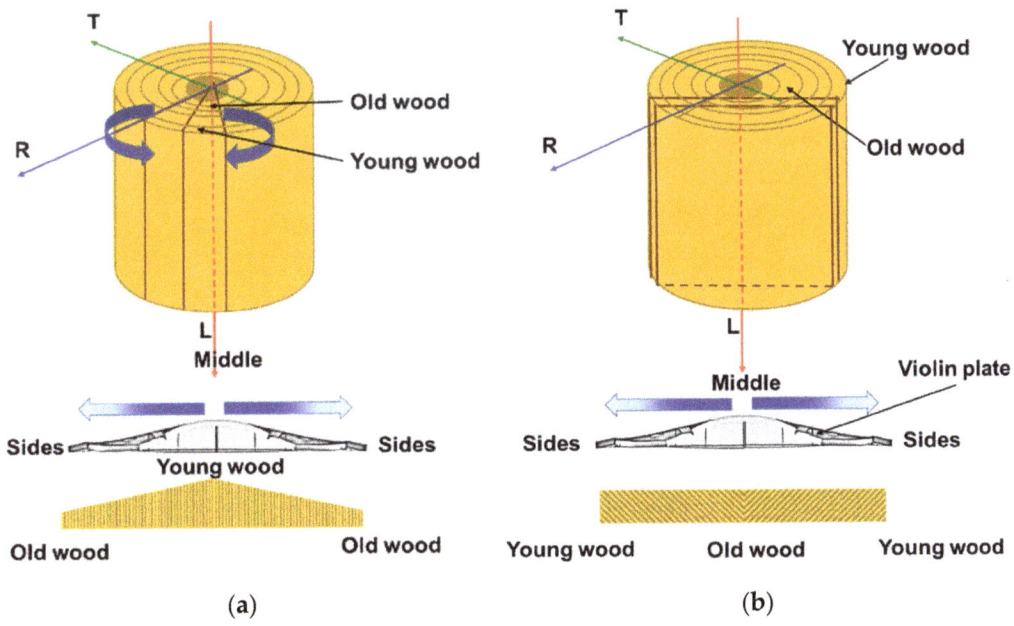

Figure 2. The main types of wood cuts for the violin: (**a**) quarter sawn; (**b**) live sawn.

Most of the wood that is used in a violin's plates construction is cut on the quarter. Nyman, 1975 [20], highlights the fact that most Cremonese violins have the wood cut into quarters (Figure 2a), compared to the style of Brescians violin makers who use live sawn, which has an end grain with growth rings of 0–90 degrees to the surface (Figure 2b). In order to obtain the arching of the plates by roughing, the initial thickness of the semi-finished products cut from logs is higher toward the middle of the plate and smaller toward the sides. Thus, the age of the wood is chronologically higher towards the outside of the violin and lower towards the inside.

The novelty of the paper consists in the comparative analysis of six violins (from XVIII–XX centuries) belonging to well-known violin schools, Stainer, Klotz, Leeb, Bela and Gliga, in order to identify their constructive particularities and the anatomical characteristics of the wood in their structure. All this information is a scientifically valuable database, especially since most studies thus far have focused on the heritage violins of the great Italian luthiers.

2. Materials and Methods

2.1. Studied Structures

In this study, six old violins and a current one were analyzed, five of them with labels containing information on the date of manufacture and belonging to a violin school: violin Jacobus Stainer, 1716; violin Johann Georg Leeb, 1742; violin Joseph Klotz, 1747; violin Babos Bela, 1920; violin Gliga Vasile Ghiorghe, 2020, and two without a label. For one of the unlabeled, the history is known (the fact that it is a copy of Jacobus Stainer coded "Jacobus Stainer Copy"), and for the other violin (coded "Unbranded"), the origin and whether it belonged to a certain school of violinists are unknown (Figure 3).

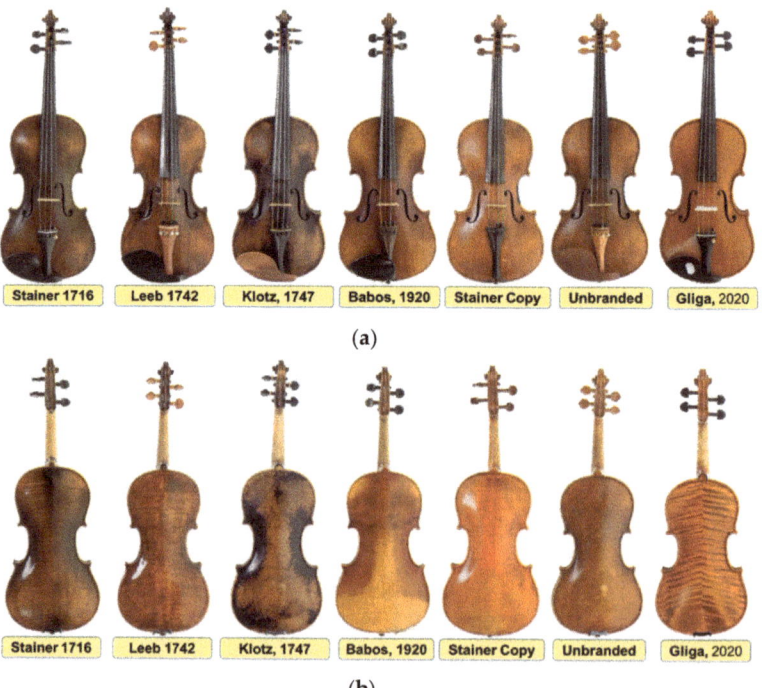

Figure 3. Old violins analyzed using non-invasive and non-destructive methods: (**a**) front view of studied violins; (**b**) back view of violins.

Figure 3 shows the seven violins studied, with images of both the faces of the violins (Figure 3a) and the back of the violins (Figure 3b). All the violins studied are constructively intact, being used in musical activities by their owners. For this reason, the methods of analysis of the constructive elements were chosen so as not to damage or harm the violins in any way. Additionally, the color of the finish of the violins and the quality of the surface in terms of clarity of wood structure at the time of investigations can be seen in Figure 3. Aspects related to the color tones, the type of finishes and the thickness of the penetration of the wood finish were studied by [3,11,12]. It was found that frequently used and aged instruments show a pattern of wear due to the degradation of the varnish after extensive manipulation and weighing by the violinist, which makes it difficult to analyze the structure of the wood [20–22].

2.2. Methods

2.2.1. Wood Structure Data Acquisition

Evaluation of the anatomical features of the wood in the construction of the top and back plates of old violins was performed using a WinDENDRO Density image analysis system (Régent Instruments Inc., Québec City, QC, Canada, 2007) from the Department of Forest Engineering, Forest Management Planning and Terrestrial Measurements, Transilvania University of Brașov, Romania (Figure 4a,b). The characteristics of the annual rings were measured in terms of the width of the annual rings denoted TRW, the width of the early wood EWW, the width of the late wood LWW and the wavelength of the curly fiber of maple (wavelength CWL) according to the method presented in previous studies [23–25]. The annual rings were measured in two or three directions, depending on the objective local difficulties in identifying the contour of the rings, especially for old violins, starting from the edge of the sides to the welding line of the face halves (Figure 4a,b). For spruce boards in the structure of violin top plates, the width of early wood and late wood was measured, the wood structure made it possible to take these data, while for maple boards in the construction of the back of the violin, only the width of each ring was measured. For verification, the resulting series of face rings were cross-dated to each other. Cross-dating was conducted within the same software, adopting a threshold of 0.60 for the Gleichläufigkeit correlation coefficient [26]. All measurements were performed without removing the metal strings and other accessories, as requested by the violins' owners.

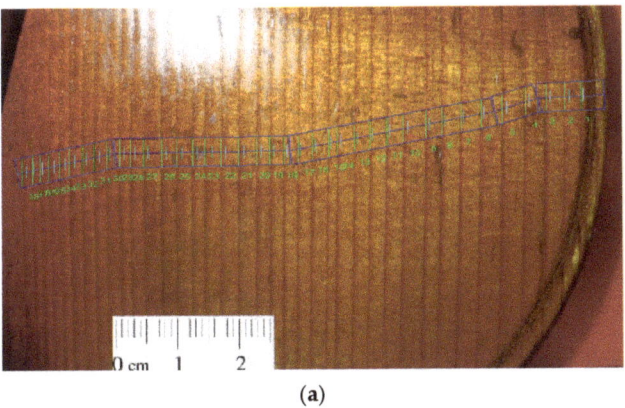
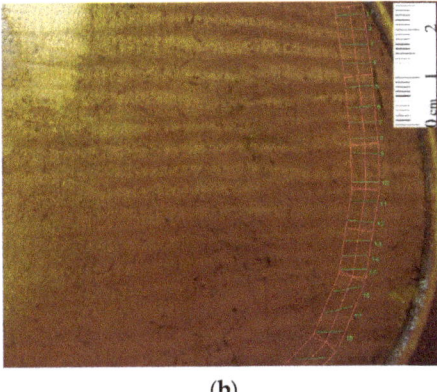

(a)　　　　　　　　　　　　　　　　　　(b)

Figure 4. The constructive principle of violin boards for obtaining the anatomical symmetry of the wood structure: (a) measurement of annual rings, early wood–late wood with Windendro system 2007 in case of violin top plate; (b) measurement of wavelength with Windendro system 2007 in case of violin back plate.

2.2.2. X-ray Imaging

In order to determine the shape and geometry of the violins, the samples were exposed to X-ray radiography at the Laboratory of Radiology and Medical Imaging, Faculty of Veterinary Medicine of Cluj-Napoca (Figure 5a). The X-ray exposures were made using a fixed radiographic device TEMCO Grx-01 (K&S Röntgenwerk Bochum GmbH&Co KG, Bochum, Germany). The exposures were made dorsoventrally, the field of view being set to cover the violin body. The parameters used to obtain the images were 50–56 kV and 13–20 mAs. The images were acquired using a DR Flat Panel detector Reyance Xmaru 1717SGC/SCC (Reyance Inc., Hwaseong-si, Gyeonggi-do, Korea) and Xmaru VetView (Reyance Inc., Hwaseong-si, Gyeonggi-do, South Korea) acquisition software.

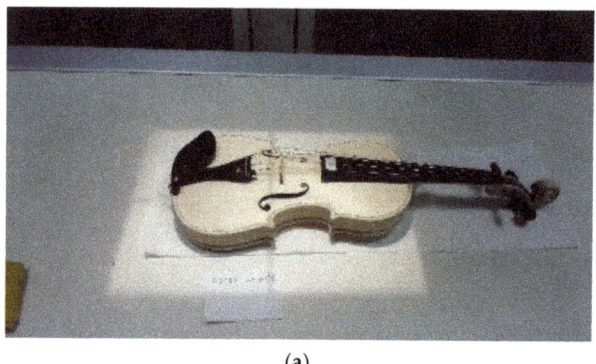
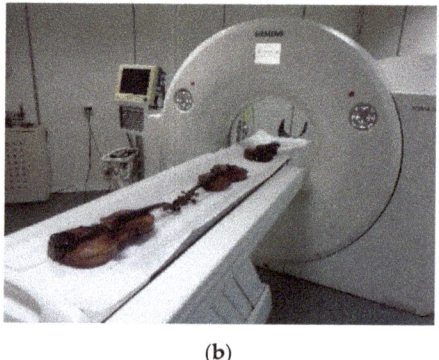

(a) (b)

Figure 5. Old violins analyzed using non-invasive and non-destructive methods: (a) X-ray of the violin; (b) Computed Tomography analysis of old violins.

2.2.3. Computed Tomography

The violins were investigated using computer tomography (denoted CT) in order to analyze the constructive elements, thicknesses and arching of the old violins (Figure 5b). The CT examinations were performed on a Siemens Somatom Scope (Siemens, Erlangen, Germany) helical CT device with 16 slices. The scans were performed using a bone reconstruction kernel. The images acquisition was conducted at 2 mm/slice and the reconstruction was performed at 0.75 mm/slice. For each violin, two axial scans were performed, one for the violin body and the second for the violin neck. The scan parameters were Nominal Total Collimation Width: 9.6 mm, Pitch Factor: 0.8 ratio, KVP: 130 kV, X-ray Tube Current: 96 mA, Exposure: 120 mA, Exposure Time per Rotation: 1 s, 512 × 512 Matrix. The images, both for X-ray and CT scan, were acquired in DICOM format; reading and post-processing of the DICOM files was performed using 3DNET PACS software and Horos DICOM viewer.

2.2.4. Data Processing

The raw data were processed, calculating the early wood proportion (EWP) and the latewood proportion (LWP). In order to assess the regularity of the rings, the following method of calculating the regularity index RI, recommended by Dinulică et al., 2015, [23] for wood, was adopted for the construction of violins:

$$RI = \frac{\max(TRW_i) - \min(TRW_i)}{\max(TRW_i)}, \qquad (1)$$

where i is a ring from the middle series of the front or back plate ($i = \overline{1 \ldots n}$) and n is the length of the series.

Then, the data were imported and processed in STATISTICS 8.0 (StatSoft 2007), following Zar's instructions (1974) [27]. To start, the variability of the experimental data was explored, and the normality was verified with the Shapiro–Wilk test. Then, the significance of the differences between the violins regarding the size of the wood structure variables was tested.

3. Results and Discussions

3.1. The Anatomical Analysis of Wood from the Construction of Old Violins

A total of 2641 front rings and 970 back rings were measured and, in Table 1 the average values and standard deviation of the main characteristics of the annual rings measured on the top and back plates of the violins are summarized.

Table 1. The anatomical features of spruce and maple wood from top and back plates of studied violins.

Variables Average Values/STDV	Studied Violins						
	Stainer 1716	Leeb 1742	Klotz 1747	Babos 1920	Stainer Copy	Unbranded	Gliga 2020
Top Plate (Spruce Wood)							
Annual rings widths (mm)	2.247 / 0.567	1.530 / 0.490	1.251 / 0.403	1.891 / 0.612	0.985 / 0.527	1.327 / 0.336	0.940 / 0.234
Early wood width (mm)	1.676 / 0.518	1.148 / 0.467	0.792 / 0.304	1.449 / 0.601	0.689 / 0.450	0.907 / 0.293	0.568 / 0.190
Latewood width (mm)	0.496 / 0.178	0.382 / 0.122	0.459 / 0.162	0.442 / 0.158	0.300 / 0.118	0.420 / 0.130	0.372 / 0.100
Early wood proportion (%)	76.184 / 9.152	73.564 / 8.507	62.635 / 8.700	74.379 / 9.942	66.127 / 11.286	67.689 / 8.921	59.766 / 8.388
Latewood proportion (%)	23.816 / 9.152	26.436 / 8.507	37.365 / 8.700	25.203 / 9.942	33.873 / 11.286	32.311 / 8.921	40.234 / 8.388
Back Plate (Maple Wood)							
Annual rings widths (mm)	1.908 / 0.531	1.246 / 0.658	1.063 / 0.902	1.026 / 0.527	1.277 / 0.297	4.563 / 1.105	1.623 / 0.666
Wavelength (mm)	4.021 / 1.577	6.421 / 2.422	NA / NA	3.946 / 1.256	4.984 / 1.589	4.585 / 1.057	6.731 / 3.371

From a statistical perspective, the measured characteristics of the wood structure of the violin sound box are continuous variables. They are not compatible with the normal law (W from Shapiro–Wilk = 0.886–0.992, $p < 0.001$), and the non-parametric Kruskal–Wallis test shows that the analyzed violins differ from each other at a very significant level in terms of all the structural characteristics (H = 257–1272, $p < 0.001$). Therefore, each violin has its structural personality, as can be seen in Figures 6 and 7. In three of the violins analyzed, the annual rings in the back plate structure are considerably finer than those in the top plate structure (Babos 1920, Leeb 1742, Stainer 1716), while in the other three violins, the annual rings of top plate are much finer than those in the back (Unbranded, Stainer Copy, Gliga 2020); for the Klotz 1747 violin, the rings have close widths in the two plates of the sound box (Figure 6a). In 40% of cases, the regularity index of the width RI (Figure 6b) is within the limits specified by Rocaboy et al. (1990) [28] for the resonant wood ($RI \leq 0.700$). It is known that the higher the RI value is, the lower the regularity of the rings is. In most violins, there are big differences between the top and back plates regarding the regularity of the rings. The rings of the spruce wood (top) are usually more regular than those of the back (maple wood) (Figure 6b). The width of the early wood in the composition of the annual ring is directly proportional to the width of the annual ring (*Spearman R rank order correlation: 0.975, $p < 0.001$*). Additionally, the width of the late wood depends to a large extent on the width of the growth ring (*Spearman R: 0.651, $p < 0.001$*). On average, late wood accounts for a third, and early wood the other two-thirds of the annual ring width. The proportions of the two components of the annual ring show a moderate level of variability (coefficient of variation: 16 and 32%, respectively) (Figure 7).

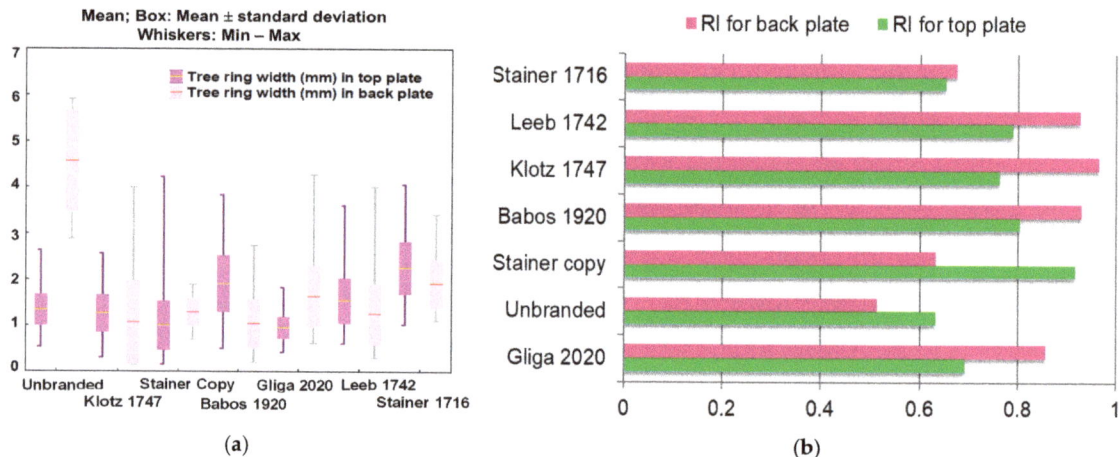

Figure 6. Comparison between anatomical descriptors of wood from old violins structures: (**a**) The variation of the annual ring width in case of top and back plates of studied violins; (**b**) The regularity level of the annual rings in the structure of the sound box for the analyzed violins.

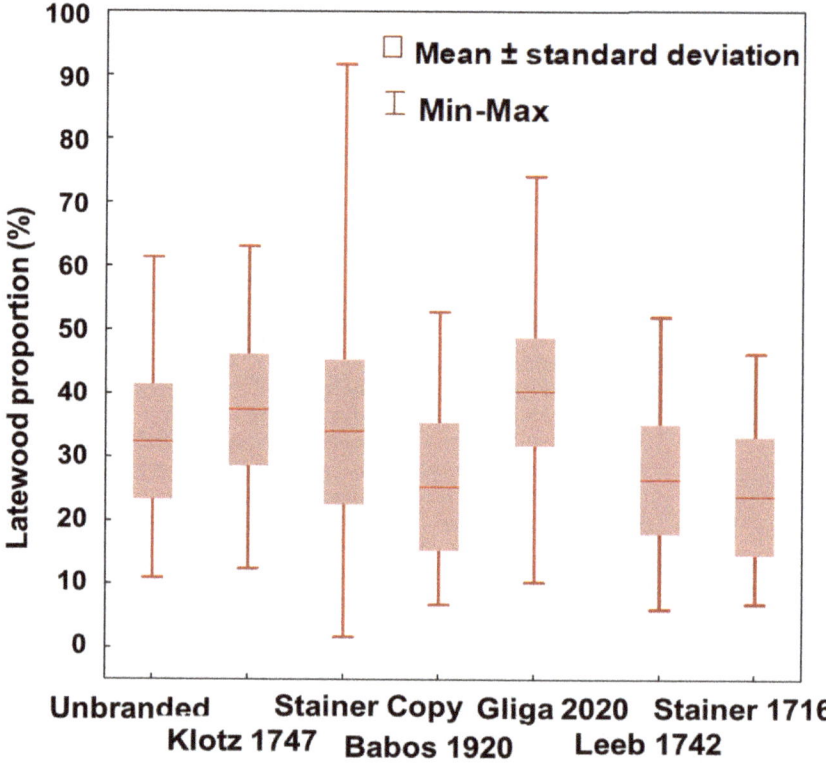

Figure 7. The variation of the proportion of late wood of top plates in the analyzed violins.

Of all the old violins investigated, 66% of the recorded values of the proportion of late wood exceed the reference level of 25% mentioned for the resonant spruce by Bucur,

2006 [29]. There are violins in which the central tendency of the late wood increases to 40% of the width of the ring, such as Klotz 1747 and Gliga 2020 (Figure 7).

It is not excluded that the result may be influenced by violin finishing techniques, which may have led to an overestimation of the width of the wood late in the imaging analysis. However, we must also take into account the fact that high values of LWP are recorded in narrow rings (the correlation coefficient between the width of the ring and the proportion of late wood is -0.623, $p < 0.001$) and rings that abound in the violins analyzed. Specifically, 38% of the total number of rings measured is less than 1 mm wide. Regarding the characteristics of curly maple wood, the wavelength of the curly fiber gravitates around 4.4 mm. The smallest value (1.35 mm) was measured on the Stainer 1716 violin, and the highest (13.11 mm) on the Gliga 2020 violin. The differences between the violins are noticeable, some have tightly created fibers, others are wide (Figure 8). It is a tendency to associate the wavelength with certain values of the annual ring width; respectively, the dense fiber appears especially in maple wood with wider rings (*Spearman R rank order correlation*: -0.156, $p = 0.04$).

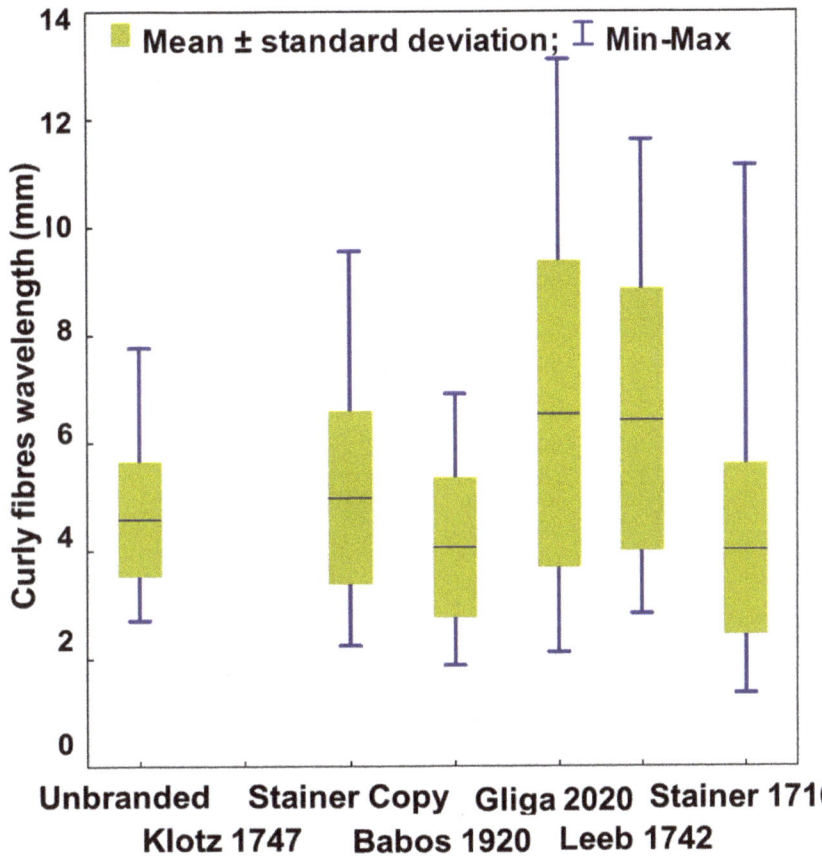

Figure 8. The variation of the curly fibers' wavelength from maple back plates of analyzed violins.

3.2. X-ray Radiography of Heritage Violins

Through the X-ray analyses within the Laboratory of Imaging Analysis and Radiography of the Faculty of Veterinary Medicine, USAMV Cluj-Napoca, it was possible to identify some constructive elements specific to the violin schools to which the violins belong. The characteristics that can be observed in X-ray radiography are the macroscopic elements of the structure, the poor resolution being, on the one hand, due to the relatively low sensitivity of X-rays to wood and, on the other hand, due to the large sample size [30]. Thus, one of the obvious constructive elements in the X-ray analysis is the corners, which have the role of strengthening the intersection between the curves of the violin as a result of changing the curvature radius, as well as increasing the gluing surface between the front plate, back plate and straps (Figure 9). As can be seen in Figure 9, the investigated violins can be grouped into the following three classes in terms of the constructive shape of the corners: violins without a corner on the inside (Stainer, 1716; Babos 1920—Figure 9a); another category of violins is the one in which the corners are stiffened with solid wood corners, cut according to the inner shape of the corners, obtaining a continuous contour inside the violin body (Figure 9b); and finally, violins with softwood slates (Klotz, 1747, the Stainer copy and the "unbranded" violin), noting that the corner reinforcing slates on the Klotz 1747 violin are found only at the corners between the central curvature and the lower curvature (Figure 9c). A few ways of joining the corners are presented in the literature [4,5].

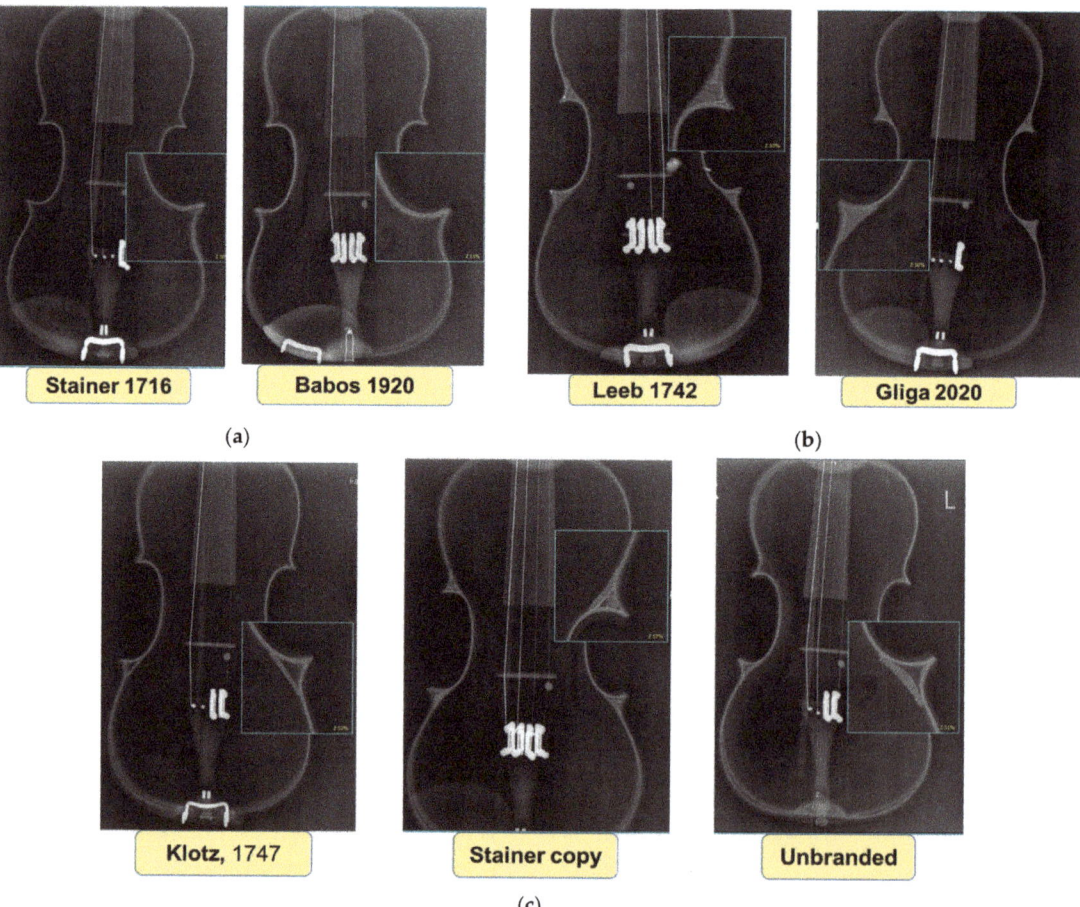

Figure 9. The constructive shape of corners visible on X-ray radiation: (**a**) Violins without corners; (**b**) Violins with solid wood corners; (**c**) Violins with softwood slates.

3.3. Imaging Analysis of Old Violins Using CT Scanning

Information on the thickness of the violin plates, the curvature of the plates, the shape of the sound bar, the dimensions of the old violins and aspects regarding the integrity or the degree of damage of the violins, all this was obtained using computed tomography of the studied violins. Computed tomography comprises a set of axial 2D images. The data volume can be reformatted and reorganized into 3D images, with the advantage of obtaining a contrast approximately 16 times higher than X-ray radiography, in order to identify some constructive elements (shapes/dimensions) inaccessible to the "naked eye". In Figure 10a,b, two cross sections are presented: one through the "unbranded" violin (Figure 10a), and one through the "Klotz, 1747" violin (Figure 10b).

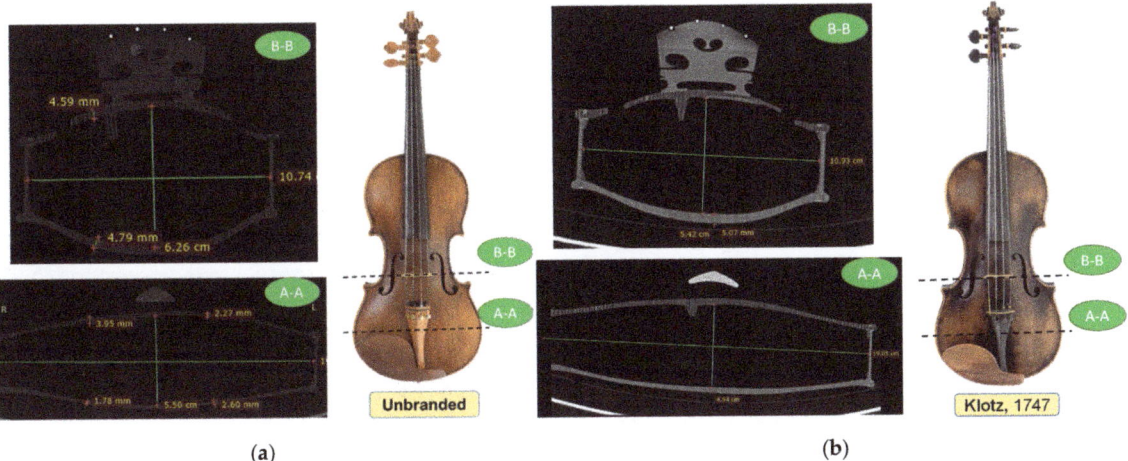

Figure 10. CT images analysis: (**a**) the Unbranded violin; (**b**) the "Klotz, 1747" violin.

Similarly, the other violins were analyzed, obtaining the dimensions of the violins in different sections, the thickness of the plates and the radius of curvature. From a constructive point of view, an interesting detail is observed in the way the resonance bar was made. The Stainer, 1716, Leeb 1742 and "unbranded" violins show the bass bar applied on the top plate (Figure 11a), compared to the violins "Stainer copy", Klotz 1747 and Bela 1920, in which the resonant bar was made by roughing the front plate, with a volumetric element on the inside of the top plate (Figure 11b).

Additionally, the arching of the violin plates and the thickness of the plates play an important role in the acoustics of the musical instrument, giving the violins the signature modes, as [31] calls them, these modes being cavity modes (A0, A1), corpus modes (CBR or C bouts rhomboidal) and main body resonance (B1+ and B1−).

Another constructive characteristic is the one related to the composition of the back plates: Unbranded and Klotz violins have a back plate made of a single wooden board, while the other violins have a back plate composed of two halves with the symmetrical and quasi-symmetrical structure of annual rings. This aspect can be observed even with the naked eye by visual analysis of the violin, but it is also confirmed by the cross-sectional views obtained on the computed tomography. The CT images offer the possibility to clearly distinguish the differences between the two wood species used for the front (spruce) and back (maple) boards, as well as eventual interventions/repairs performed on the violins, the degree of wood wear and biological attacks of the wood, as can be seen in Figure 12, in the case of the Babos 1920 violin, and Figure 13, for the Stainer 1716 violin. In the highlighted areas in Figure 12, the trajectories of the voids produced by larvae of coleoptera can be observed, the color contrast and shape of the voids being specific to the biological attack. Taking into account the way the violin plates were assembled, where the young wood is found in the joint area, we assume that the holes observed at CT were produced by *Anobium pertinax* or *Anobium punctatum*, which laid eggs under the rhytidome, and the larvae bear irregular galleries of maximum 3 mm and filled them with sawdust. The wood can be attacked inside without being noticed on the outside [32,33].

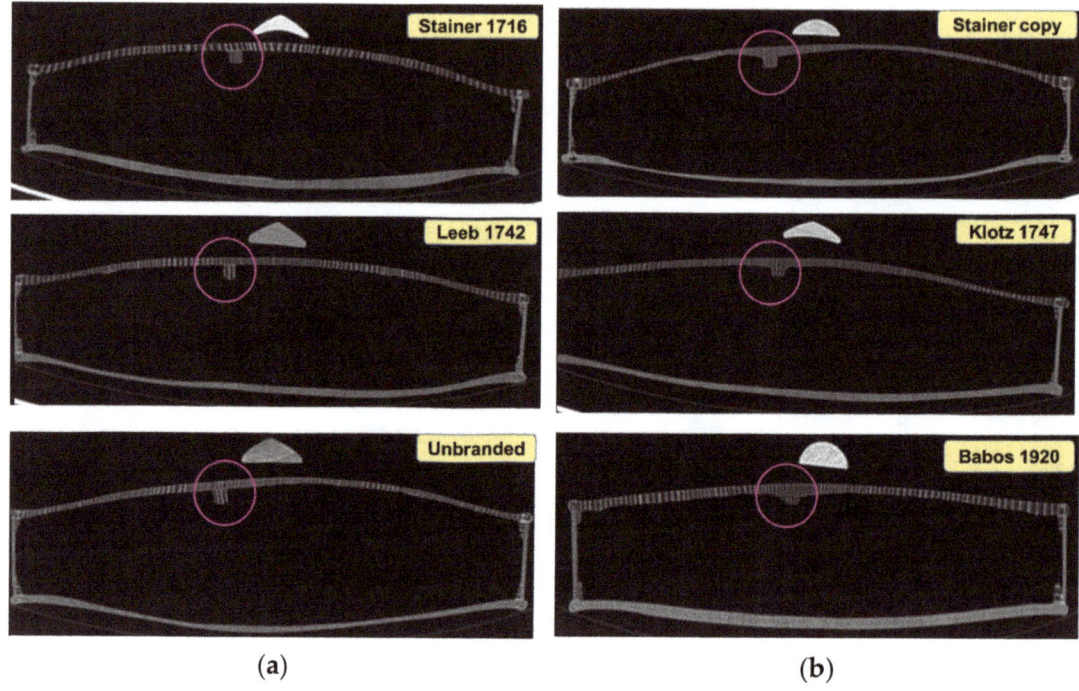

Figure 11. Cross section through CT-scanned violins: (**a**) violins with the applied resonance bar; (**b**) violins with the resonant bar processed from the top plate thickness.

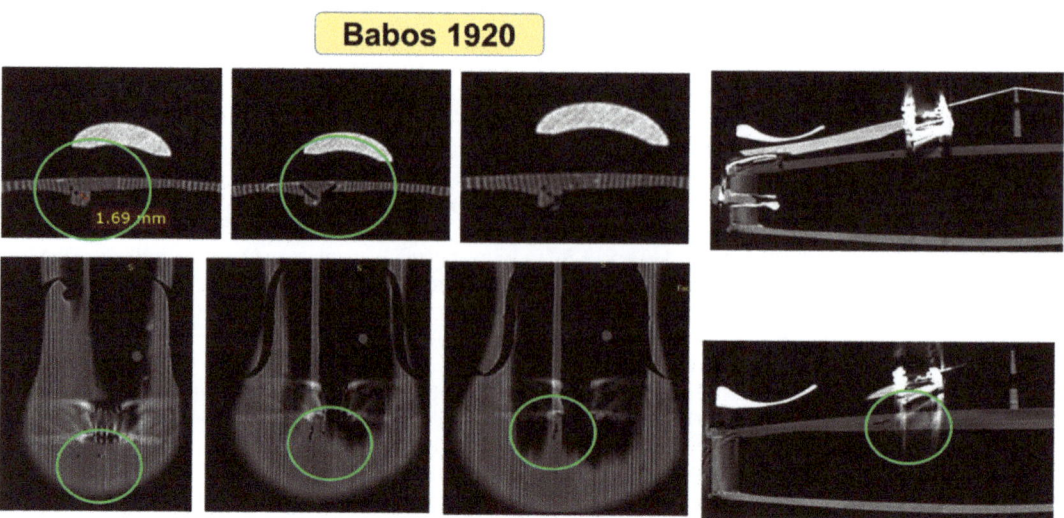

Figure 12. The trajectory of voids produced by coleoptera in case of the violin "Babos, 1920".

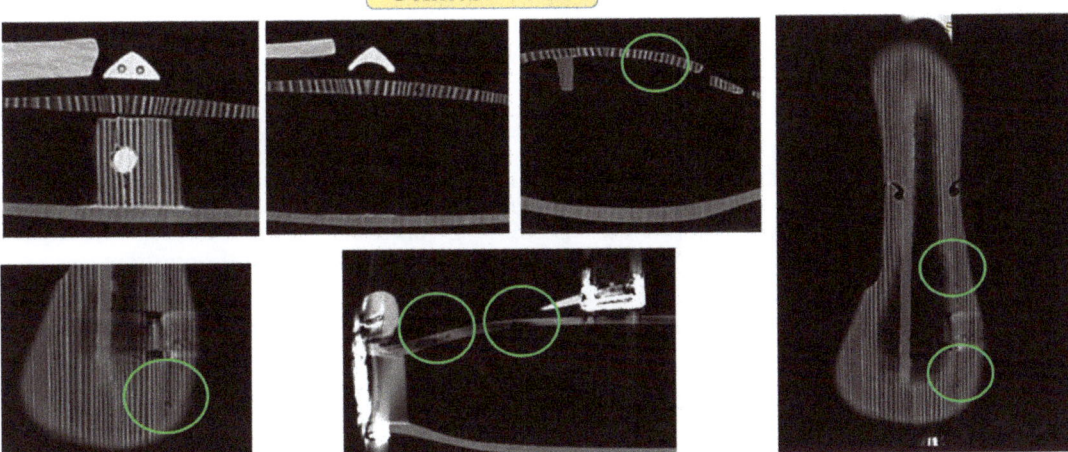

Figure 13. The trajectory of voids produced by coleoptera in case of the violin "Stainer, 1716".

4. Conclusions

The study presents the non-invasive spectroscopic approaches related to anatomical patterns of wood from ancient and modern violin plates, as well as the morphological and geometrical features, such as the shape of the body, the arching, the corners and the f-holes. Integrating and comparing the results, it was possible to characterize the studied violin, as can be seen in Table 2. Finally, it can be appreciated that the integration of imaging techniques with information on wood processing and its properties provides a useful database for luthiers and musicians, and in perspective, the authors of this paper aim to both date unknown violins (Unbranded and Stainer copy) and conduct acoustic analysis on these violins, compared to the Stradivarius Elder Voicu 1702 violin, which is part of Romania's cultural heritage.

Table 2. The features of studied violins.

Features	Stainer 1716	Leeb 1742	Klotz 1747	Babos 1920	Stainer Copy	Unbranded	Gliga 2020
				Studied Violins			
Types of wood cuts for the violin	Two pieces	Two pieces	One piece	Two pieces	Two pieces	Two pieces	Two pieces
Annual rings width	Wide	Narrow	Narrow	Wide	Narrow	Narrow	Narrow
Regularity of annual rings	Medium	Weak	Good	Medium	Weak	High	High
Symmetry of top plate	Medium	Weak	Medium	Weak	Weak	Medium	High
Types of wood cuts for the violin	Two pieces	Two pieces	One piece	Two pieces	Two pieces	Two pieces	Two pieces
Annual rings width	Wide	Narrow	Medium	Narrow	Narrow	Wide	Wide
Wavelength	Medium curly fibers	Low curly fibers	No curly fiber	Low curly fibers	Medium curly fibers	Medium curly fibers	High curly fibers
Regularity of annual rings	Medium	Weak	Weak	Weak	Medium	Medium	Weak
Constructive Elements							
Type of bass bar	Applied	Applied	From top plate	From top plate	From top plate	Applied	Applied
Type of corner	No corners	Solid wood	Softwood slats	No corners	Softwood slats	Softwood slats	Solid wood
Coleoptera voids	Yes	No	No	Yes	No	No	No

In further work, the correlation between the anatomical structure of wood, especially the values of the indicators for characterizing the symmetry and the frequencies spectra, dominant frequencies, quality factor and damping of old and new violins from A dynamic test will be presented. For each violin, the signature mode will be identified and quantified in terms of eigenvalues.

Author Contributions: Conceptualization, M.D.S. and M.M.; methodology, F.D., R.P.; R.L. and A.M.N.; software, M.D.S., F.D., R.P. and R.L.; validation, G.V.G. and A.M.N.; formal analysis, M.D.S. and M.M.; investigation, F.D., R.P. and R.L.; resources, M.D.S. and G.V.G.; data curation, M.D.S.; writing—original draft preparation, M.M. and A.M.N.; writing—review and editing, M.D.S.; visualization, M.D.S. and G.V.G.; supervision, F.D.; project administration, M.D.S.; funding acquisition, M.D.S. All authors have read and agreed to the published version of the manuscript.

Funding: This research was funded by a grant from the Ministry of Research, Innovation and Digitization, CNCS/CCCDI, UEFISCDI, PN-III-P2-2.1-PED-2019-2148, project number. 568PED/2020 MI-NOVIS, Innovative models of violins acoustically and aesthetically comparable to heritage violins, within PNCDI III.

Institutional Review Board Statement: Not applicable.

Informed Consent Statement: Not applicable.

Data Availability Statement: The data presented in this study are available on request from the corresponding author.

Acknowledgments: We thank Mădălina Mădălina Gyorke, instrumental artist from the Brașov Philharmonic for the support offered, as well as the company AP Studio Brașov for the professional photographs of the violins.

Conflicts of Interest: The authors declare no conflict of interest.

References

1. Tai, H.-C.; Shen, Y.-P.; Lin, J.-H.; Chung, D.-T. Acoustic evolution of old Italian violins from Amati to Stradivari. *Proc. Natl. Acad. Sci. USA* **2018**, *115*, 5926–5931. [CrossRef]
2. Sandu, I.C.A.; Negru, I.C.; Sandu, A.V.; Vasilache, V. Authentication of an Old Violin by Multianalytical Methods. *Appl. Sci.* **2020**, *10*, 306. [CrossRef]
3. Lämmlein, S.L.; Mannes, D.; Van Damme, B.; Schwarze, F.W.; Burgert, I. The influence of multi-layered varnishes on moisture protection and vibrational properties of violin wood. *Sci. Rep.* **2019**, *9*, 18611. [CrossRef]
4. Rigon, L.; Vallazza, E.; Arfelli, F.; Longo, R.; Dreossi, D.; Bergamaschi, A.; Schmitt, B.; Chen, R.; Cova, M.A.; Perabò, R.; et al. Synchrotron-radiation microtomography for the non-destructive structural evaluation of bowed stringed instruments. *E-Preserv. Sci.* **2010**, *7*, 71–77.
5. Erdélyi, S. *Hungarian Treasures of Violinists*; DRESKULT: Budapest, Hungary, 1997. (In English and Hungary)
6. Schwarze, F.W.M.R.; Morris, H. Banishing the myths and dogmas surrounding the biotech Stradivarius. *Plants People Planet* **2020**, *2*, 237–243. [CrossRef]
7. Čufar, K.; Beuting, M.; Demšar, B.; Merela, M. Dating of violins—The interpretation of dendrochronological reports. *J. Cult. Herit.* **2017**, *27*, S44–S54. [CrossRef]
8. Klein, P.; Mehringer, H.; Bauch, J. Dendrochronological and wood biological investigations on string instruments. *Holzforsch. Int. J. Biol. Chem. Phys. Technol. Wood* **1986**, 197–203, 197–203. [CrossRef]
9. Klein, P.; Pollens, S. *The Technique of Dendrochronology as Applied to Violins Made by Giuseppe Guarneri del Gesù*; Biddulph Publisher, Giuseppe Guarneridel Gesù: London, UK, 1998.
10. Topham, J.; McCormick, D. A dendrochronological investigation of British stringed instruments of the violin family. *J. Archaeol. Sci.* **1998**, *25*, 1149–1157. [CrossRef]
11. Topham, J.; McCormick, D. A dendrochronological investigation of stringed instruments of the cremonese school (1666–1757) including "The Messiah" violin attributed to Antonio Stradivari. *J. Archaeol. Sci.* **2000**, *27*, 183–192. [CrossRef]
12. Le Conte, S.; Le Moyne, S.; Ollivier, F. Modal analysis comparison of two violins made by A. Stradivari. In Proceedings of the Acoustics 2012 Nantes Conference, Nantes, France, 23–27 April 2012, hal-00811010.
13. Fiocco, G.; Gonzalez, S.; Invernizzi, C.; Rovetta, T.; Albano, M.; Dondi, P.; Licchelli, M.; Antonacci, F.; Malagodi, M. Compositional and Morphological Comparison among Three Coeval Violins Made by Giuseppe Guarneri "del Gesù" in 1734. *Coatings* **2021**, *11*, 884. [CrossRef]
14. Lehmann, E.H.; Mannes, D. Wood investigations by means of radiation transmission techniques. *J. Cult. Herit.* **2012**, *13*, S35–S43. [CrossRef]

15. Stoel, B.C.; Borman, T.M. A Comparison of Wood Density between Classical Cremonese and Modern Violins. *PLoS ONE* **2008**, *3*, e2554. [CrossRef]
16. Nia, H.T.; Jain, A.D.; Liu, Y.; Alam, M.R.; Barnas, R.; Makris, N.C. The evolution of air resonance power efficiency in the violin and its ancestors. *Proc. Math. Phys. Eng. Sci.* **2015**, *471*, 20140905. [CrossRef]
17. Cattani, C.; Dunbar, R.L.M.; Shapira, Z. Value Creation and Knowledge Loss: The Case of Cremonese Stringed Instruments. *Organ. Sci.* **2012**, *24*, 813–830. [CrossRef]
18. Marcon, B.; Goli, G.; Fioravanti, M. Modelling wooden cultural heritage. The need to consider each artefact as unique as illustrated by the Cannone violin. *Herit. Sci.* **2020**, *8*, 24. [CrossRef]
19. Sodini, N.; Dreossi, D.; Chen, R.; Fioravanti, M. Non-invasive microstructural analysis of bowed stringed instruments with synchrotron radiation X-ray microtomography. *J. Cult. Herit.* **2012**, *13*, S44–S49. [CrossRef]
20. Nyman, C.D. 1. History of the Development of the Violin 2. Construction of the Violin 3. Repairs of the Violin String Instruments. Master's Thesis, Utah State University, Logan, UT, USA, 1975; p. 750. Available online: https://digitalcommons.usu.edu/gradreports/750 (accessed on 15 June 2021).
21. Blümich, B.; Baias, M.; Rehorn, C.; Gabrielli, V.; Jaschtschuk, D.; Harrison, C.; Invernizzi, C.; Malagodi, M. Comparison of historical violins by non-destructive MRI depth profiling. *Microchem. J.* **2020**, *158*, 105219. [CrossRef]
22. Hargrave, R. A Violin by Jacobus Stainer. 1679. Available online: https://www.roger-hargrave.de/PDF/Artikel/Strad/Artikel_1987_09_Jacobus_Stainer_1697_PDF.pdf (accessed on 15 June 2021).
23. Dinulică, F.; Albu, C.; Borz, S.A.; Vasilescu, M.M.; Petrițan, C. Specific structural indexes for resonance Norway spruce wood used for violin manufacturing. *Bioresources* **2015**, *10*, 7525–7543. [CrossRef]
24. Dinulică, F.; Albu, C.T.; Vasilescu, M.M.; Stanciu, M.D. Bark features for identifying resonance spruce standing timber. *Forests* **2019**, *10*, 799. [CrossRef]
25. Dinulică, F.; Stanciu, M.D.; Savin, A. Correlation between anatomical grading and acoustic–elastic properties of resonant spruce wood used for musical instruments. *Forests* **2021**, *12*, 1122. [CrossRef]
26. Pilcher, J.R. Sample preparation, cross-dating and measurement. In *Methods of Dendrochronology*; Cook, E.R., Kairiukstis, L.A., Eds.; Kluwer Academis Publishing: Dordrecht, The Netherlands, 1990; pp. 40–51.
27. Zar, J.H. *Biostatistical Analysis*; Prentice-Hall Inc.: Englewood Cliffs, UK, 1974.
28. Rocaboy, F.; Bucur, V. About the physical properties of wood of twentieth century violins. *J. Acoust. Soc. Am.* **1990**, *1*, 21–28.
29. Bucur, V. *Acoustics of Wood*, 2nd ed.; Springer: Berlin/Heidelberg, Germany, 2006; pp. 173–196.
30. Mannes, D.; Lehmann, E.; Niemz, P. Tomographic investigations of wood from macroscopic to microscopic scale. *Wood Res.* **2009**, *54*, 33–44.
31. Bissinger, G. Structural Acoustics of Good and Bad Violins. *J. Acoust. Soc. Am.* **2008**, *124*, 1764–1773. [CrossRef]
32. Kasal, B.; Adams, A.; Drdácký, M. New applications in radiographic evaluation of structural components. *Eur. J. Environ. Civ. Eng.* **2010**, *14*, 395–410. [CrossRef]
33. Hýsek, Š.; Löwe, R.; Turcáni, M. What Happens to Wood after a Tree Is Attacked by a Bark Beetle? *Forests* **2021**, *12*, 1163. [CrossRef]

MDPI
St. Alban-Anlage 66
4052 Basel
Switzerland
Tel. +41 61 683 77 34
Fax +41 61 302 89 18
www.mdpi.com

Materials Editorial Office
E-mail: materials@mdpi.com
www.mdpi.com/journal/materials

www.ingramcontent.com/pod-product-compliance
Lightning Source LLC
LaVergne TN
LVHW070737100526
838202LV00013B/1260